P. Patrick MacDaniel

BEACH AND NEARSHORE SEDIMENTATION

Based on a Symposium
Sponsored by the Society of
Economic Paleontologists and Mineralogists

Edited by

Richard A. Davis, Jr., University of South Florida, Tampa
and
R. L. Ethington, University of Missouri, Columbia

SOCIETY OF ECONOMIC PALEONTOLOGISTS AND MINERALOGISTS

Special Publication No. 24

Tulsa, Oklahoma, U.S.A. October, 1976

A Publication of

The Society of Economic Paleontologists and Mineralogists

a division of

The American Association of Petroleum Geologists

PREFACE

Sedimentology, like many other disciplines within the geological sciences, has experienced tremendous advances during the past two decades. Prior to the early and mid-1950's, most sedimentologic research was directed primarily toward descriptive aspects of sediments and sedimentary rocks with most emphasis on textural parameters. Sedimentary structures commonly were included in these descriptive studies primarily for purposes of characterizing rock units or for correlation.

During the early 1950's some new avenues of approach to sedimentology were initiated. Prominent among these was the work of W. C. Krumbein and his colleagues and students at Northwestern University who used statistics extensively in their research. Concurrently the late P. D. Krynine and his students at Pennsylvania State University made significant advances in the use of petrography to interpret sedimentary depositional and diagenetic environments.

These activities along with numerous others provided some of the impetus for an explosion of sedimentological research devoted to the interpretation of depositional environments. It is important to note that until about 20 years ago detailed knowledge of modern depositional sedimentary environments was lacking. This, of course, makes it very difficult for the field geologist who is studying the rock record to apply the "Principle of Uniformitarianism" in any detail.

Large-scale efforts by the petroleum company research groups at Shell Development Company and the Humble (now Exxon) Research laboratories during the fifties provided much of the basic data and technology for the study of modern sedimentary environments. This led to similar efforts by other petroleum research groups, state and federal agencies and academic institutions. The majority of these studies dealt with coastal and shallow marine environments. Prominent among these efforts were the studies of the Mississippi Delta complex by H. N. Fisk and his colleagues and the studies of barrier island complexes of the Gulf Coast by H. A. Bernard and colleagues. Numerous other noteworthy activities of a similar nature paralleled and followed these studies.

Such studies were devoted largely to detailed investigation of the sedimentary facies patterns utilizing textural parameters and sedimentary structures with some attention also paid to the faunal relationships with respect to these sediment characteristics. To a large extent these studies also were descriptive in that the processes operating in these environments were treated in only a superficial manner or not at all. The results of these efforts have been invaluable however, in that they have provided the geologist with the information needed to make rather good interpretations as to the sedimentary environments represented in the rock record.

These research efforts also served to demonstrate the necessity of geologists, whether they be stratigraphers, sedimentologists or paleontologists, having some training and practical experience working in modern sedimentary environments. Since the late fifties it has been common practice for many, if not most, geologists in the petroleum industry to spend some time in various modern environments in order to gain this insight into their interpretations of the rock record.

Although the aforementioned efforts represented considerable progress in our understanding of the rock record, they were primarily descriptive approaches to the analysis of sedimentary environments. Little detailed work was done to relate the sediment parameters and sedimentary structures to the processes by which they were distributed or generated. In other words quantitative process-response data were lacking.

At this point proper credit should be given to those groups who were engaged in such quantitative efforts at an early date. The old Beach Erosion Board which envolved into the Coastal Engineering Research Center of the U.S. Army Corps of Engineers has been sponsoring this type of research at least since the early 1930's. Although most of their effort has been in the area of engineering, marine geologists such as F. P. Shepard and D. L. Inman of the Scripps Institution of Oceanography have been associated with this group for some time.

A good bit of the initial process-response type of data necessary for the understanding of sediment transport and formation of sedimentary structures was conducted in flumes and wave tanks by engineers and fluvial sedimentologists. The results of this laboratory research and related field studies were drawn together in the annual SEPM Research Symposium for the 1964 meetings in Toronto. This research symposium, organized by G. V. Middleton of McMaster University, resulted in a landmark publication entitled *Primary Sedimentary Structures and their Hydrodynamic Interpretation* (SEPM Special Publication 14). The title perhaps suggests a broader coverage than is actually included, with most of the papers devoted to sedimentary structures generated by unidirectional flow, e.g. streams and flumes. This publication and numerous of its contained papers has become a classic contribution to the discipline of sedimentary processes.

During the ten years that have passed since publication of the above described symposium, there has been considerable research into quantitative aspects of sedimentary processes. A significant portion of this effort has been supported by the Geography Programs of the Office of Naval Research although the Coastal Engineering Research Center and the Marine Geology Branch of the U.S. Geological Survey have also been prominent in their support of this type of sedimentological research.

This decade has resulted in the formulation of numerous process and process-response models for a wide variety of modern sedimentary environments. Many of these studies and their resulting models are in the area of beach and nearshore sedimentation. As a result of the rapidly expanding interest and significant rise in the level of understanding of beach and nearshore sedimentation, the Society of Economic Paleontologists and Mineralogists elected to convene its annual research symposium under this topic. The topic is timely and valuable from two primary areas of application. Quantitative process-response models for these coastal sedimentary environments provide the data necessary to more completely comprehend these environments and thereby extrapolate to the rock record in order to interpret ancient depositional environments. This is the obvious and rather traditional application of such geologic research. Another application for geologic research in the beach and nearshore environments is directly involved with man's occupation of the coastal zone. Although engineers have long dominated this type of research effort, the past decade or so has seen much increase in the efforts of geologists, thereby providing a comprehensive approach to beach and nearshore research. With increased concern for proper utilization and zoning of the coast as well as the need for understanding coastal erosion processes, coastal sedimentologists have provided much insight to understanding the process-response mechanisms in the beach and nearshore.

The 1975 SEPM Research Symposium was organized around topical considerations of *Beach and Nearshore Sedimentation—Physical and Biological.* The intent was to cover the topic from the generation of processes through the mechanics of the processes, interaction of processes with sediment and culminating in distribution of sediment and structures across the environment in question. During the organization of topics and speakers for this symposium it became clear that the biological aspect of the topic, that is animal-sediment relationships in the beach and surf zone, is being neglected by both biologists and geologists. Hopefully there will be considerable expansion of the research effort in this zone by both groups.

The present volume contains all papers presented at the Dallas meeting with two exceptions; the paper on coastal currents by P. D. Komar and one on animal-sediment relationships in coastal environments by J. Dörjes and J. D. Howard. The Komar paper is presented here in condensed form and appears in its entirety elsewhere. Two papers which were not a part of the Dallas symposium have been added in order to provide a more comprehensive volume.

The first paper by W. T. Fox and R. A. Davis represents a rather unique perspective from which to approach process-response mechanisms in beach and nearshore sedimentation. These investigators have conducted detailed time-series field studies at numerous sites on all coasts of the United States. The paper presented here, *Weather Patterns and Coastal Processes*, represents a general summary of the empirically derived conceptual models for the relationships between meteorological parameters and coastal processes. The fundamental variable in these models is barometric pressure. As cyclonic or frontal systems approach and pass over a coast there is a predictable pattern to such processes and process-related parameters as wind velocity and direction, wave height and direction, and longshore current direction and velocity. The predictable nature of these processes provides important capabilities to coastal engineers, planners and researchers who are involved in the investigations of process-response phenomena along the coast.

The paper by V. Goldsmith entitled *Wave Climate Models for the Continental Shelf: Critical Links Between Shelf Hydraulics and Shoreline Processes* represents the other topic in which remotely generated processes and their effects on the coast are considered. The complex patterns in wave orientation that are developed as waves move across the shoreface and into the breaker zone are critical to a proper understanding of energy distribution along the beach. Wave refraction studies have long been a part of coastal engineering and sedimentation research but only recently have they become truly quantified through the use of the computer. Goldsmith has developed a very detailed model for the Virginia Sea which shows the interrelationships between wave parameters and inner shelf topography. The resulting concentrations and dissipations of energy can be related to textural and erosional trends which have been observed along this coast and also to inlet hyraulics.

Paul D. Komar's paper on the *Evaluation of Longshore Current Velocities and Sand Transport Rates Produced by Oblique Wave Approach* is a brief summary of his complete study which is published elsewhere. This paper represents a "state of the art" summary of Komar's work on longshore sediment transport, and provides theoretical and observed data on the prediction of sediment transport rates in this complex and dynamic environment.

In order to provide sedimentologists with basic principles of wave mechanics and provide background information for some of the succeeding papers the paper entitled *Wave Modeling and Hydromechanics* by J. I Collins was solicited for inclusion in the volume. Collins is not a geologist but an engineer. He has considerable experience in modeling coastal processes in the breaker and surf zones. This paper represents a summary of basic wave mechanics. It provides background information for the four succeeding papers, all of which deal with some aspects of hydromechanics.

R. L. Miller's paper, *Role of Vortices in Surf Zone Prediction: Sedimentation and Wave Forces,* represents a landmark contribution to our knowledge of the relationships between waves and sediment in the breaker zone. Although some of Miller's research has been in the field, most was accomplished in the wave tank. Through high speed photographic techniques, breaker types have been characterized in detail. Analysis of the vortices and their relationship to breaker types has enabled Miller to establish predictable changes in bedforms and wave-foreshore interactions. Further refinement of these predictive modeling techniques will provide the coastal engineer and sedimentologist with the ability to predict surf and breaker zone changes from breaker type.

Because of the rigor of the breaker zone there is great difficulty in monitoring the processes taking place. Using rather sophisticated instrumentation, B. M. Brenninkmeyer has investigated *Sand Fountains in the Surf Zone.* Breaking of waves causes explosions of sediment from the bottom. Little or no suspension of sediment occurs in the outer breaker zone with the maximum suspension at the still water level (plunge zone). The monitoring studies of Brenninkmeyer provide some detailed data on the nature and magnitude of sediment entrainment by breaking waves. Because of the obvious interrelationships between this study and that of other authors in this volume (especially Miller) this paper was included although it was not a part of the research symposium.

The swash zone is also an area of the beach and nearshore environment that has received little detailed attention with respect to its process-response mechanisms. This is a very complex environment because it represents a barrier to wave action, it contains opposing water motions and it interacts with the groundwater system of the beach. Evans Waddell's paper on *Swash-Groundwater-Beach Profile Interactions* considers just such phenomena. His detailed measurements provide considerable insight into the complex interactions of swash processes and also those between swash and wave processes.

Virtually all of the preceding papers deal primarily with those aspects of coastal sedimentation which are primarily directed toward interpretation of modern coastal environments with little application to interpretation of the rock record. The final three papers in this volume provide data to enable geologists to make detailed environmental interpretations of the rock record. H. E. Clifton's paper entitled *Wave Formed Sedimentary Structures—A Conceptual Model* provides a wealth of quantitative process-response data on nearshore sedimentary structures. Clifton's conceptual model has evolved from considerable field study on numerous coasts. He has provided a rather simple, yet comprehensive set of relationships between wave parameters, sediment textures and bedforms.

Detailed analysis of box cores by R. G. Davidson-Arnott and Brian Greenwood enabled them to be able to establish *Facies Relationships on a Barred Coast, Kouchibouguac Bay, New Brunswick, Canada.* This study represents one of the most comprehensive treatments of sedimentary structures from any modern sedimentary environment. The authors have been able to characterize numerous subenvironments within the nearshore zone. They have also related bedforms in this environment to the concept of flow regime.

A similar study to the above was carried out by G. W. Hill and R. E. Hunter along the barrier island coast of Texas. Their paper, entitled *Interaction of Biological and Geological Processes in the Beach and Nearshore Environments, Northern Padre Island, Texas,* concentrates on the detailed zonation of these environments utilizing both physical and biogenic sedimentary structures. Like the paper by Davidson-Arnott and Greenwood, this paper provides the geologist with the necessary detail to make explicit environmental interpretations of rocks which represent beach and nearshore environments.

In summary, this volume provides a wealth of data on the details of the generation, hydromechanics and sediment interaction of processes operating in beach and nearshore environments. The subject is treated in such a way as to be of considerable benefit to coastal engineers, sedimentologists and field geologists. The organization of the volume is such that the latter types of scientists will probably find it convenient to start their reading at the end of the book and progress toward the beginning. In this way they will be able to recognize features in the modern environment that compare with those observed in the rock record and then proceed to probe into conditions of their generation. The book was organized according to the interests of sedimentologists and coastal engineers who are primarily concerned with the conditions under which sediment is entrained, transported and incorporated into various structures. It is hoped that this volume will provide the interested reader with considerable insight into the complexities of sedimentation in the beach and nearshore environment.

ACKNOWLEDGEMENTS

This volume represents the combined talents of many people with quite diverse backgrounds and interests. The authors deserve much thanks for their cooperation and quality efforts. Special thanks is extended to J. Ian Collins for his paper which was solicited after the research symposium was held. Many of the authors were of great assistance in reading and editing manuscripts of other contributors. Robert L. Miller was particularly helpful in this capacity and his efforts are thoroughly appreciated. Thanks are also due to William L. Wood and Donn S. Gorsline for their editorial assistance. Virtually all retyping of manuscripts at various stages was done by the editor's secretary, Wanda B. Evans.

The editor is especially grateful to Ruth Tener of the SEPM headquarters office for her valuable insight and suggestions, and to Raymond Ethington, Chairman of the SEPM Publications Committee, for his cooperation and editorial assistance.

RICHARD A. DAVIS, JR.
Convenor
February 1, 1976

CONTENTS

WEATHER PATTERNS AND COASTAL PROCESSES

WILLIAM T. FOX AND RICHARD A. DAVIS, JR.
Williams College, Williamstown, Massachusetts 01267, and
University of South Florida, Tampa, 33620

ABSTRACT

Waves and longshore currents, which are responsible for deposition and erosion on sand beaches, are closely tied to storm patterns. Three coastal areas, the eastern shore of Lake Michigan, Mustang Island, Texas and the central Oregon coast, were studied to develop coastal process-response models. Time-series analysis of weather, waves and current data was coupled with frequent beach surveys to construct the models.

Thirty-day studies were completed at Holland and Stevensville, Michigan, during the summers of 1969 and 1970. Longshore current was correlated with the first derivative of barometric pressure, and breaker height was correlated with the second derivative. During storms, longshore currents were shunted offshore by beach protuberances and cut through nearshore bars. After the storms, the bars reformed and migrated toward the beach.

At Mustang Island, Texas, beach studies were undertaken during the fall and winter of 1971-1972. Weather patterns were dominated by intense, offshore-moving cold fronts or northers marked by a rapid rise in pressure, drop in temperature and sudden reversal in wind direction. Broad beaches are relatively straight with a low seaward slope and two or three nearshore bars. Saddles in the bars are not related to protuberances on the beach.

Waves and currents on the Oregon coast are controlled by the East Pacific High in the summer and the North Pacific storm track in the winter. In summer of 1973, strong north winds set up southward flowing longshore currents and frequent upwelling. The summer beaches are characterized by broad intertidal sand bars separated by longshore troughs and rip channels. During winter storms of 1973-74, sand was removed from the beach and stored offshore on subtidal bars.

The three areas differ in their response to coastal processes because of their position relative to storm tracks, orientation of the shoreline, tidal range and nearshore topography.

INTRODUCTION

Weather patterns and storm tracks play a major role in the development and destruction of coastal features. Large coastal land forms such as continental shelves, submarine canyons, deltas and barrier islands are influenced by global tectonics and eustatic changes in sea level. Smaller features including beaches, nearshore bars and rip channels are created and removed by the local waves, swells and longshore currents which are generated by surface winds and storms. Therefore, the existing coastal configuration of an area is a combination of large erosional or depositional features influenced by tectonic controls or sea level changes upon which smaller features are superimposed as a function of the local wave regime.

Several coastal areas in the United States, including the east and west coasts of Lake Michigan, Mustang Island, Texas on the Gulf of Mexico, and the central Oregon coast, were studied to develop a model for predicting the relationships between weather patterns and coastal processes. The east coast of Lake Michigan was studied from 1969 through 1974. It provided a simple model for the rapid response of beaches, and adjoining nearshore bars provide ideal conditions for developing and testing a coastal process-response model. The model developed for the eastern shore of Lake Michigan was also tested on the western shore of the lake where the weather systems move offshore.

Two time-series studies were conducted during the fall of 1971 and the winter of 1972 at Mustang Island on the central Texas coast. The fall and winter weather pattern for the Texas coast is dominated by cold fronts or northers which move seaward across the coast. A broad gently sloping beach with two or three nearshore bars characterizes the Texas coast. Tides of 1.0 to 1.5 meters spread out the effects of waves and currents over the broad, flat beaches. Reversals of wind direction and longshore currents recorded on Lake Michigan were also evident on the Texas coast. Although the cold fronts moved offshore, the general model developed for Lake Michigan accounted for the succession of waves, longshore currents and beach erosion observed on the Texas coast.

The Oregon coast has frequent winter storms which produce huge waves and also has a large tidal range which provides a marked contrast with Lake Michigan and the Texas coast. Weather and wave conditions were studied for one year, from June, 1973 through May, 1974. During the summer, the Oregon coast was dominated by the East Pacific High, which resulted in strong north winds and upwelling, with waves from 1 to 2 meters high. A few small low pressure systems which moved through the area in the summer caused short reversals in wind direction and currents along the coast. During the fall and winter, a series of large storms moved across the Oregon coast with heavy rains and high winds. Sand, which built up as a series of intertidal bars in the summer, was removed by waves and longshore currents during the winter storms. The waves during the winter storms were more than 8 meters high and caused considerable erosion along the coast. In the spring, the sand returned to the beach in the form of large broad sand bars.

The relationships between weather patterns and coastal processes were studied by time-series analysis of waves and currents in conjuction with maps of the beach and nearshore area. The time-series studies of coastal processes ranged from 15 days to one year. The beaches studied ranged from 244 to 488 meters in length and ranged in width from a few meters on Lake Michigan to several hundred meters on the Oregon coast. Although the sizes of the areas and the energy conditions encountered on the coasts varied considerably, definite similarities emerged for the relationship between weather patterns and coastal processes.

COASTAL PROCESSES

Scale of coastal processes.—Coastal features are constructed and destroyed by processes which operate at different scales in time and space. Large coastal features such as mountain ranges, deltas and continental shelves are strongly influenced by crustal plate movements and develop over millions of years. Intermediate scale features including estuaries, spits and barrier islands are more closely related to changes in sea level caused by tectonic processes or glaciation and may be formed in hundreds or thousands of years. Small scale features including beaches, nearshore bars and ridge and runnel topography are controlled by waves, longshore currents and tidal currents. The waves and currents which control beach and nearshore topography are strongly influenced by local weather patterns and major storm tracks. Small scale features are often formed in a few days or destroyed by a single storm. In unravelling the geomorphologic and sedimentologic history of a coast, it is necessary to consider the processes operating at each of the different scales with the smaller features superimposed on the larger features.

Plate tectonics and coastal processes.—The major geomorphic configuration of a coastal area is closely related to the movement of crustal plates away from spreading centers, toward subduction zones or along transform faults. Inman and Nordstrom (1971) have classified coasts into three groups; trailing edge coasts which face a spreading center, collision coasts which abut a subduction zone or transform fault, and marginal coasts which adjoin an island arc. As seafloor spreading progresses and continents move away from a spreading center, a full cycle of shoreline development occurs. Near the spreading center or ridge, the trailing edge coasts are bouyed up by the thermally elevated spreading centers and form steep, tectonically active coasts. As the spreading continues and the continental margin moves away from the spreading center, the coastline subsides and the gradient of the continental shelf decreases. At the same time, sediment eroded from the interior of the continent is dumped along the continental margin forming a broad flat continental shelf. Barrier islands frequently develop on the trailing edge coastlines where the slope is very low. At collision coasts, on the other hand, rugged mountains form where one plate is subducted beneath another, or fault blocks form where transform faults delineate the continental margin.

Sea level changes.—Eustatic and tectonic changes in sea level have an influence on the formation of coastal features such as barrier islands, cliffs, wave-cut terraces and submerged shorelines. Eustatic changes in sea level can be directly related to tectonic forces which cause the vertical movement of a portion of the coast. Changes in sea level can also be created by the horizontal movement of crustal plates. With increased spreading rates, the volume of elevated material in the mid-oceanic ridges is increased (Sclater and others, 1971) and the continental margins are flooded by a widespread marine transgression. When the spreading rate decreases, the elevation of the ridges drops and the marginal seas withdraw resulting in a margin regression. Because the volume of the ridges adjusts slowly to changes in spreading rates, the transgressions and regressions endure for hundreds of thousands or millions of years. With slow transgressions or regressions, the coastal processes are able to adjust to the changes in sea level and the products of sedimentation are spread over a wide area of the sea floor.

Valentin (1952) recognized the nonequilibrium conditions along coastlines related to changes in sea level and glaciation. When sea level is dropping or the coast building out by accretion, there is a gain of land or an advancing coast. Under weak tidal action, lagoon-barrier and dune-ridge coasts form, while under strong tidal action, tidal flat and barrier-islands coasts develop. With a rapid rise in sea level on a coast which has been excavated by glaciers or rivers, a fjord or embayed coastline forms. If the submergence takes place over a broad coastal plain, there is development of barrier islands and spits, typical of an emergent coastline. Shepard (1963) shows a rapid rise in sea level between 20,000 and 5,000 years ago, with a fairly slow rise in sea level for the past 5,000 years.

Waves and currents.—Weather patterns which influence wave and current conditions are responsible for small scale coastal features. Davies (1964) developed a world wide dynamic classification of coastal environments based on the distribution of wave types. He distinguished four major types of wave climate; first, the storm wave environment, second, the west coast swell environment, third, the east coast swell environment and fourth, the low energy environment. The coastal features formed within each environment are related to the waves impinging on the coast. The storm wave environment is generally restricted to the higher latitudes where storm tracks are concentrated and a belt of strong and variable winds generates storm waves (Fig. 1). In the northern hemisphere, the storm belt shifts from 46°N in the winter to 62°N in the summer with the occurrence of larger storms more frequent in the winter. On the North American continent, the zone of storm wave activity extends northward from California on the west coast and from New York and New England on the East coast. Because the storms approach from the west in the mid-latitudes, the wave activity on the west coast is generally much greater than on the east coast. In Europe, the storm wave zone extends from northern France to northern Norway and in Asia, from Japan northward. In the southern hemisphere, the storm belt varies from 54°S in the winter to 56°S in the summer. Because only the tip of South America extends into the storm belt, storm wave coasts are much more restricted in the southern hemisphere.

In addition to waves, longshore currents are important in the formation of small scale coastal features. Where storm paths move directly across a coast, the reversal of wind direction during the passage of a storm will result in a change in angle of wave approach and a shift in longshore current direction. If the storm track is roughly parallel to the coast, the winds may shift from onshore to offshore, but the dominant direction of longshore transport remains the same. A constant angle of wave approach or longshore current direction causes the enhancement of longshore coastal features such as spits, bars and rhythmic topography. Where storms are less frequent and long period swells dominate the wave climate, the waves are refracted so that they approach the coast at a smaller angle and generate slow longshore currents. Although long period swells do not generate strong longshore currents, they

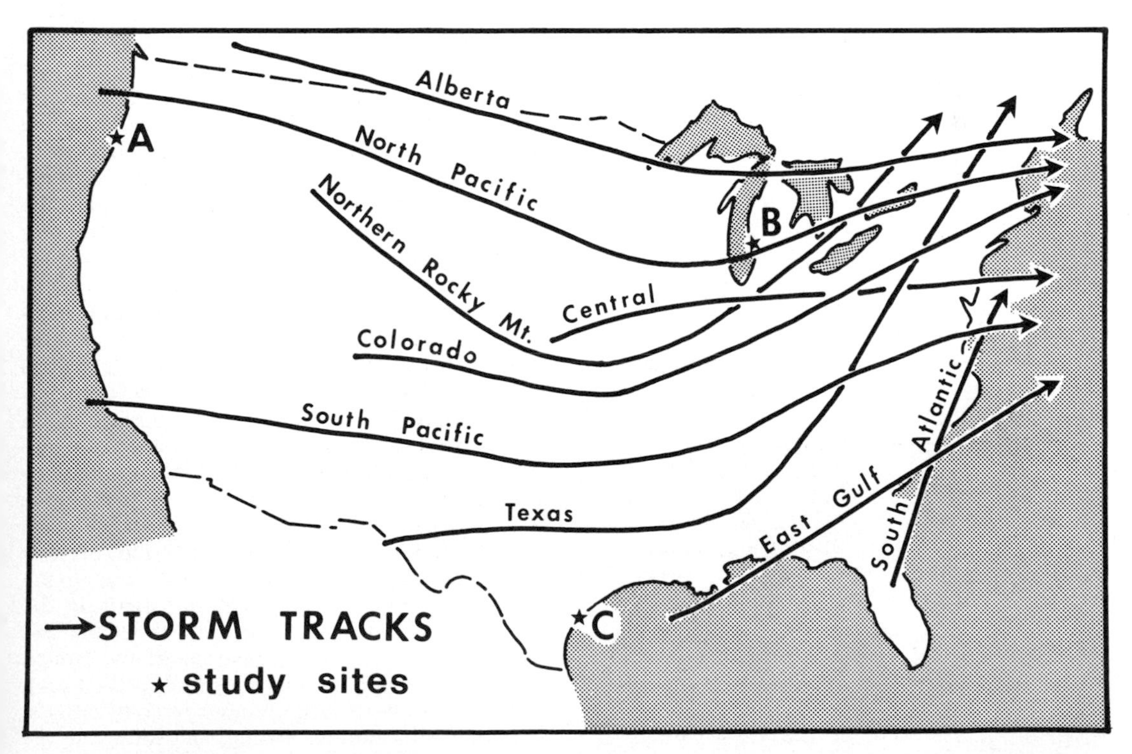

Fig. 1.—Typical paths of cyclones appearing in various regions of the United States and location of study sites at *A*, central Oregon coast; *B*, eastern shore of Lake Michigan; and *C*, Mustang Island, Texas.

often set up standing waves or edge waves which may form cusps or rhymthic topography along the coast.

LAKE MICHIGAN

Lake Michigan study area.—The relationship between weather patterns and coastal processes has been studied at several locations on Lake Michigan by the authors (Fig. 2). In July, 1968, a two-week pilot study was completed at Stevensville, Michigan, to test field procedures and make preliminary environmental observations. A 30-day field study was conducted at Stevensville, Michigan, during July and August, 1969, with weather and wave observations taken at 2 hour intervals (Fox and Davis, 1970a, 1970b). Daily topographic profiles were surveyed across the beach and nearshore bars to measure erosion and deposition under varying wave conditions. In July, 1970, a similar 30-day study was conducted at Holland, Michigan, about 95 kilometers north of Stevensville (Fox and Davis, 1971a; Davis and Fox 1971, 1972b). During the summer of 1972, a field study was completed at Sheboygan, Wisconsin, on the western shore of Lake Michigan (Fox and Davis, 1973a). A pair of simultaneous field studies were conducted at Zion, Illinois, and South Haven, Michigan, during July, 1974 to determine the effects of a single storm on the eastern and western shores of the lake (Davis and Fox, 1974).

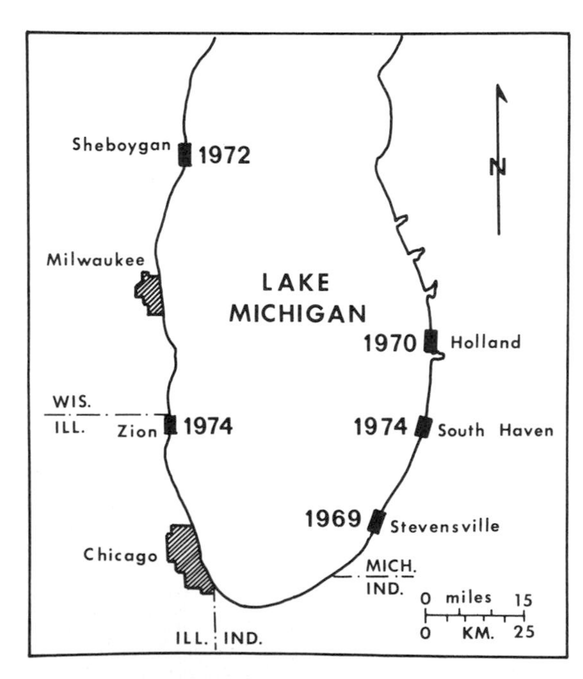

FIG. 2.—Location map showing Lake Michigan Study site for Stevensville, Michigan (1969), Holland, Michigan (1970), Sheboygan, Wisconsin (1972), Zion, Illinois (1974) and South Haven, Michigan (1974).

Lake Michigan lies across and to the south of the major North Pacific and Alberta storm tracks (Fig. 1). Summer storms which originate in the Pacific northwest proceed from west to east across the northern United States and southern Canada and generally pass over or somewhat to the north of Lake Michigan. During a typical summer, a storm moves through the area every 7 to 10 days. The general storm pattern during the summer is quite regular with a predictable shift in wind direction and angle of wave approach as the storms move across the Great Lakes region.

Stevensville, Michigan.—The typical wind, wave, and current pattern generated by a storm on the eastern shore of Lake Michigan can be shown by studying the effects of a fairly large storm which passed north of Stevensville, Michigan, in July, 1969 (Fox and Davis, 1970b). On July 26, 1969, a large circular low pressure system moved across the northern United States and the center of the low passed over Lake Michigan (Fig. 3). The surface winds circulated around the low in a counterclockwise direction and produced large waves on the lake. As the low pressure system moved across to the north of the study area, waves approached the shore out of the southwest. After the passage of the low, the wind shifted around to the northwest and large waves built up from that direction. The waves moved across Lake Michigan and resulted in extensive erosion and coastal damage along the southeastern shore of the lake. In many places the beaches were eroded back to the base of the dunes; at some places the dunes were cut back from 6 to 10 meters.

When the low pressure system passed over the lake, a 30 day time-series study was in progress at Stevensville, Michigan. Observations on 17 variables were collected at 2 hour intervals to obtain a detailed plot of weather and coastal processes during the passage of the storm. Weather related variables include barometric pressure, wind speed and direction, air temperature and cloud cover. Wave measurements were made offshore on a staff 180 meters from the beach in a depth of 5.5 meters, in the surf zone at a depth of one meter, and in the plunge zone at a depth of 0.4 meters. Wave measurements include offshore wave height and period; breaker height, period, angle, distance and type in the surf zone; and breaker height and depth in the plunge zone. The direction and speed of the longshore current was measured in the trough between the sand bar and the shore. Daily beach profiles were used to determine rates of erosion and deposition before, during, and following the storm.

Weather and wave measurements taken at 2-

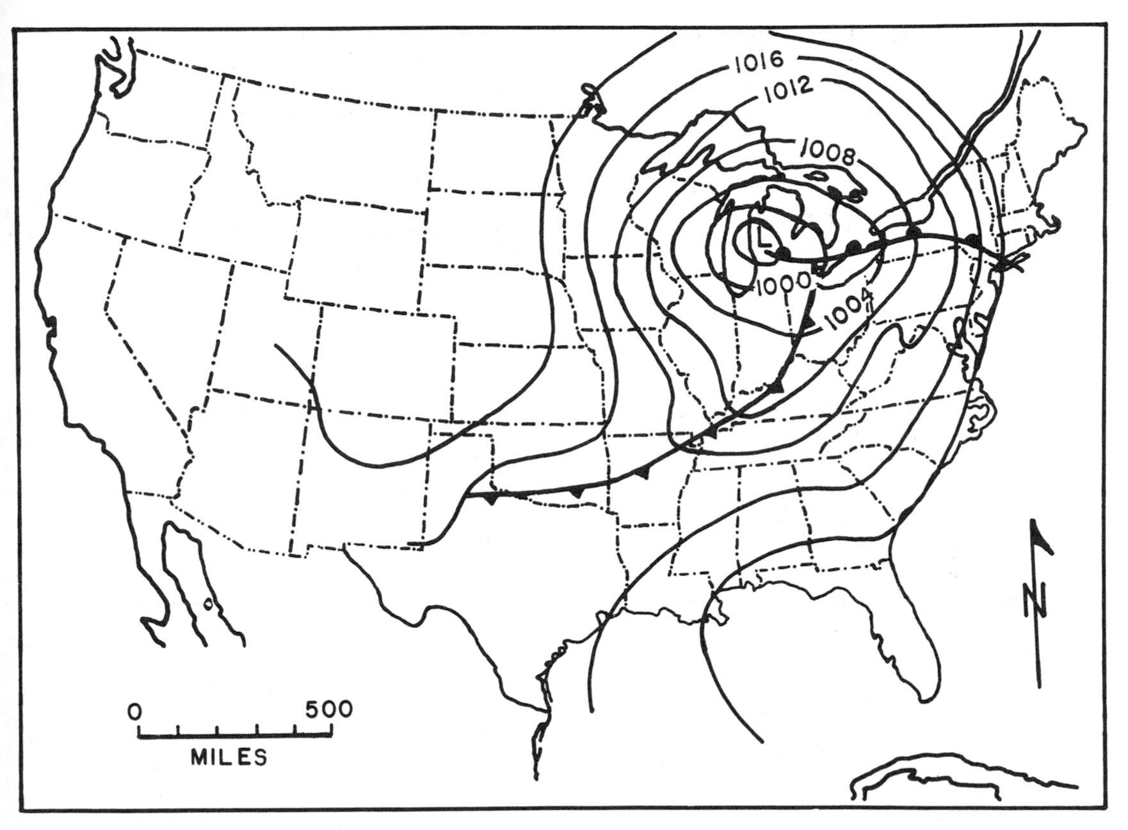

FIG. 3.—Map showing position of intense low pressure area over Lake Michigan on July 28, 1969.

hour intervals provide a time series with 360 observations which can be smoothed using Fourier analysis. The period, phase and amplitude of the first 15 Fourier components were computed for each of the variables. In order to study the effects of the passage of the storm, the cumulative curves for the first 15 harmonics for barometric pressure, longshore wind, longshore current and breaker height were plotted for a 15-day period from 8 a.m. on July 21 through 6 a.m. on August 4, 1969 (Fig. 4). The curve for barometric pressure shows a sharp drop when the low passed through the area reaching a minimum of 1000.4 millibars (29.54 inches) on July 27. Following the storm, the barometric pressure rose to 1017.3 millibars (30.04 inches) on July 30 and 1020.0 millibars (30.12 inches) on August 3. There is a sharp rise in breaker height with the drop in barometric pressure, reaching 1.8 meters on July 28. The peak in the breaker height curve is displaced somewhat to the right (later) relative to the low point in barometric pressure.

As the storm passed over Stevensville, a cold front moved in on July 28 with a shift in wind direction from the southwest to the northwest and an increase in wind velocity from 9.5 to 12 m/sec. The southwest winds preceding the storm produced waves 0.5 meters high from the southwest forming an angle of 21 degrees with the nearshore bar and generating a northward flowing longshore current with a speed of 120 cm/sec. Following the passage of the cold front, the wind shifted to the northwest and the wave height increased to 1.8 meters. The northwest waves formed an angle of 23 degrees as they crossed the bar and generated a longshore current of 210 cm/sec to the south. The highest winds, waves and longshore current occurred after the low pressure system had passed and the wind had shifted over to the northwest. Strong northwest winds following the passage of the cold front accounted for the high waves and strong longshore current to the south.

On the eastern shore of Lake Michigan where the shoreline is almost perpendicular to the storm path, breaker height and longshore current velocity can be roughly predicted directly from the curve for barometric pressure (Fox and Davis, 1971b, 1973a). The curve for longshore current closely resembles the curve for a constant multiplied by the first derivative for barometric pressure. Since the cosine is the derivative of a sine, and the

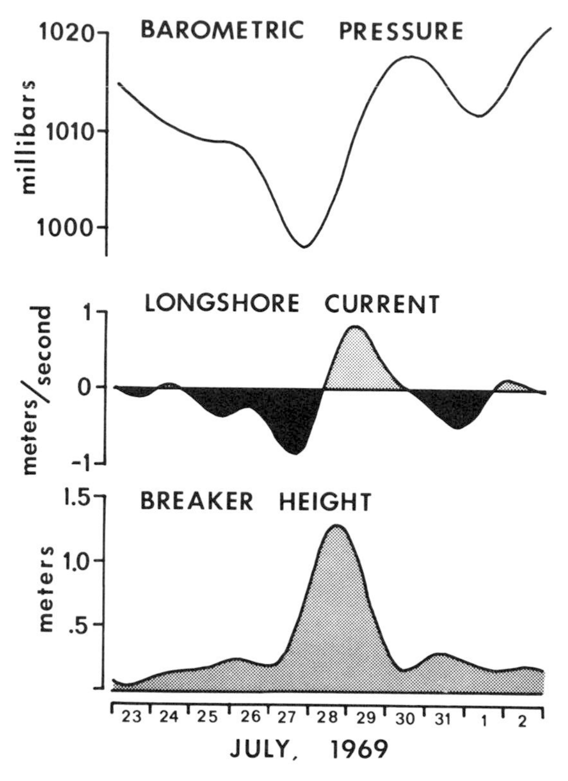

FIG. 4.—Smoothed curves for barometric pressure, longshore current and breaker height at Stevensville, Michigan, July, 1969.

Fourier curves are formed from a summation of a series of sine curves, it is possible to take the derivative of barometric pressure and generate a curve for longshore current. The curve for breaker height resembles the second derivative of barometric pressure. If a time lag of 3 to 6 hours is incorporated in the equations for predicting longshore current and breaker height, it is possible to reproduce the curves for longshore current and breaker height with a fair degree of accuracy.

Holland, Michigan.—During July, 1970, a series of summer storms passed over Lake Michigan at approximately one week intervals. In that month a 30-day time series study was in progress at Holland, Michigan (Fox and Davis, 1971a). Holland is located on the eastern shore of Lake Michigan about 95 kilometers north of Stevensville and 150 kilometers northeast of Chicago (Fig. 2). The beach at the Holland study site trends in a north-south direction and is bounded on its landward side by a low cliff cut into grass-covered sand dunes.

Observations were made on weather and wave conditions at Holland, Michigan for 2-hour intervals from 8 a.m. on June 29 through 6 a.m. on July 29, 1970. During the 30-day study, the barometric pressure varied from a low of 1002.7 millibars (29.61 inches) to a high of 1023.4 millibars (30.22 inches) with a mean of 1011.8 millibars (29.88 inches). Four distinct storms, indicated by low points in the barometric pressure curve, occurred on July 4, 9, 15 and 19 (Fig. 5).

The longshore component of the wind reversed from south to north as each low pressure system passed over the study area. Because the storm paths typically are located to the north of the study area and the surface winds circulate in a counterclockwise direction around the center of the low, the winds which preceed each storm are from the southwest.

For the first storm which moved through on July 4th, the northern component of the longshore wind following the reversal was much stronger than the southern component. The reversal in wind direction for July 4 was accompanied by a rapid rise in barometric pressure. For the second storm

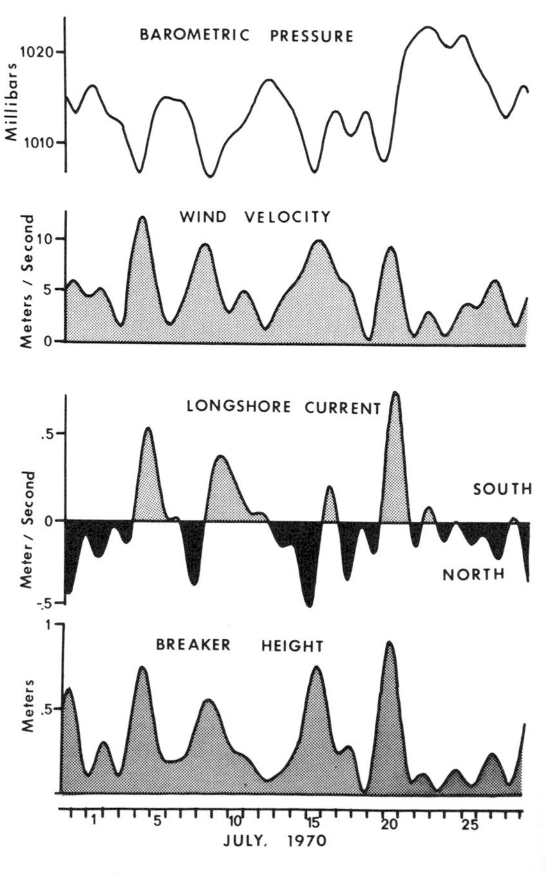

FIG. 5.—Smoothed curves for barometric pressure, wind velocity, longshore current and breaker height at Holland, Michigan, July, 1970.

on July 8, the north and south components of the longshore wind are approximately equal, and the depression in the barometric pressure for July 8 is nearly symmetrical. The curve for the third storm which occurred on July 15 shows a stronger southern wind component followed by a weak north wind (Fig. 5). The barometric pressure curve for July 15 shows a long gradual drop in pressure followed by a rapid rise in pressure. The fourth storm which moved through on July 19 had a short drop in pressure followed by a large increase as a strong cold front and high pressure system moved into the area. The high pressure system was accompanied by strong northwest winds on July 19 and 20. The different patterns displayed by the curves for barometric pressure and longshore winds are determined by the locations, sizes and paths of the individual storms and cold fronts. When a small intense storm moves directly across the study area, the barometric pressure curve is steep and the curve for longshore wind is symmetrical. When a broad, less intense storm system moves diagonally across the shore, an asymmetrical curve is produced for the longshore wind and current.

The peaks and valleys in the curve for longshore current velocity closely match the highs and lows for longshore wind (Fig. 5). The longshore current measured in the trough between the nearshore bar and the shore is roughly parallel to the shore. On the curve for longshore current velocity, the positive longshore current flows to the south and corresponds to the north and northwest wind, whereas the negative longshore current is flowing to the north and related to the southwest winds. Longshore current velocity is a function of breaker angle, wave period, breaker height and bottom topography. Because breaker angle is controlled by wind direction and wave refraction in the nearshore zone, one would expect a close correspondence between the longshore component of the wind and longshore current velocity. There is a close alignment between the smoothed curves for longshore wind and longshore current with minor differences in the amplitude of the peaks. The longshore wind reached a maximum of 15 meters/second on July 4, while the maximum longshore current of 115 centimeters/second was recorded on July 20 (Fig. 5). Although the longshore component of the wind was lower on July 20 than on July 4, the onshore component, which influences breaker height, was greater.

The curve for breaker height has four peaks which occur shortly after the troughs in the barometric pressure curve (Fig. 5). The highest breakers were recorded on July 4 and 20, with lower breakers associated with the storms on July 9 and 15. Lake Michigan waves are generated by local winds with a maximum fetch of about 150 kilometers. Therefore, there is a very rapid response of the waves to increase in wind speed and changes in wind direction. As the low pressure center approaches, the wind picks up out of the southwest generating waves from the same direction. As the waves break on the coast, a northward flowing longshore current is produced. After the storm passes, there is a rapid shift in wind direction which is followed a few hours later by a change in angle of wave approach and a reversal in longshore current direction. On Lake Michigan, the time lag in this shift is usually 3 to 6 hours after the change in wind direction.

Conceptual model for Lake Michigan.—A conceptual model was formulated to study the relationship between storm cycles and beach erosion on Lake Michigan (Davis and Fox, 1972b; Fox and Davis, 1973b). The model is based on field studies conducted at Stevensville and Holland, Michigan, during the summers of 1969 and 1970. Generalized curves are plotted for barometric pressure, wind speed, longshore current and breaker height (Fig. 6). In the model, two low

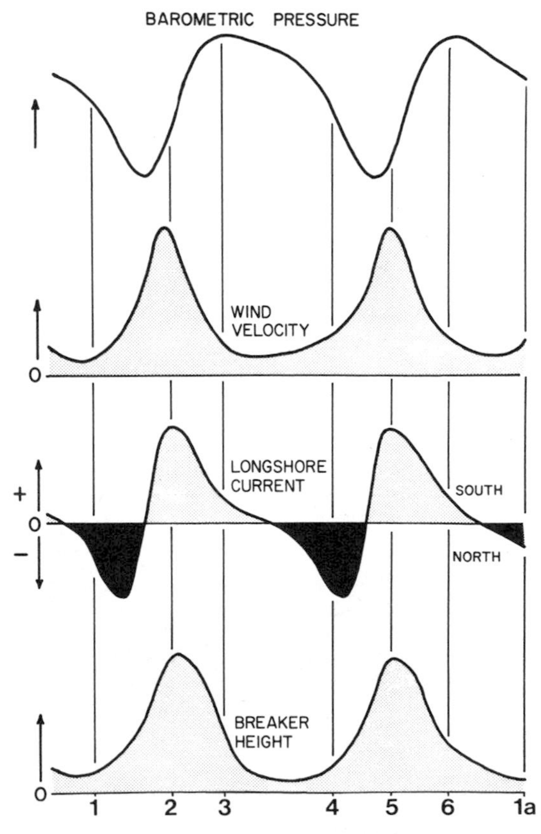

Fig. 6.—Conceptual model for barometric pressure, wind velocity, longshore current and breaker height for 2 storm cycles on Lake Michigan. Numbered vertical lines refer to stages in the model shown in Figure 7.

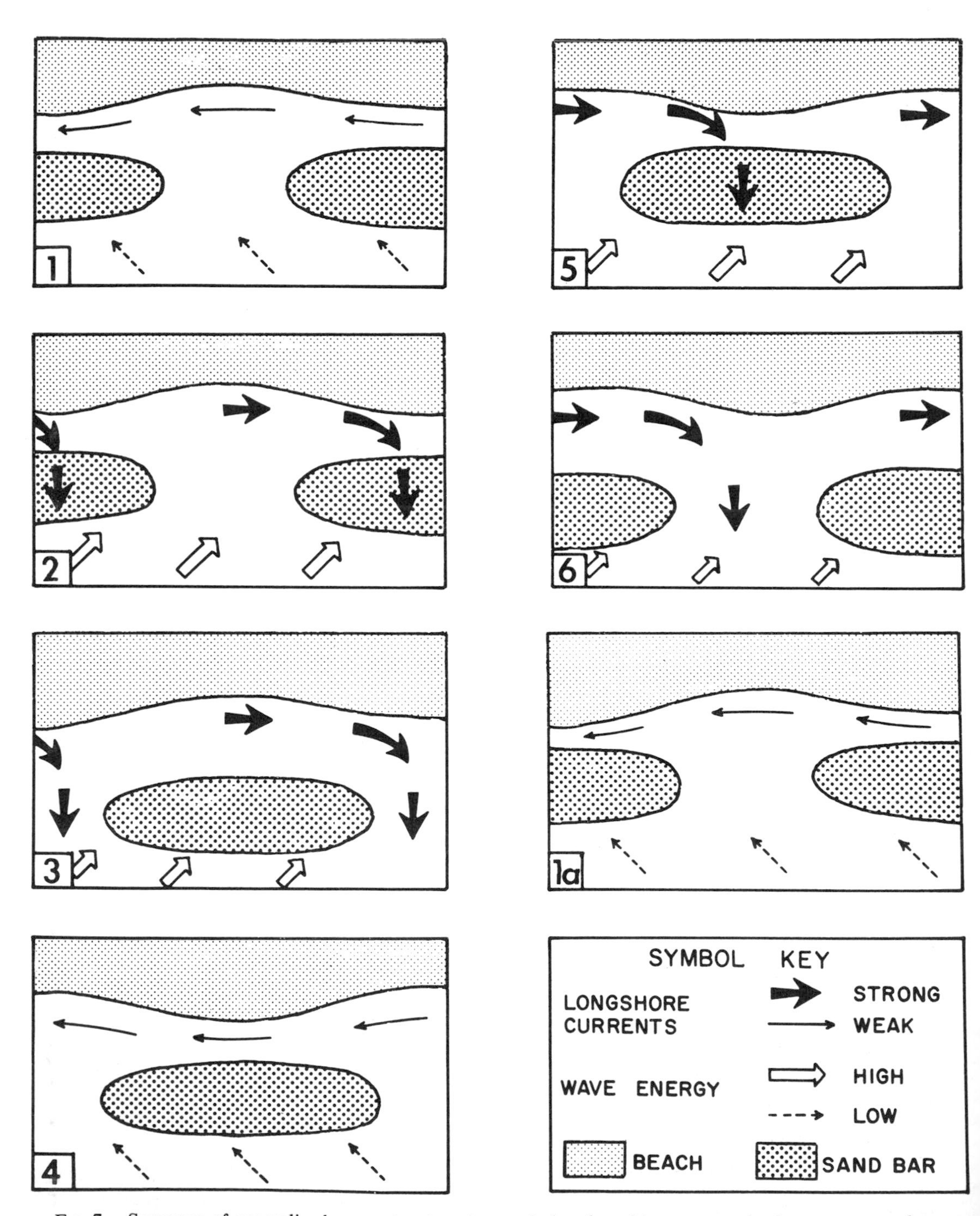

FIG. 7.—Sequence of generalized maps showing changes in beach and bar topography in response to changes in magnitude and direction of the waves and currents on Lake Michigan.

pressure systems pass through the area forming two storm cycles which are separated by a post-storm recovery period.

Six stages in the model (Fig. 6) are numbered to correspond to maps of the beach and nearshore area (Fig. 7). The first storm cycle is marked by a sharp drop followed by a rise in barometric pressure between stages 1 and 3. The interval between 3 and 4 represents the post-storm recovery. The second storm cycle occurs between stages 4 and 6 with a post-storm recovery between stages 6 and 1a. In the model, stage 1 represents falling barometric pressure, low wind speed, weak longshore current to the north and low breaker height as a storm approaches. Between stages 1 and 2, the low passes over and the longshore wind and current reverse direction. During stage 2 the cold front or high pressure ridge moves in. As the barometric pressure rises, wind speed, breaker height and longshore current reach a maximum. At stage 3, barometric pressure is at its highest while wind speed, longshore current and breaker height die down. In the post-storm recovery between stages 3 and 4, the wind slowly shifts over to the southwest, generating low waves and weak currents to the north.

Seven generalized maps show changes in beach and bar topography as the storms pass over the coastline (Fig. 7). The initial shoreline has a rhythmic pattern with two nearshore bars separated by a saddle or rip channel (Evans, 1940). As the low waves approach from the southwest, a weak longshore current flows to the north. At stage 2, high waves build up out of the northwest with strong southward flowing longshore currents between the bars and the shore. The longshore currents are shunted offshore by the protuberance on the beach and cut a channel across the nearshore bar. At stage 3, a new bar is formed in the old saddle and currents cut across the position of the earlier bar. During the post-storm recovery, between stages 3 and 4, small waves move across the bar which advances toward the beach (Olson, 1958). At the same time, a protuberance builds out behind the bar and sets the stage for the next storm cycle. In stages 5 and 6, the storm cycle is repeated again and the bar returns to its original position (Fig.7).

In the conceptual model, the bar oscillates back and forth during storms exchanging positions with the rip channels or saddles between bars. The post-storm recovery with the shoreward migration of the bars and the growth of the protuberances behind the bars, plays an important role in the model. During the post-storm recovery, erosion takes place on the beach at the head of the rip channels. The sand is transported along the beach by the weak longshore currents forming a protuberance behind the bar. During the next storm cycle, the longshore current is directed offshore by the protuberance and cuts across the nearshore bar forming a rip current. A new bar is formed between the pairs of rip currents near the center of the nearshore circulation cell (Fig. 7). If two storm cycles are closely spaced in time, the protuberance will not have an opportunity to form and the bars will not shift position. On the other hand, if the time interval between storm cycles is long enough, the bar will migrate onto the shore and become welded to the beach (Davis and others, 1972). In a tidal environment, the bar migration would correspond to the ridge and runnel formation described by Hayes and others (1969).

MUSTANG ISLAND, TEXAS

Texas study area.—Two time-series studies were conducted on the Texas coast during October and November, 1971, and January and February, 1972 (Davis and Fox, 1972a). The study area is located on Mustang Island near the middle of the extensive Texas barrier island system (Fig. 8). The barrier islands are approximately 2 to 3 kilometers wide with the major shoreward portion covered by vegetated barrier flats and wind tidal flats. A narrow foredune complex and a gently sloping beach extend along the seaward margin of the islands. The dunes are generally 5 to 8 meters high, but may reach 15 meters locally. In front of the foredunes, the backbeach slopes toward the sea slightly and often is covered with small wind-shadow dunes.

During the fall and winter of 1971-72 when the studies were being made, a prominent berm was not present but there was a noticeable slope difference between the foreshore and backshore. The berm which is normally present on the beach was destroyed by tropical storm Fern in September, 1971, and was not reconstructed because of the relatively high wave conditions during the fall and early winter of 1971-72 (Davis, 1972). The foreshore zone is typically covered by low antidunes generated by swash runoff on the low angle foreshore. During most of the study, three bars were present in the nearshore zone spaced at about 80, 170, and 260 meters from the mean high tide line. Rip currents frequently cut across low saddles in the bars. Although the bars shifted slightly with varying wave conditions, they did not migrate onto the beach and remained relatively stable in position.

The beach study site was located about 5.6 kilometers southeast of the jetty at Aransas Pass. The study site extends approximately 460 meters (1500 feet) southeast of Access Road #1.

Field observations.—The fall time-series study was conducted for 30 days from 8 a.m. on October 18, 1971, through 8 p.m. on November 16. The

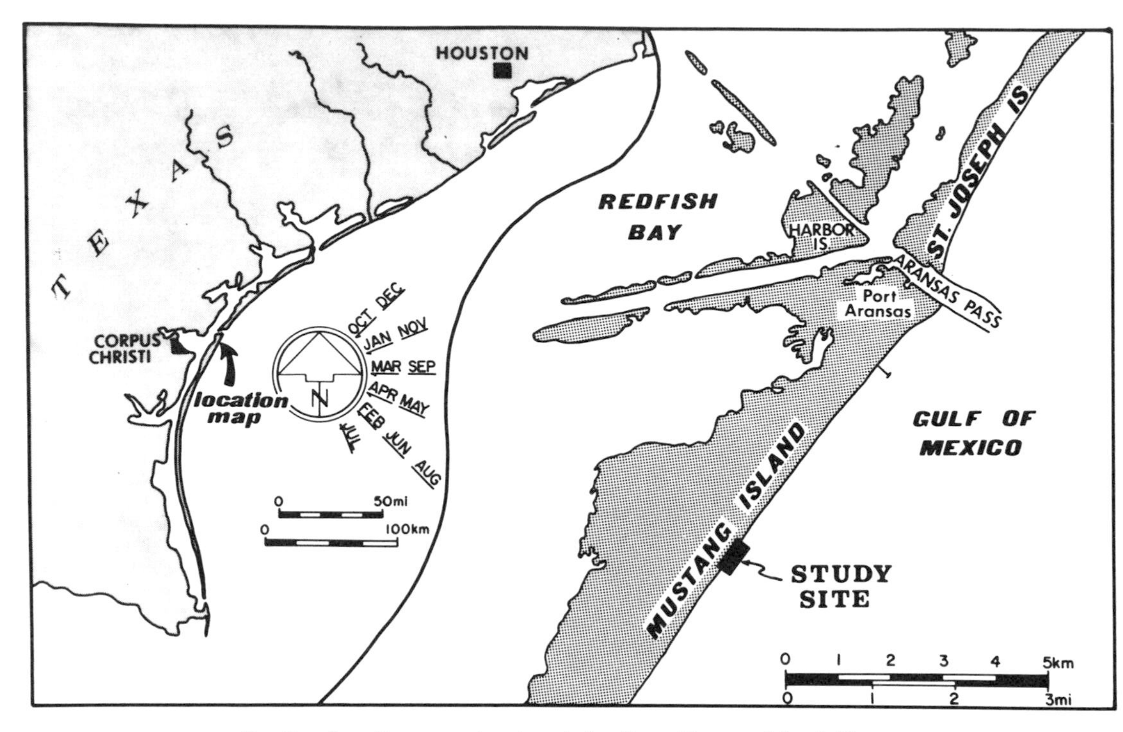

FIG. 8.—Location map showing study site at Mustang Island, Texas.

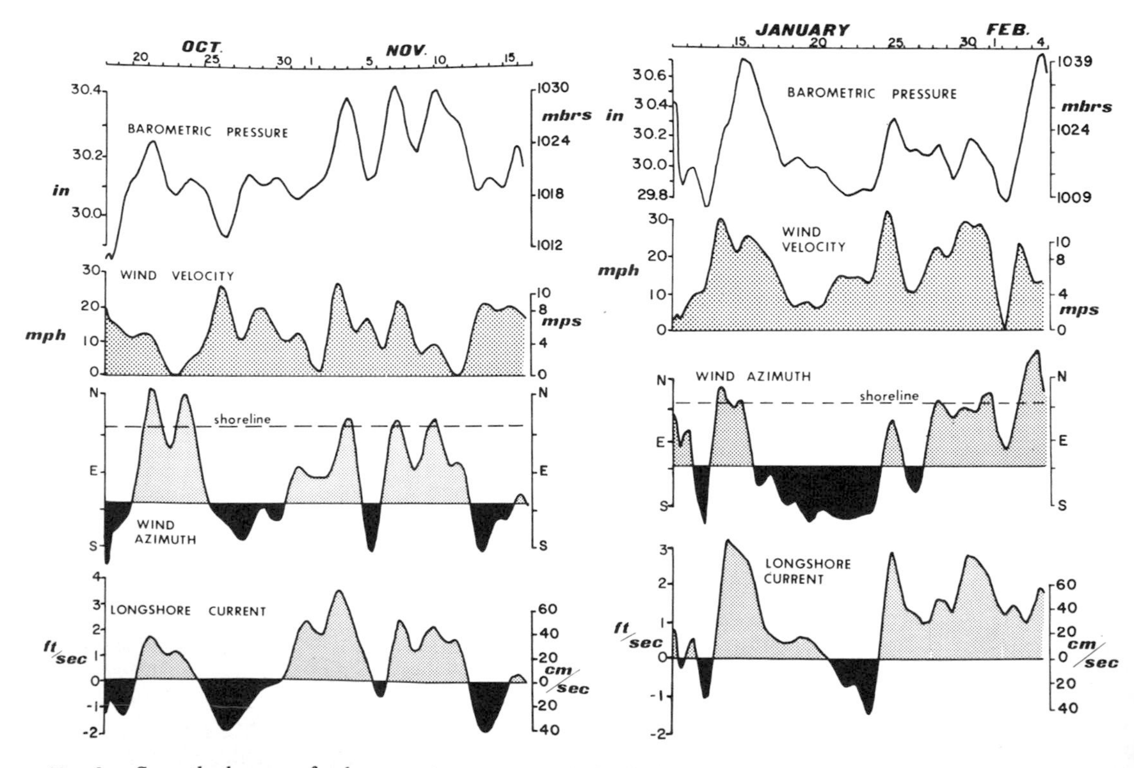

FIG. 9.—Smoothed curves for barometric pressure, wind velocity, wind azimuth and longshore current during fall 1971 and winter 1972 at Mustang Island, Texas.

winter study which lasted from 8 a.m. on January 11, 1972, through 8 p.m. on February 4, was terminated after 25 days because of cold weather and high winds. Wave period, offshore wave height, breaker height, distance, angle and type, and longshore current velocity were measured in the surf at the beach study site or from wave staffs mounted on the Horace Caldwell Pier about 3 kilometers northeast of the beach study site. The wave and current data were taken at 8 a.m., 2 p.m. and 8 p.m. each day. Barometric pressure, tides, wind speed and direction, and water temperature were recorded continuously at the University of Texas Marine Science Institute at Port Aransas, Texas. Readings were extracted at 2-hour intervals for the time-series study in order that they could be compared directly with the wave and current data.

During each of the field studies on the Texas coast, several cold fronts passed over the study area. During the fall of 1971, the range in barometric pressure was 1010.5 to 1031.8 millibars with a mean of 1019.0, while in the winter of 1972, the range was 1007.8 to 1039.6 millibars with a mean of 1019.3 (Fig. 9). The greater range in barometric pressure reflects the more intense cold fronts which passed through the area during the winter.

Wind data were recorded on a two channel recorder, with the directional vane and anemometer mounted approximately 15 meters above sea level at the Marine Science Institute pier. A rapid increase in wind velocity and sudden shift in wind direction are associated with the passage of cold fronts. During the fall, the range in wind velocity was from 0 to 16 m/sec with a mean of 7.5, while in the winter, the range was 0 to 14 m/sec with a mean of 8.2. In both the fall and winter, the wind velocity increased from zero to the maximum in less than one hour when a norther passed over the coast. In the fall, high wind velocities occurred on October 26 and November 3 and 7. On October 26, a storm moved onshore from the Gulf of Mexico and hence was not a typical norther and did not display a rapid rise in barometric pressure and reversal in wind direction (Fig. 9). The events on November 3 and 7 are typical of northers with a sudden increase in wind speed and reversal in wind direction as the front passed through the study area.

Three distinct northers passed over the Texas coast during January and early February of 1972 (Fig 9). During the first norther on January 13, the wind increased from 4 to 14 m/sec and reversed direction from south to north. For the second norther on January 25, the wind increased from 7 to 17 m/sec with a similar reversal in wind direction. In the third norther of the winter on February 3, the wind increased from 2.3 to 15 m/sec, but shifted more to the northwest instead of northeast. The cold front extended in a northwest-southeast direction, so that the south to north reversal in longshore wind was not as prominent and the offshore component of the wind was larger. The strong offshore wind subdued the incoming waves resulting in a decrease in breaker height, wave period and the longshore current velocity. The tides were also almost 0.3 meters lower than predicted because water was forced out of the estuary by the offshore wind.

In the fall study, wave height reached a maximum of 1.2 meters on October 26, and in winter, the wave height reached about 1.3 meters on January 17. The mean breaker height in the winter was the same as in the fall, both were about 0.65 meters. The maximum breaker heights occurred shortly after the maximum wind velocity with a slight time lag. Because the strongest winds were blowing along the shore or slightly offshore, a long fetch was not available to build up large waves.

Longshore current velocity was measured just shoreward of the inner bar using a whiffle ball attached to a 15 meter (50 foot) nylon line. The morning and evening measurements (8 a.m. and 8 p.m.) were made at the Caldwell fishing pier, and the mid-afternoon readings (2 p.m.) were taken at the beach study site. During the fall, the longshore current velocity ranged from −90 cm/sec to the northeast to 120 cm/sec to the southwest (Fig. 9). During the winter, the velocity varied between −50 and 110 cm/sec. The mean current velocities for the fall and winter of 1971-72 were 12 and 24 cm/sec respectively. The lows in the barometric pressure curves are aligned with negative longshore currents and high barometric pressure corresponds to positive longshore currents (Fig. 9). The curve for longshore current during the winter shows a greater asymmetry with positive currents from the northeast being both dominant and prevailing.

The cold front or norther which moves across the Texas coast in an offshore direction creates a pattern of coastal processes which is quite similar to that observed on Lake Michigan. In both areas, the center of the low pressure cell moves from west to east, but passes several hundred kilometers north of the study sites. On Lake Michigan, the counterclockwise circulation around the center of the low pressure cell in conjuction with weak cold fronts produces a shift from southwest to northwest winds as the storm passes. Along the Texas coast, intense cold fronts extending as a southwest line from the low pressure center, exert a strong influence on the coastal processes. In Texas, the shift from southwest to northeast winds is much more abrupt as the cold front passes. Therefore, the curves for barometric pressure, wind velocity and longshore current are more asymmetrical on the Texas coast where strong cold fronts dominate the fall and winter weather pattern and coastal processes.

Conceptual model for Texas coast.—A conceptual model showing the environmental variables on the Texas coast (Fig. 10) was plotted to contrast with the conceptual model developed for Lake Michigan (Figs. 6, and 7). As a cold front or norther approaches the Texas coast from the northwest, barometric pressure drops steadily and the wind velocity is moderate at 8 to 10 m/sec. The waves are generally low, ranging between 0.5 to 0.8 meters, and longshore currents are slow, between 35 to 50 cm/sec to the southwest. Immediately after the passage of the cold front, there

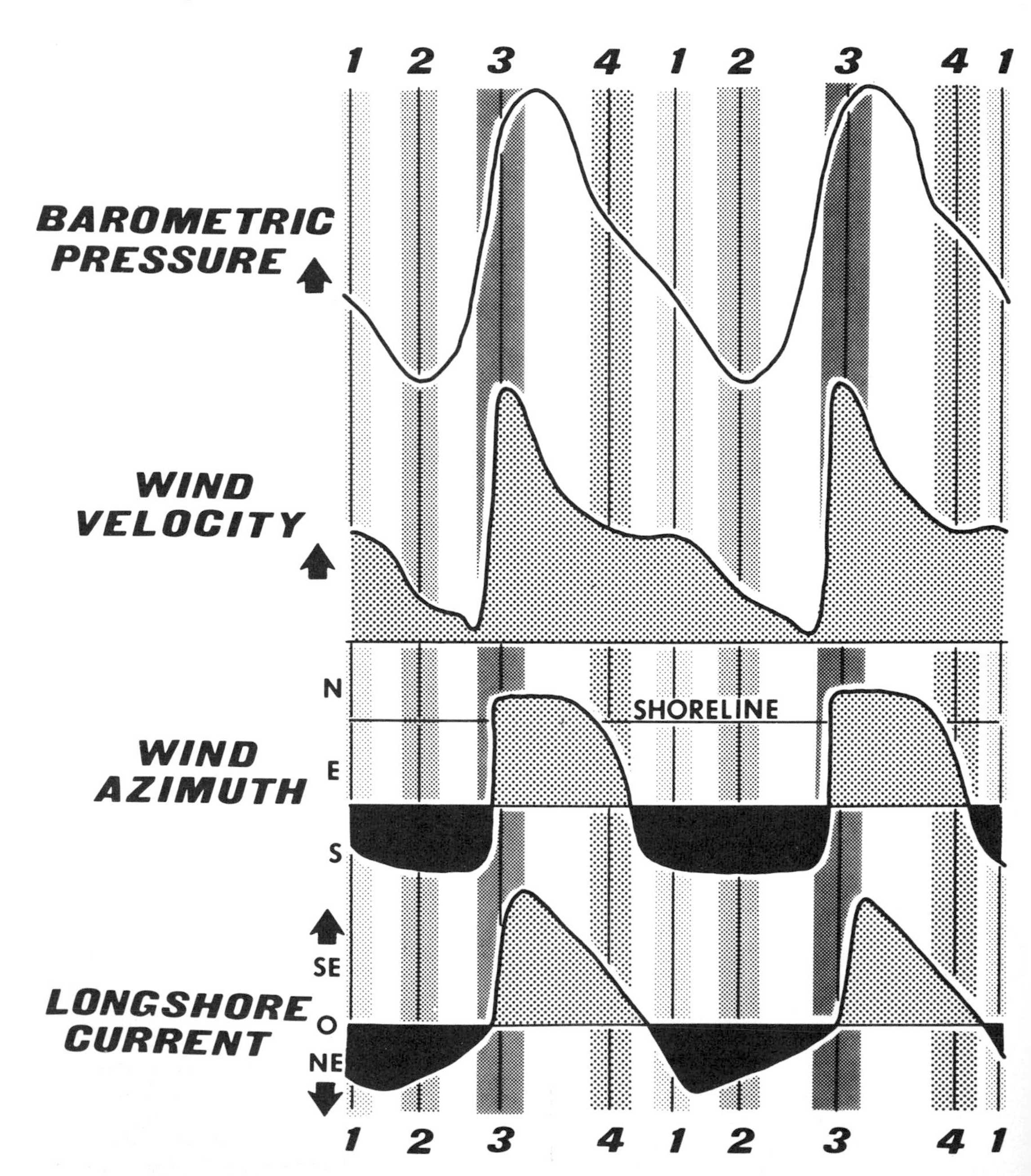

FIG. 10.—Conceptual model for barometric pressure, wind velocity, wind azimuth and longshore current on the Texas coast.

is an abrupt rise in barometric pressure and a sudden drop in air temperature. Wind direction changes to the northeast and the wind velocity increases to about 15 m/sec. The waves approach the shore at an angle open to the southwest and generate longshore currents of 60 cm/sec to the southwest. After the front passes seaward of the coast, the barometric pressure falls, wind shifts toward the southeast and slackens, and the longshore current decreases.

Because the strongest components of the wind are blowing along the shore or slightly offshore, the wave height on the Texas coast does not build up as rapidly as on Lake Michigan. The curves depicted on the model for the Texas coast (Fig. 10) are more asymmetrical than those for Lake Michigan because of the rapid change in wind direction with the passage of cold fronts. The wind speed curve on the Texas model shows an extended hump behind the initial rise as the front passes. The abrupt changes in wind speed, wind direction and longshore current occur somewhat later with respect to barometric pressure on the Texas coast as compared with Lake Michigan (Figs. 6, 10). In general configuration, the models for Lake Michigan and Texas are quite similar, with differences in symmetry for longshore wind and current and a significant difference in wave height with offshore versus onshore wind.

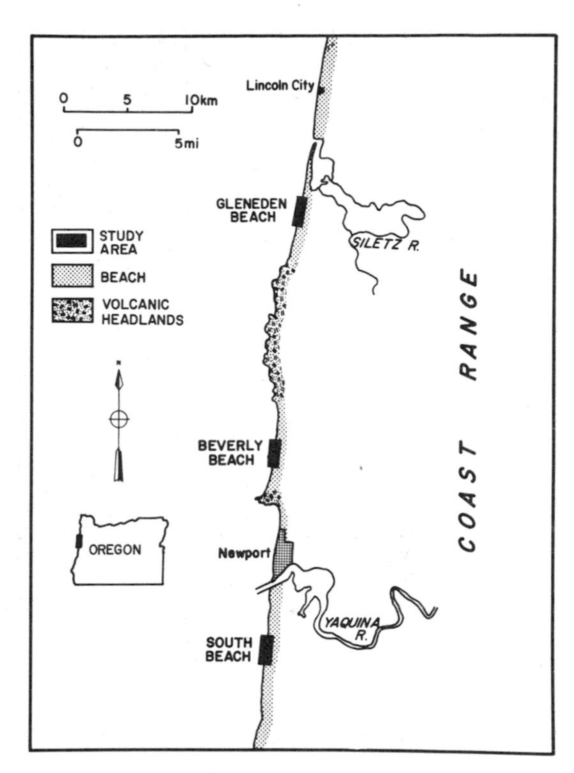

FIG. 11.—Location map of 1973–74 study sites at Gleneden Beach, Beverly Beach and South Beach, Oregon.

OREGON COAST

Oregon study site.—During 1973 and 1974, a summer study of 45 days and a full year study were conducted on the Oregon coast in the vicinity of Newport, Oregon. In the summer of 1973, a 45-day study extended from July 1 through August 14, when wave conditions were low and longshore currents were relatively slow (Fox and Davis, 1974). Wave and current conditions were measured three times each day in the surf zone. Hourly readings for barometric pressure, wind speed and direction, air temperature, humidity and tide level were taken from the continuously recording instruments at the U. S. Weather Station located in the Oregon State University Marine Science Center at Newport, Oregon. Significant wave height and period were interpreted four times each day from microseismograph records which were correlated directly with pressure sensor observations (Longuet-Higgins, 1950). For the full year study, the continuously recording weather instruments and microseismograph were used exclusively to record the weather and wave data (Creech, 1973). Direct surf observations of waves and currents were not attempted in the winter because of the inherent danger and possible loss of men and equipment in the heavy surf.

During the summer months, three beach areas were surveyed on a rotating basis with one area mapped each day. The primary study site at South Beach, Oregon, was located about 3 kilometers south of the jetty at the mouth of Yaquina Bay in Newport, Oregon (Fig. 11). The study area extended 488 meters along the coast and about 300 meters seaward from the base of the cliff. Additional beaches were also monitored at the south end of Beverly Beach State Park about 12 kilometers north of Newport, and at Gleneden Beach about 35 kilometers north of Newport (Snavely and MacLeod, 1971). The study site at South Beach was mapped once every two weeks at spring low tide for the full year study. The Oregon coast has a mixed, diurnal-semidiurnal tide with a spring tide range of 4.0 meters and a neap tide range of 2.0 meters (Cox, 1973).

During the summer months, the weather pattern on the Oregon coast is dominated by the East Pacific High (Frye, Pond and Elliot, 1972). The high, which is frequently situated a few hundred kilometers off the Oregon or Washington coast, influences the flow of wind and weather patterns along the coast (Fig. 12). In the fall, the high is displaced to the south and the North Pacific storm track moves across the Oregon coast (Cramer, 1973).

The smoothed curve for barometric pressure

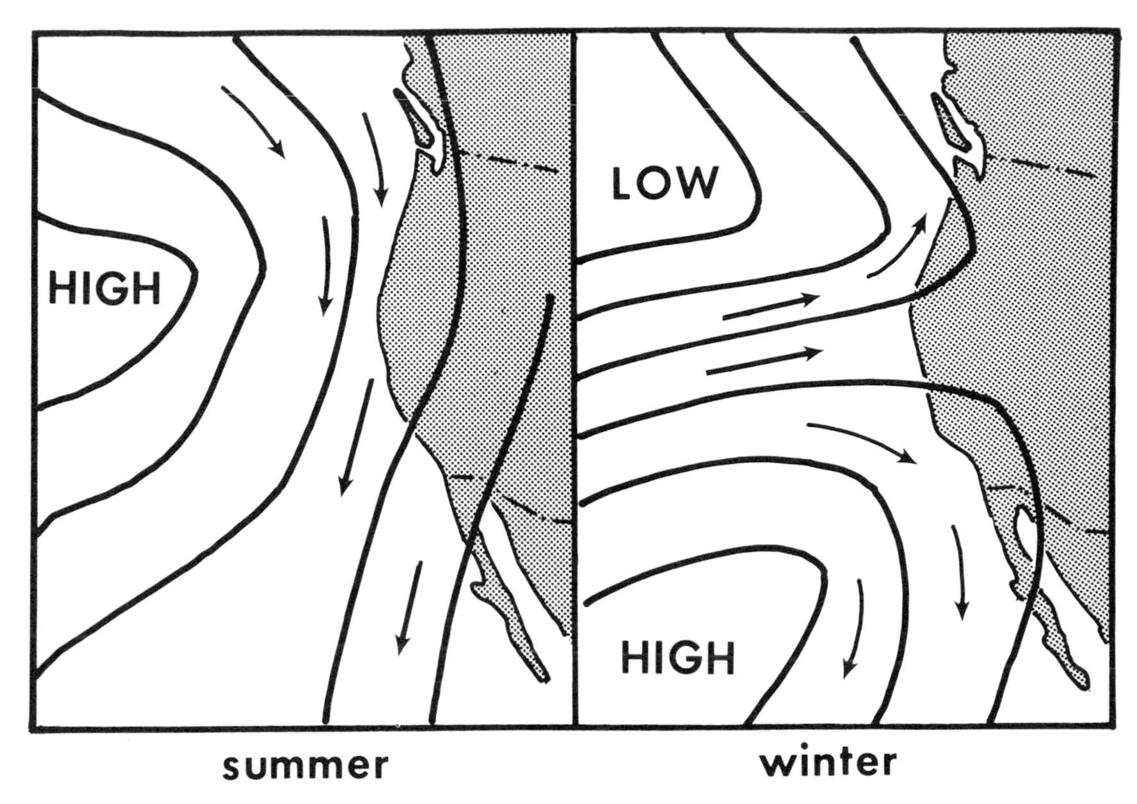

FIG. 12.—Schematic diagram of the seasonal shift of the East Pacific High and associated wind patterns.

has several low points which correspond to weak fronts which moved in from the Pacific Ocean or up from California (Fig. 13). The major frontal systems which had an effect on the coastal area moved across on July 7, 14, 16 and 28 and August 6 and 10, 1973 (Fox and Davis, 1974). The barometric pressure ranged from a low of 1013.2 millibars on July 16 to a high of 1026.2 millibars on July 11. Under the influence of the East Pacific High, north to northwest winds dominate the summer wind pattern and wind-induced coastal upwelling is frequent and long lasting (Smith, 1972). During the upwelling which started on July 17, the water temperature in the surf zone dropped from 14.4°C to 6.6°C (58°F to 44.4°F) in two days.

During July and August, the East Pacific High generally produced strong north to northwest surface winds, roughly parallel to the coast. The surface wind pattern is shown by the longshore component of the wind (Fig. 13). During the study, the longshore wind blew out of the north for 35 days reaching a maximum of about 15 m/sec on July 12, and out of the south for 10 days reaching a peak of 5 m/sec on July 8. On July 1, a high pressure system dominated the weather with winds about 12.5 m/sec out of the north. On July 2 through 8, a weak front stalled offshore with southwest winds reaching 6 m/sec. From July 9 through 16, north winds swept along the coast reaching 15 m/sec on July 12. A low pressure system moved up the coast from California on July 16 and displaced the East Pacific High seaward until July 21, producing southwest winds of 3 to 4 m/sec. The longest continuous period of north winds and upwelling occurred for 18 days between July 22 and August 8, 1973. A weak low started to move up from California on July 28 and decreased the north component of the longshore wind, but did not push the subtropical high offshore. Low pressure systems moved in from the Pacific on August 9 and 13, producing weak southwest winds on the 9th and stronger winds, up to 5 m/sec, on August 12th and 13th.

The longshore winds on the Oregon coast have a definite diurnal component which was smoothed out on the plotted curves (Fig. 13) but shows up in the observed data (Fox and Davis, 1974). The longshore wind reaches a peak in the afternoon and drops off considerably at night. The sea breeze pattern is deflected along the coast by the large difference in air and water temperature. The onshore component of the wind shows a fairly regular pattern with an onshore peak of about 5 m/sec in the mid-afternoon, and an

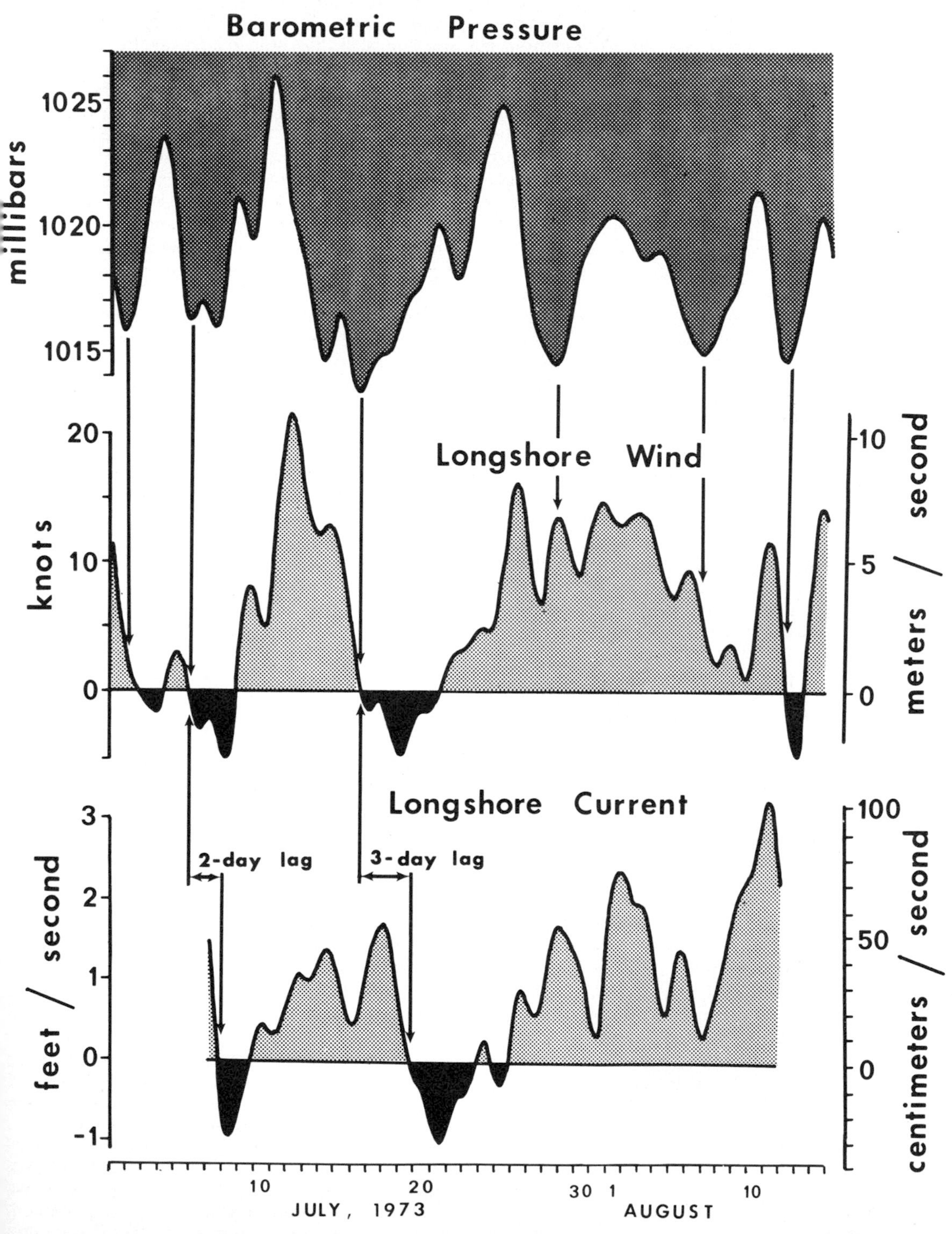

FIG. 13.—Smoothed curves for barometric pressure, longshore wind and longshore current at Newport, Oregon.

offshore breeze of about 2 m/sec at night. Normally, one would expect a sea and land breeze to show a 12-hour pattern with the onshore breeze picking up in the afternoon and an offshore breeze during the late evening. However, along the northwest Pacific coast of the United States, the diurnal wind field is superimposed on the subtropical high and the result is a strong 24-hour, instead of 12-hour, periodicity in the wind speed (Frye, Pond and Elliot, 1972). The cold surface water which reaches a low of about 6°C during upwelling, tends to inhibit the development of a land breeze at night.

The nearshore current speed and direction was measured at least twice each day at the South Beach study site (Fox and Davis, 1974). When the South Beach site was being surveyed in the mid-afternoon, longshore current was also measured at 2 p.m. However, when surveys were being made at Beverly Beach or Gleneden Beach, the afternoon current measurements were not taken. Because rip currents were present in the intertidal zone at different times in the tidal cycle, the azimuth as well as speed of the nearshore current was recorded. Therefore, current speed and azimuth measurements were taken in the center of the rip channel, as well as 30 meters north and 30 meters south of the channel. The average longshore component of the nearshore current was recorded as the longshore current velocity. The offshore component, which represents the rip current, was strongest on July 25 when it reached 25 cm/sec. The longshore component of the current was generally much stronger than the offshore component and reached a maximum of 90 cm/sec on August 11.

There is a close relationship between barometric pressure, longshore wind and longshore current on the Oregon coast (Fig. 13). For the first half of the study, the lows in barometric pressure corresponded quite closely to reversals in wind direction from north to south. The lows in pressure on July 2, 5 and 17 were coincident with changes in wind direction from north to south. A similar pattern of reversals in longshore current direction was observed, but the current showed a 2 to 3 day lag in the reversal. Because the longshore currents are closely related to rip currents on the Oregon coast, the lag time in reversal of current direction may be related to the shifting of nearshore bars and rip channels.

Ocean wave periods and heights were derived from microseismograph records at the Marine Science Center, which were verified twice each day by visual observations (Creech, 1973). During the 45-day study, the significant breaker height varied from a minimum of 0.76 meters on July 2 to a maximum of 2.60 meters on July 16 (Fig. 14). The plot of breaker height shows a general increase for the first 16 days, then low waves for the remainder of the study with small peaks of about 1.2 meters on July 26 and 1.6 meters on August 8. The low points on the barometric pressure curve correspond with the peaks on the breaker height curve similar to the general model developed for Lake Michigan. During the latter half of the study, from about July 24 through August 5, strong winds were blowing along the coast with speeds up to 15 m/sec, but they did not produce high waves. Because the high waves correspond to the lows in the barometric pressure curve and not the local wind conditions, it can be assumed that the waves were produced by the winds offshore as the low approached the coast. At the coast, the winds are predominantly along the shore and do not correspond closely with local wave conditions.

Topographic changes.—The primary study site at South Beach, Oregon was located about 3 kilometers south of the jetty at the mouth of the Yaquina River in Newport, Oregon (Fig. 11). The study area extends for 488 meters along the coast and about 350 meters seaward from the base of the cliff. The beach covers a wave-cut terrace in the Nye Mudstone which is overlain unconformably by Pleistocene marine terrace deposits which crop out in the upper half of the cliff (Lund, 1972). The layer of sand on the beach is up to two meters thick during the summer with portions of the wave-cut terrace exposed during the winter.

Two lines of bars were present in the intertidal zone on July 4, an inner set (bars A and E) about 140 meters from the cliff and an outer set (bars B, D and C) about 230 meters from the cliff (Fig. 15a). Between July 4 and 10, bars A and B advanced toward the shore at about 5 m/day, and to the south at about 16 m/day (Fig. 15h). Bar C advanced toward the beach at 4 m/day and bars E and F welded onto the beach. Some rip channel excavation took place in front of the bars as they advanced toward the beach.

Between July 10 and 18, significant changes took place on the beach and in the intertidal zone (Fig. 15c). Bars A and C merged and advanced across the beach and to the south. On July 16, strong longshore currents due to high waves from the north excavated a trough more than a meter deep and 15 meters wide in front of bar C. Large megaripples in the trough provided evidence of strong longshore currents at high tide. A new bar, G, developed about 80 meters seaward from bar A and started to migrate toward the beach.

Between July 18 and 25, low wave conditions persisted on the Oregon coast and the intertidal sand bars advanced shoreward across the beach (Fig. 15d). Bar G moved toward the shore at 14 m/day and built up in elevation. A rip channel developed between the south end of bar G and

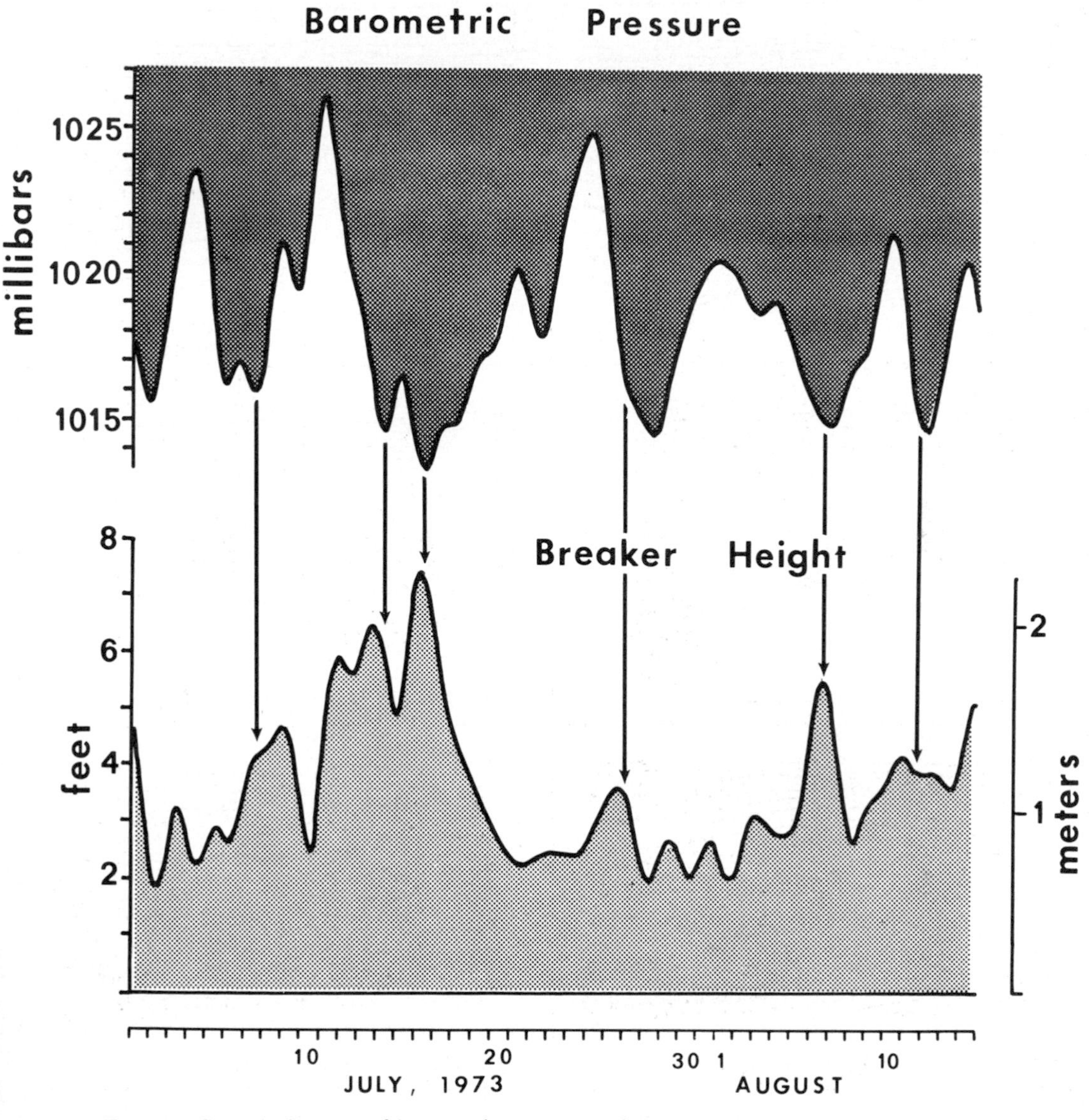

FIG. 14.—Smoothed curves of barometric pressure and breaker height at Newport, Oregon.

bar C. At the same time, bar C advanced toward the shore at about 2 m/day and closed the southern exit from the longshore trough. Therefore, the water carried over bar C by the waves was no longer able to drain out to the south and increased the discharge out the rip channel between bars G and C (Fig. 15d). From July 25 through August 4, the south end of bar G continued to enlarge southward forcing the rip channel between bars G and C to the south. As the current emerged from the rip channel, the northwest waves directed the current southward along the shore. The seaward face of bar C was eroded by wave action and longshore currents from the rip channel. A portion of the sand from the seaward side of bar C was transported across the bar and deposited on the longshore trough as the bar advanced up the beach.

Between August 4 and August 12, bar G continued to build out to the south until it extended beyond bar C. It then proceeded to advance shoreward and merge with bar C forming a continuous bar the full length of the study area. At the same time, two new bars, H and I, formed seaward of bars G and C and started their advance across the beach. At the end of the study, one long continuous bar extended along the entire beach with a longshore trough which drained to

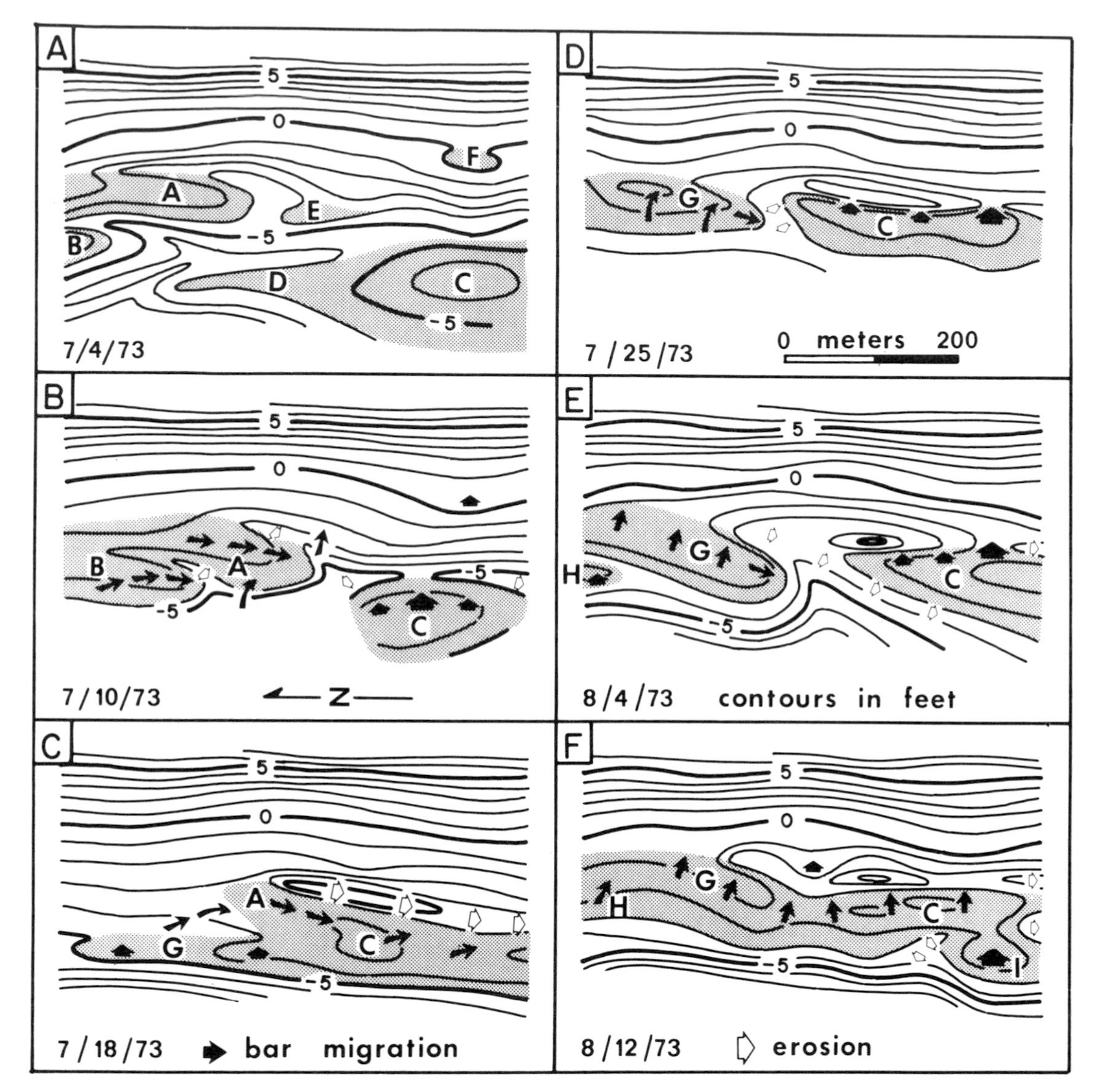

FIG. 15.—Sequence of maps showing bar migration and beach erosion during July and August, 1973, at South Beach, Oregon.

the south. Two smaller bars were forming seaward of the long bar and were moving across the beach.

In summary, the major mode of sand transport across the Oregon beaches in the summer is in the form of large intertidal sand bars. The bars form at a depth of 1 to 2 meters and advance toward the cliff at 1 to 5 m/day. During the summer, the wind is generally out of the north to northwest at 5 to 15 m/sec with waves 1 to 3 meters high. The northwest winds and waves generate longshore currents to the south resulting in a southward migration of the bars. The rate of bar movement to the south varies from 10 to 15 m/day. During storms or high waves longshore troughs and rip channels are excavated by the longshore currents. During times of low wave activity, the bars advance toward the cliff and bury the longshore trough.

Seasonal trends on the Oregon coast.—Very definite seasonal trends are present in the weather and wave conditions on the Oregon Coast. In order to separate the seasonal trends from the shore period fluctuations due to individual storms, the first 18 Fourier harmonics were computed from the data, and the cummulative curves were plotted for barometric pressure, longshore wind,

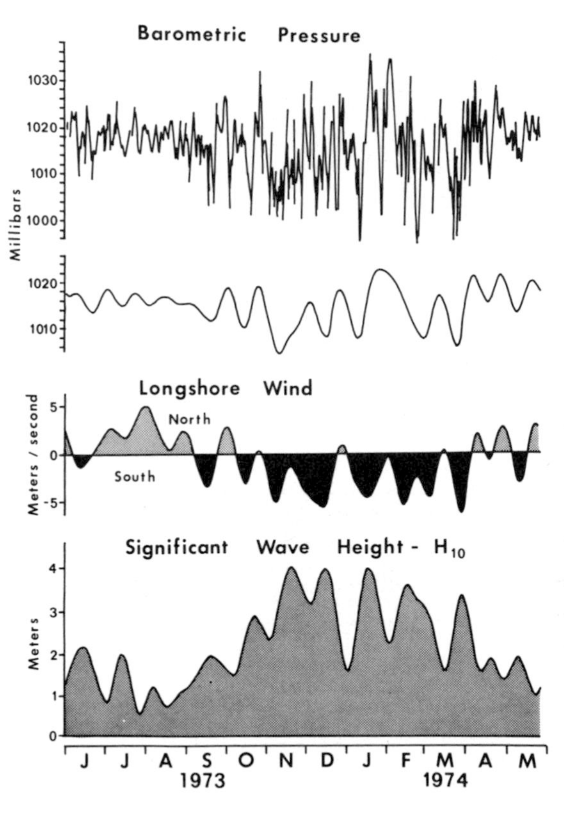

FIG. 16.—Observed curve for barometric pressure and smoothed curves for barometric pressure, longshore wind and wave height from June, 1973, through May, 1974, at Newport, Oregon.

and significant wave height (Fig. 16). The first 18 Fourier components filter out any fluctuations with a period of less than 20 days.

The smoothed curve for barometric pressure shows minor broad oscillations from June through October, 1973, with the first major low in mid-November. The remaining low points on the smoothed curve for barometric pressure occur in mid-December, mid-January, late March.

The curve for significant wave height reaches a low point in late August and builds up to several broad peaks in the late fall and winter. Offshore waves of more than 7 meters in height were recorded during storms in November and February. The peaks in the wave height curve for mid-November, mid-December, mid-January, late February and late March correspond to the low points on the pressure curve. Each broad peak on the wave height curve consists of several distinct storms, but the groups of storms are separated by 8 to 10 days of relative calm with low wave activity. The longest break in the weather occurred in early January when it was clear for 10 days and the wave heights dropped below one meter. The range in oscillation of barometric pressure varied from a minimum of 8 millibars during the summer to a maximum of 43 millibars in the midwinter storms. The wave height reached maximum of 2 meters during late July and early August compared with about 7 meters in November and February.

No attempt was made to measure longshore current velocity during the winter months, but the longshore component of the wind was recorded (Fig. 16). Because there is a close correlation between longshore current velocity and the longshore component of the wind during the summer (Fig. 13), it is assumed that the longshore current and longshore wind would also be correlative during the winter. This assumption may not be justified because the larger waves in the winter would be refracted and approach the shore more closely parallel to the beach. However, based on the deep excavation of longshore troughs, and the general build up of sand on the south side of jetties during the winter, it is assumed that the northward flowing longshore currents play an important role in coastal erosion and sediment transport during the winter months. The smoothed curve for longshore wind shows the dominant north wind under the influence of the East Pacific High during the summer months (Figs. 12, 16). The dominant wind direction shifts to the southwest in September and remains from the south until the following April. The peaks in the longshore wind from the south also line up with the peaks for the smoothed curve for breaker height. Although they are masked in the smoothed curve, reversals in wind direction occur within each individual storm, but the south component of the wind generally is stronger than that from the north. The onshore component of the wind also increases during the winter from a minimum of about 2 to 3 m/sec during the summer to maximum of almost 15 m/sec in the winter. The strong onshore winds significantly build up the winter waves. Offshore winds of up to 10 m/sec reduce the wave height in early January.

During the summer, intertidal sand bars were formed on the lower foreshore and advanced across the beach under low wave conditions. In September and October, the intertidal bars were removed from the beach and the beach surface was concave upward. In November, more than 0.6 meters of sand was stripped from the entire surface beach. During the winter storms, portions of the beach were eroded down to bedrock. In the calm intervals between major storm periods, large bars formed and started to migrate across the beach. Deep rip channels and longshore troughs were cut on the shoreward margin of the bars as they advanced across the beach. Therefore, the sand on the backbeach was removed when the longshore trough was excavated, and more sand was deposited as the bar advanced across the trough. Because a deep trough is frequently excavated on the shoreward side of the sand bar and filled in as the bar progresses up the beach, a new layer of sand is accumulated by each successive set of bars. Therefore, the new sand is not simply laid down as a carpet of sand over the existing sand on the beach, but the pre-existing sand is eroded on the shoreward edge of the trough and the bar fills in the trough as the bar and trough migrate together across the beach.

DISCUSSION

Three coastal areas, the eastern shore of Lake Michigan, Mustang Island, Texas, and the central Oregon Coast were selected for detailed analysis of the relationship between weather patterns and coastal processes. The shoreline on the eastern coast of Lake Michigan extended perpendicular to the storm paths and the storms moved onshore to the north of the study area. With a tidal range of less than 10 centimenters and fetch of about 150 kilometers, the waves respond quickly to changes in weather conditions, and swells and tidal currents do not influence beach erosion. A simple storm cycle model was developed for Lake Michigan relating breaker height and longshore current velocity to barometric pressure. The general shapes for the barometric pressure, longshore current and breaker height curves for Lake Michigan are given in Figure 17. The curve for barometric pressure is symmetrical with the rate of pressure drop about equal to the rate of pressure rise. The curve for longshore current is again symmetrical with a somewhat faster longshore current to the south than to the north. The maximum longshore current of the three areas was 210 cm/sec at Stevensville, Michigan, in July 1969. Wave heights reached a maximum of 1.8 meters during summer storms, but were generally less than 1.0 meter on Lake Michigan.

On the Texas coast, Mustang Island is oriented in a northeast-southwest direction with the Gulf

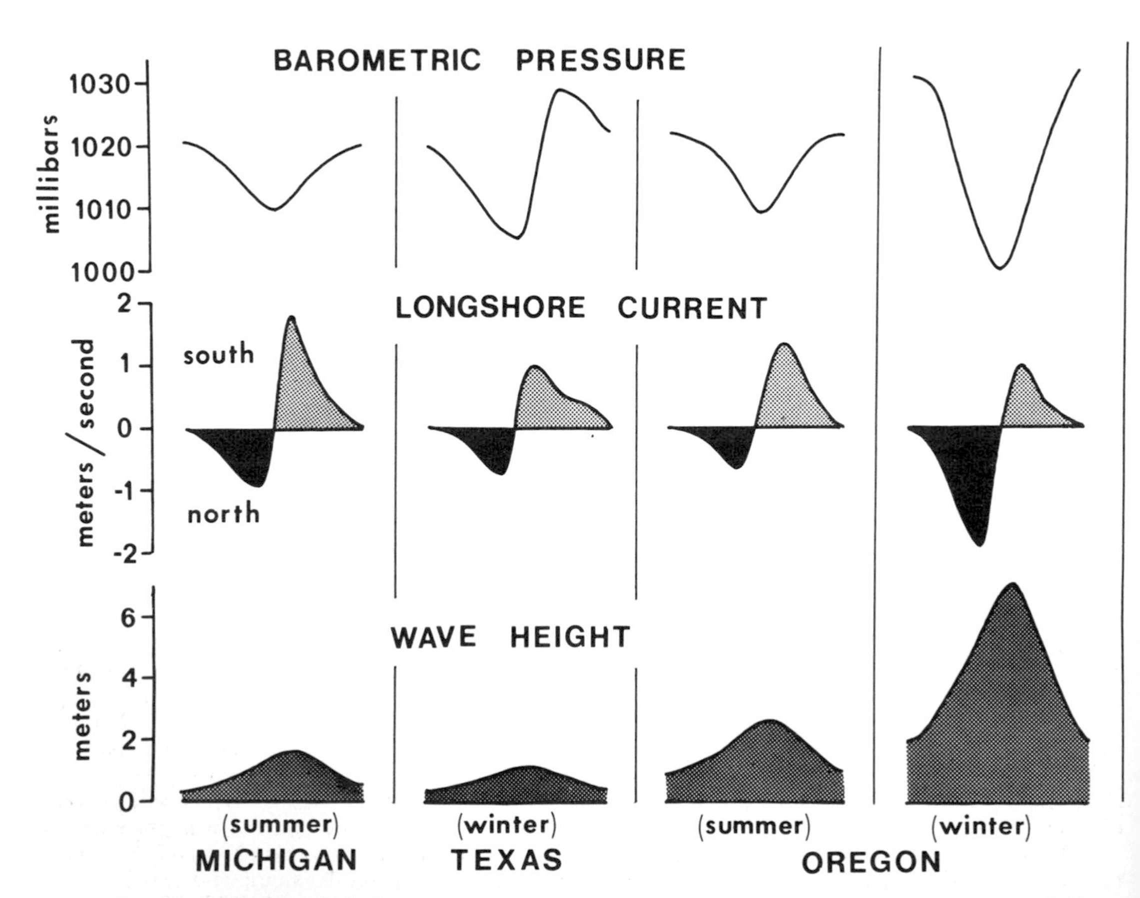

FIG. 17.—Generalized curves for barometric pressure, longshore current and breaker height at Holland, Michigan; Mustang Island, Texas; and Newport, Oregon.

of Mexico to the east. In the fall and winter, low pressure systems move to the north of Mustang Island with intense cold fronts trailing off to the southwest. As a low pressure center moves to the east, a cold front sweeps across the coast accompanied by a sharp rise in barometric pressure and rapid reversal in longshore current direction. Because winds blow along shore or offshore, the fetch is short and breaker height does not increase very much. Maximum breaker height during the fall and winter studies at Mustang Island was about 1.2 meters, whereas the breaker height reached 1.8 meters on Lake Michigan during the July 1969 storm. Although the fetch is less on Lake Michigan, onshore winds generally produced larger waves and faster longshore currents on Lake Michigan than on the Texas coast (Fig. 17). The warm gulf air contrasted sharply with the polar air masses in the late fall and winter producing intense cold fronts or northers which crossed the Texas coast. Cold fronts connected with the low pressure systems that passed over Lake Michigan during the summer were not as intense because of the smaller difference in temperature. The asymmetry of curves for barometric pressure and longshore current, and the smaller breaker heights are plotted at the same scale as the curves for Lake Michigan (Fig. 17).

The summer weather pattern on the Oregon coast is dominated by the East Pacific High. The high pressure system is situated a few hundred kilometers off the Oregon coast. North winds circulating around the high blow along the coast from April through September. Weak low pressure systems move over the coast during the summer, causing a reversal in wind direction and longshore current (Fig. 17). The south component of the longshore wind reaches only 5 m/sec, while the north component is frequently over 15 m/sec. Breakers on the Oregon coast reached heights of 2.5 meters and longshore currents approached 100 cm/sec. Usually the high waves were refracted so that the wave crests were almost paralled to the coast, and longshore currents were more strongly influenced by the topography of the intertidal sand bars. The tide range on the Oregon coast is up to 4 meters in spring tides which contrasts with the 1 meter spring tides on the Texas coast and 10 centimenter tide on Lake Michigan.

During the winter, the East Pacific High moves to the south and the major Pacific storm track shifts over the Oregon coast. During 1973–74, storms passed with regular frequency during the fall and winter months boosting the rainfall on the Oregon coast to 92 inches. Wave height during the winter storms reached 8 meters with wave periods of 8 to 14 seconds. Most of the storm activity was concentrated between the months of November and March with breaks of 8 to 10 days between sets of 3 to 5 storms. The onshore wind generally dominated over the offshore wind with speeds reaching 15 m/sec. During the storms, the south component of the longshore winds reached 20 m/sec, while the north component was only 10 m/sec. Although it was not possible to measure the longshore currents during the winter, there was a positive correlation between longshore wind and longshore current during the summer. It can be assumed that the strong south winds during the winter were also responsible for strong longshore currents to the north.

Regular patterns of topographic change were recorded in each of the three study areas. On Lake Michigan, the beach and nearshore bars responded quickly to waves and longshore currents within each storm cycle. In the intervals between storms, protuberances built out along the shore, and the bars migrated toward the beach. During storms, the waves approached the shore at a large angle and generated strong longshore currents. The currents were shunted offshore by the protuberances and eroded a channel through the sand bars. Thus, the position of the bars shifted with each storm cycle under the influence of waves and longshore currents.

Along the Texas coast, the beach is broad with 2 to 3 well developed longshore bars. Although low amplitude cusps were present on the foreshore during most of the study, the rhythmic shoreline pattern found on Lake Michigan was not as pronounced on the Texas coast. With the straight shoreline, longshore currents were confined to the nearshore trough and rip currents were not regularly spaced. However, the shoreward migration of bars during low energy conditions, and seaward expansion of the trough during high energy were observed on both the Michigan and Texas coasts. Saddles are present on the Texas sand bars, but they are not related to protuberances along the beach.

On the Oregon coast, the 4 meter tidal range exposed an intertidal beach almost 300 meters wide. Sand bars migrated across the intertidal beach with large rip channels developed between the bars. The rip channels are affected by rising and falling tides which flood and drain the beach. During the summer, the bars advanced across the beach at 1 to 5 m/day, and new bars formed at the seaward margin of the intertidal zone. During the winter, high waves and strong currents stripped a large portion of the sand off the beach. During occasional quiet conditions in midwinter, however, large bars would form and move across the beach. Most of the sand which was removed from the beach by large winter storms apparently was stored below mean low tide on large sand bars. An equivalent volume of sand was returned to the intertidal beaches during the spring and summer.

CONCLUSIONS

The following conclusions can be drawn concerning weather patterns and coastal processes on Lake Michigan, Mustang Island, Texas and the central Oregon Coast.

1. On the eastern shore of Lake Michigan, longshore current velocity is a function of the first derivative of barometric pressure, and breaker height is a function of the second derivative.
2. On Lake Michigan, rip channels between bars occur offshore from protuberances on the beach. During storms, longshore currents are shunted offshore by protuberances and form rip channels between the bars.
3. On the Texas coast, fall and winter weather patterns are dominated by intense cold fronts or northers marked by a rapid rise in pressure, drop in temperature and sudden shift in wind from northeast to southwest.
4. Beaches along the Texas coast are relatively straight with 2 to 3 offshore bars. Rip channels forming saddles between the bars are not related to protuberances along the coast.
5. During the summer, the weather pattern on the Oregon coast is dominated by north winds circulating around the East Pacific High. Waves are 1 to 3 meters high with strong longshore currents to the south.
6. During the winter on the Oregon coast, the East Pacific High shifts to the south and the Pacific storm track passes over the Oregon coast. Wave heights reach 6 to 8 meters during the storms and southwest winds predominate forming strong northward flowing longshore currents.
7. The 4-meter tides on the Oregon coast play an important role in spreading the influence of waves and currents over a 300 meter wide intertidal beach.
8. Broad intertidal sand bars migrate across the Oregon beaches during the summer at a rate of 1 to 5 m/day. During the winter storms, between November and March, sand is removed from the beach and stored in subtidal sand bars. In the spring the bars migrate onto the beach returning the sand lost by the winter storms. Large sand bars also form on the beach in the winter during the quiet intervals between major storms.
9. Beach erosion and bar migration observed in the three study areas are closely related to waves and longshore currents during storms. The areas differ in their response to wave and currents because of their position relative to storm tracks, orientation of the shoreline, tidal range, and nearshore topography.

ACKNOWLEDGEMENTS

This research was carried out under the sponsorship of the Geography Branch of the Office of Naval Research under Research Contract N00014-69-C-0151 Task No. NR 388-092. Several student assistants including: E. Cartnick, C. Collinson III, J. Cunningham, H. Flint, C. Foster, T. Getz, R. Kerhin, D. Lehman, R. LoPiccolo, D. McTigue. P. Murphy, W. Murphy, K. Murray, D. Rosen and E. Thorp helped with various aspects of the field work, laboratory analysis and computer programming for the project. Computer facilities at Williams College, Western Michigan University and Oregon State University were used for plotting the curves. Wind and weather data were made available from the Marine Science Institute at the University of Texas and from Oregon State University.

REFERENCES

Cox, W. S., 1973. Oregon estuaries: Oregon State Land Board, Salem, Oregon, 48 p.

Cramer, O. P., 1973, Mesosystem weather in the Pacific Northwest, a summer case study: Monthly Weather Rev., v. 101, p. 13–23.

Creech, C., 1973. Wave climatology of the central Oregon coast: Oregon State Univ., Marine Sci. Center Tech. Rep. NOAA Sea Grant Project 04-3-158-4, 19p.

Davies, J. L., 1964, A morphogenetic approach to world shorelines: Zeitschr. Geomorphologie, v. 8, Sonderheft, p. 127–142.

Davies, R. A., Jr., 1972, Beach changes on the central Texas coast associated with Hurricane Fern, September 1971: Contr. Marine Sci., v. 16, p. 89–98.

—— and Fox, W. T., 1971, Beach and nearshore dynamics in eastern Lake Michigan: Williams College, Tech. Rep. 4, ONR Contract 388-092, 145 p.

—— and ——, 1972a, Coastal dynamics along Mustang Island, Texas: Williams College, Tech. Rep. 9, ONR Contract 388-092, 68 p.

—— and ——, 1972b, Coastal processes and nearshore bars: Jour. Sed. Petrology, v. 42, p. 401–412.

—— and ——, 1974, Simultaneous process-responses study on the east and west coasts of Lake Michigan: Williams College, Tech. Rep. 13, ONR Contract 388-092, 61 p.

——, ——, Hayes, M. O., and Boothroyd, J. C., 1972, Comparison of ridge and runnel systems in marine and non-marine environments. Jour. Sed. Petrology, v. 42, p. 413–421.

Evans, O. F., 1940, Low and ball of the eastern shore of Lake Michigan: Jour. Sed. Petrology, v. 48, p. 476–511.

Fox, W. T., and Davis, R. A., Jr., 1970a, Fourier analysis of weather and wave data from Lake Michigan: Williams College, Tech. Rep. 1, ONR Contract 388-092, 47 p.
—— and ——, 1970b, Profile of a storm-wind, waves and erosion on the southeastern shore of Lake Michigan: Internat. Assoc. Great Lakes Res., Proc. 13th Conf. on Great Lakes Research, p. 233-241.
—— and ——, 1971a, Fourier analysis of weather and wave data from Holland, Michigan, July 1970: Williams College, Tech. Rep. 3, ONR Contract 388-092, 79 p.
—— and ——, 1971b, Computer simulation model of coastal processes on eastern Lake Michigan: Williams College, Tech. Rep. 5, ONR Contract 388-092, 114p.
—— and ——, 1973a, Coastal processes and beach dynamics at Sheboygan, Wisconsin, July 1972: Williams College, Tech. Rep. 10, ONR Contract 388-092, 94 p.
—— and ——, 1973b, Simulation model for storm cycles and beach erosion on Lake Michigan: Geol. Soc. America Bull., v. 84, p. 1769-1790.
—— and ——, 1974, Beach processes on the Oregon coast: Williams College, Tech. Rep. 12, ONR Contract 388-092, 86 p.
Frye, E. E., Pond, S., and Elliott, W. P., 1973, Note on the kinetic energy spectrum of coastal winds: Monthly Weather Rev., v. 100, p. 671-673.
Hayes, M. O., et al., 1969, Coastal environments, northeastern Massachusetts and New Hampshire: Univ. Massachusetts Coastal Res. Group, Contr. 1, 462 p.
Inman, D. L., and Nordstrom, C. E., 1971, On the tectonic and morphologic classification of coasts: Jour. Geology, v. 79, p. 1-21.
Lund, E. H., 1972, Coastal landforms between Yachats and Newport, Oregon: The Ore Bin, v. 34, p. 73-91.
Longuet-Higgins, M. S., 1950, A theory of the origin of microseisms: Royal Soc. London Philos. Trans., Ser. A, v. 243, p. 1-35.
Olson, J. S., 1958, Lake Michigan dune development; 3. Lake level beach and dune oscillations: Jour. Geology, v. 66, p. 473-483.
Sclater, J. G., Anderson, R. N., and Bell, L. M., 1971, Elevation of ridges and evolution of the central eastern Pacific: Jour. Geophys. Res., v. 76, p. 7888-7915.
Shepard, F. P., 1963, Thirty-five thousand years of sea level: *In* R. L. Miller (ed.), Essays in marine geology in honor of K. O. Emery: Univ. Southern California Press, p. 1-10.
Smith, R. L., 1972, A description of current, wind, and sea level variations during coastal upwellings off the Oregon coast, July-August, 1972: Jour. Geophys. Res., v. 79, p. 435-443.
Snavely, P. D., Jr., and Macleod, N. S., 1971, Visitors guide to the geology of the coastal area near Veverly Beach State Park, Oregon: The Ore Bin, v. 33, p. 85-105.
Valentin, H., 1952, Die Küsten der Erd: Petermanns Geog. Mitt. Ergänzunashelf, 246 p.

WAVE CLIMATE MODELS FOR THE CONTINENTAL SHELF: CRITICAL LINKS BETWEEN SHELF HYDRAULICS AND SHORELINE PROCESSES[1]

VICTOR GOLDSMITH
Virginia Institute of Marine Science, Gloucester Point, 23062

ABSTRACT

In 1947, Munk and Traylor's classic paper clearly showed the importance of shelf bathymetry upon surface wave processes, and linked these processes to shoreline changes to the extent ". . . that wave refraction is the primary mechanism controlling changes in wave height along a beach . . ." (Munk and Traylor, 1947, p. 1). With the application of high speed digital computers in the 1960's, wave refraction diagrams have become commonplace in shoreline and nearshore studies.

The *Virginian Sea Wave Climate Model* (Goldsmith and others, 1974) represents a significant advance in the computation and application of "wave refraction diagrams" through the use of new and more sophisticated techniques such as: (1) use of a regional approach in which 52,000 km^2 of continental shelf (out to depths of 300 m), and 160 km of shoreline, are incorporated into one wave ray diagram; (2) voluminous depths are chosen from numerous original hydrographic sounding sheets and interpolated depths are avoided: e.g., 100,000 depths were acquired for the Virginian Sea Model; (3) these depths are transferred to a common grid using a specially computed transverse Mercator projection "centered" on the study area in order to minimize distortion caused by the earth's curvature (i.e., waves travel great circle paths); (4) 19 different ray parameters are computed along each ray including surface wave heights and bottom orbital velocities; (5) an improved understanding of wave behavior in the area of crossed wave rays (Chao and Pierson, 1972) has been applied to the interpretation of such wave phenomena as curved caustics (over the shelf-edge canyons and ridge and swale bathymetry) and straight caustics (over deep channels off the mouths of Delaware and Chesapeake Bays); (6) this information is then used to delineate areas of "confused seas" and bottom "scour" for specific wave and tidal conditions.

Wave ray diagrams, shoreline histograms and shelf contour maps of various wave parameters for various combinations of 122 distinct wave conditions, as computed in the Virginian Sea Wave Climate Model, are being used to increase our understanding of shelf sedimentology, historical shoreline changes, and inlet hydraulics.

INTRODUCTION

Wave refraction is essentially the bending of wave crests caused by their slowing down as depth decreases and as waves pass through shallow water in their approach to the shore. Because of the large variation in depths along most coastal areas, the slowing down of the waves occurs differentially. Such wave bending, or refraction, is the major determinant of the shoreline wave energy distribution along the East and Gulf Coasts of the United States because of the wide, shallow continental shelves. Refraction is also critical in understanding the local processes and resulting geomorphology along the United States West Coast.

When a series of wave refraction diagrams is computed with the particular wave conditions chosen because they are considered to be the important wave conditions for the area, we may refer to this total series of wave computations as a wave climate model.

In this discussion we will first briefly review the historical development of the "art" of wave refraction, some of the basic behavior patterns of refracted waves, and the consequences to the adjacent shoreline. Secondly, the latest developments and applications of wave climate models will be reviewed, with emphasis on the Virginian Sea Wave Climate Model (VSWCM). Finally, the present state of the art will be briefly reviewed, and some problems necessitating solution will be enumerated.

This review is not meant to be all inclusive, but merely to highlight the state of the art—both the accomplishments and the problems—as it relates to the study of nearshore processes. For a more exhaustive treatment of the subject the interested reader is referred to Goldsmith and others (1974), Goldsmith and others (in preparation) and to the standard texts.

HISTORICAL DEVELOPMENT

The study of wave refraction is thousands of years old. Early wave analysts included the Polynesians who navigated their way around the south Pacific by using the crossed wave patterns resulting from waves bending around the numerous islands (Lewis, 1974). The Polynesians constructed the oldest known wave refraction diagrams using bent twigs.

Modern wave refraction studies began in the 1930's and largely owe their origin to M. P. O'Brien and his colleagues at the University of California (Berkeley) who applied Snell's Law to the process of wave slowing (i.e., wave "re-

[1]VIMS Contribution No. 708

fraction''). Some of this early work, discussed in O'Brien (1942, 1947), Anonymous (1950), Horrer (1950) and Arthur (1951), relates to breakwater problems of stone displacement, to harbor shoaling at Long Beach and Santa Barbara, California, respectively, and to other similar West Coast studies.

Emphasis in the early 1940's shifted to wartime applications—especially surf prediction on proposed Allied landing beaches (Bates, 1949). Experience gained in these applications is summarized in Johnson (1948). The manual construction of wave refraction diagrams, pioneered by O'Brien, is detailed in Johnson and others (1948), a basic reference for those seeking instruction on the manual construction of wave refraction diagrams, a practice now discontinued with the development of computers.

Basic relationships between offshore wave refraction, the resulting shoreline wave energy patterns and shoreline processes were most clearly detailed and verified in Munk and Traylor's (1947) classic investigation. Much of the basic wave refraction relationships affecting the beach and nearshore were delineated in their work, and are still valid. Subsequent efforts have developed more sophisticated approaches, but have added little to the basic relationships outlined in Munk and Traylor's (1947) study.

Basic wave refraction patterns.—Because early wave refraction studies were mostly on the West Coast, emphasis in these studies was on the effects of the canyons which approached quite close to shore, in some cases within less than a kilometer of the beach. Munk and Traylor (1948) noted that for a three kilometer stretch of shore, wave rays tended to diverge opposite and down-wave from the canyon, whereas along the shoreline opposite from an inter-canyon ridge the wave rays converged (Fig. 1). Many basic wave refraction patterns were delineated within their study: (1) larger measured wave heights occurred along the shoreline near computed wave ray convergences, and lower measured wave heights occurred in shoreline areas of computed wave ray divergences for specific wave conditions (Fig. 2); (2) short wave periods gave smaller variations in shoreline wave heights (this is because the longer waves ''feel'' the bottom sooner, and are refracted more than the shorter waves); (3) different wave approach directions for the same wave periods changed the shoreline locations, but not the spacing of areas of higher and smaller calculated wave heights; (4) crossed-wave fronts, which developed on the downwave margins of steeply decreasing bathymetry (e.g., canyon rims), were detected in the refraction diagrams and verified in photographic studies.

Such commonly occurring crossed-wave patterns, termed caustics from the application of geometrical optics to wave studies, were further described by Pierson (1951). Wave caustics are

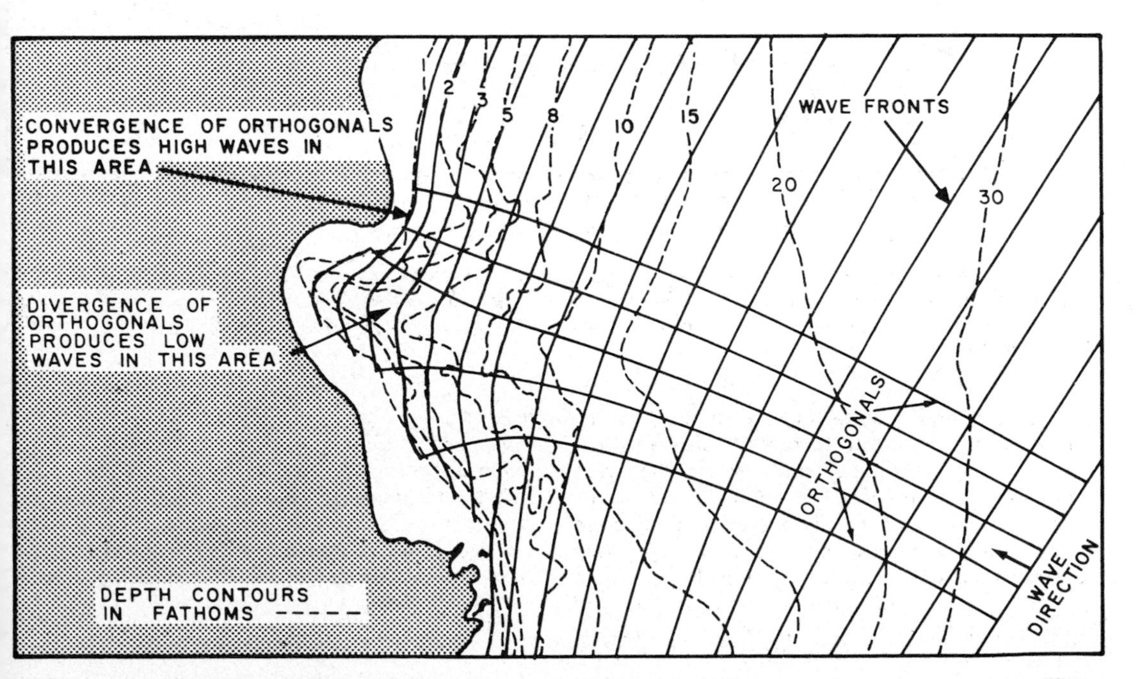

Fig. 1.—Schematic diagram illustrating the link between continental shelf waves and nearshore processes (from Goldsmith and others, 1974, p. 8).

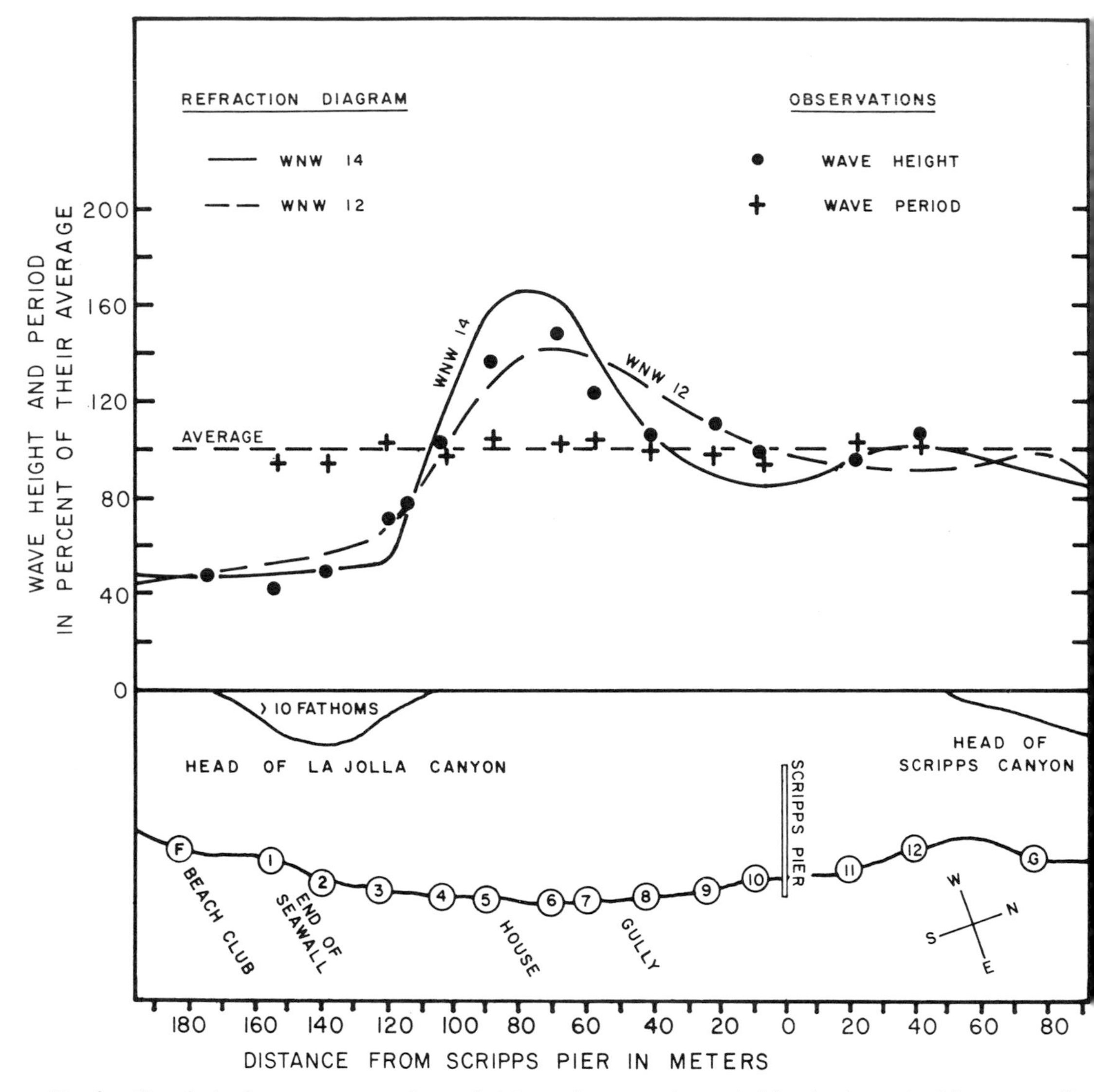

FIG. 2.—Correlation between measured wave heights and computed wave heights (as determined from manually drawn refraction diagrams) for specific wave conditions. Note that low wave heights occurred in areas of wave ray divergence and larger wave heights in areas of wave ray convergence (from Munk and Traylor, 1947, Fig. 16).

one of the major areas of controversy in the interpretation of wave refraction diagrams because of the inability of linear wave theory to mathematically describe the wave caustic.

Many investigators of nearshore processes were quick to apply these aspects of wave refraction to their studies. Shepard and Inman (1950) related nearshore circulation such as rip currents and geomorphology to areas of wave ray divergence caused by wave refraction. Bascom (1954) used wave refraction diagrams to help explain how this longshore variation in wave heights, due to wave refraction, controlled the location of stream outlets. That is, in shoreline areas of wave ray divergence, the resulting beach berms were at lower elevations due to the lower impinging wave heights, thereby encouraging streams to enter the ocean at these areas. Bascom (1954) found these relationships to hold for examples of wave refraction around tombolos, headlands and opposite marine canyons.

With the advent and application of high-speed computers, wave refraction really came of age (Lepetit, 1964; Harrison and Wilson, 1964). There was one change in the theory employed in most of the computer-drawn wave refraction diagrams, however, which is often overlooked by those interpreting these diagrams. This relates to the variation in the spacing between the wave rays, which is used as an indication of wave heights along the wave fronts. In the older, manual-drawn diagrams a simple ratio of the distance between adjacent rays in deep water relative to shallow water was used to calculate the shallow water wave heights, wave energy, and other parameters. In the computer-drawn diagrams a method suggested by Munk and Arthur (1951) has been adopted. This method assumes that a second ray is spaced an infinitesimal distance from the first ray, and the mathematical expressions relating to "wave intensity" proposed by Munk and Arthur (1951) are used to calculate this ray spacing, and consequently the wave height. Thus, in the wave refraction diagrams employing this technique, wave heights and other related wave parameters are calculated along each wave ray, and each ray is "unaware" of the presence of the other wave rays. Partly for this reason, Chao (1974) suggested reverting back to a variation of the manual method for the proper interpretation of crossed-waves, even for computer-drawn diagrams. However, Chao and others (1975) in a more recent paper have returned to the Munk and Arthur (1951) wave intensity method, with some modification within the wave caustic area. Additional aspects of the interpretation of crossed-wave fronts will be discussed in a later section of this paper.

Recent applications.—Wave refraction diagrams have been used to trace the paths of seismic sea waves across the Pacific (Wilson and Torum, 1968; Keulegan and Harrison, 1970) and in particular, to interpret the high destruction at Crescent City, California, relative to adjacent coastal areas (Roberts and Kauper, 1964).

More commonly, wave refraction has been used to understand dramatic longshore variations in shoreline erosion and accretion (Goldsmith and Colonell, 1970), nearshore bottom sediment distribution (Farrell and others, 1971), the role of wave climate in river delta morphology (Coleman and Wright, 1971), the mysterious loss of two British trawlers in the North Sea (Pierson, 1972), the development and maintenance of nearshore sediment and morphology cells (May and Tanner, 1973), reef design for recreational surfing (Walker and others, 1972), the development of offset inlets (Goldsmith and others, 1973), and many other applications.

VIRGINIAN SEA WAVE CLIMATE MODEL (VSWCM)

The Virginian Sea Wave Climate Model differs from previous models in the following important elements:

1. The model covers a very large geographic area of the continental shelf and shoreline, Cape Henlopen, Delaware, to Cape Hatteras, North Carolina, an area of 52,000 km^2 within a single large grid (Fig. 3). The importance of this approach is that the resulting graphical display allows the investigator to visually integrate patterns of wave behavior which would escape detection when smaller areas are used. As a result, regional differences in behavior within the grid stand out. More detailed studies can then be made on a finer-mesh grid in specific subareas by using the wave information from the large grid as input to the smaller grid.

2. Distortions due to flat representations of the spherical earth and problems resulting from the fact that waves travel great circle paths were overcome by constructing a transverse Mercator map projection tangent to the earth along the center of the grid.

3. An improved understanding of wave behavior in the area of crossed-wave rays is now available from the theoretical studies of Chao (1972) and Pierson (1972). These studies have been applied to the interpretation of such wave phenomena as curved caustics which occur over continental shelf ridge and swale bathymetry and straight caustics which occur directly over the margins of the deeply incised channels off the mouths of the Delaware and Chesapeake Bays.

The depth grid utilized an input of 84,420 depths with a unit cell of 0.5 nautical miles on a side. The specified wave input conditions considered nine initial directions for six different wave frequencies, two wave heights, and two tidal conditions for three approach directions. In all, 122 separate wave conditions were used with 19 different wave parameters computed as output for the entire shelf and adjacent shoreline.

These aspects are thoroughly described in Goldsmith and others (1974) from which much of the discussion in this section is taken. Our exhaustive studies have shown that one of the major weaknesses in such an effort is the horizontal and vertical accuracy of depth information on the original hydrographic soundings sheets available for much of the United States East Coast. Furthermore, the depth information is considered to be a far more important problem in the efficacy of the methodology than any weaknesses in the wave theory discussed here and elsewhere.

DATA INPUT

Depths.—Despite the wide usage of original sounding sheets, few sources of written information exist on the accuracy criteria desired and met in these surveys as well as the corrections employed or not employed and their justifications. In order to fill this critical information gap a study on the accuracy of the depth and navigational positioning has been made by Sallenger and others

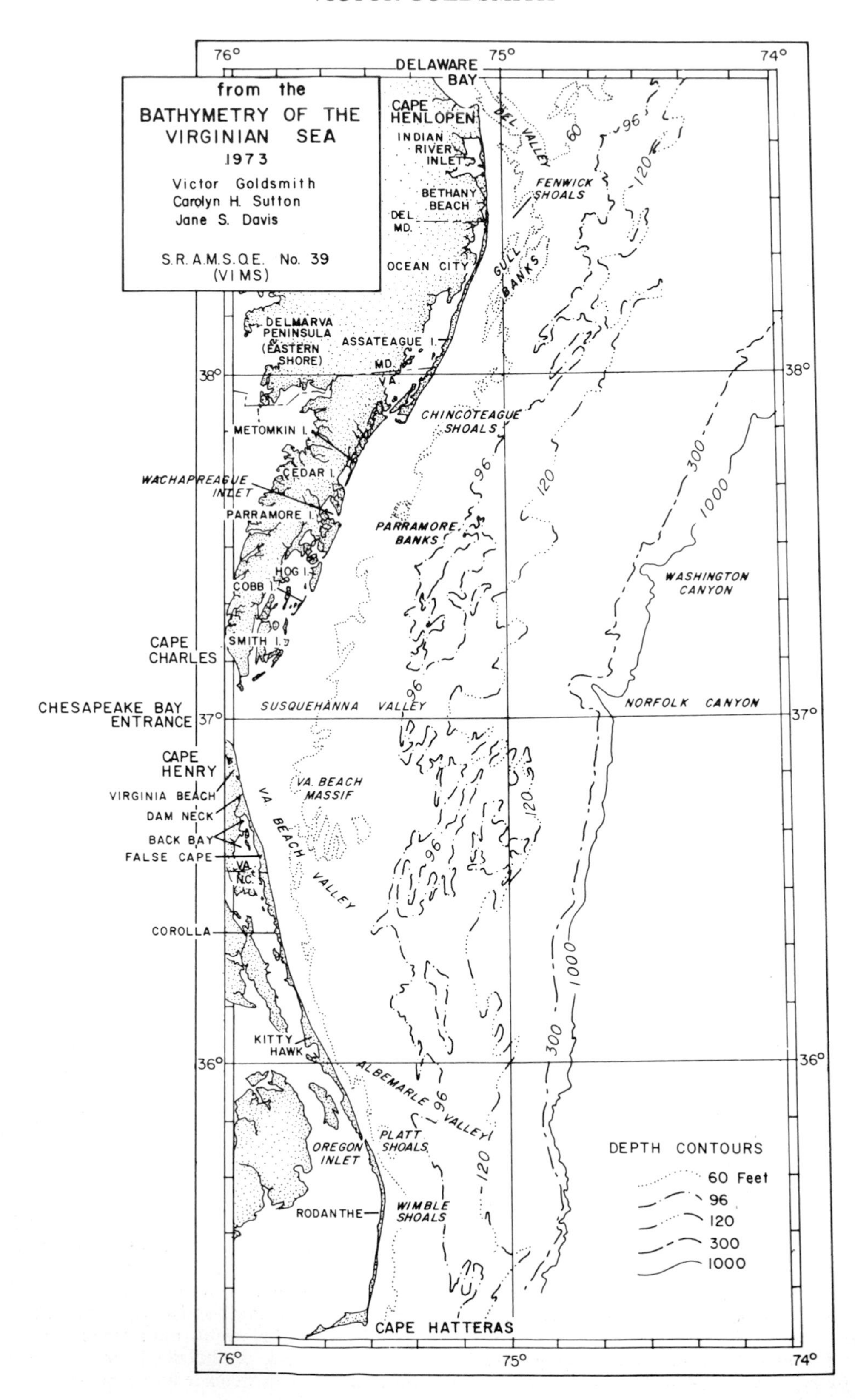
from the
BATHYMETRY OF THE
VIRGINIAN SEA
1973
Victor Goldsmith
Carolyn H. Sutton
Jane S. Davis
S.R.A.M.S.O.E. No. 39
(VIMS)
DELAWARE BAY
CAPE HENLOPEN
INDIAN RIVER INLET
BETHANY BEACH
DEL. MD.
OCEAN CITY
DEL VALLEY
FENWICK SHOALS
GULL BANKS
DELMARVA PENINSULA (EASTERN SHORE)
ASSATEAGUE I.
MD. VA.
CHINCOTEAGUE SHOALS
METOMKIN I.
CEDAR I.
WACHAPREAGUE INLET
PARRAMORE I.
PARRAMORE BANKS
HOG I.
COBB I.
SMITH I.
WASHINGTON CANYON
CAPE CHARLES
CHESAPEAKE BAY ENTRANCE
SUSQUEHANNA VALLEY
NORFOLK CANYON
CAPE HENRY
VIRGINIA BEACH
DAM NECK
BACK BAY
FALSE CAPE
VA. NC.
VA. BEACH MASSIF
VA. BEACH VALLEY
COROLLA
KITTY HAWK
ALBEMARLE VALLEY
PLATT SHOALS
OREGON INLET
RODANTHE
WIMBLE SHOALS
CAPE HATTERAS
DEPTH CONTOURS
60 Feet
96
120
300
1000
76°
75°
74°
38°
37°
36°

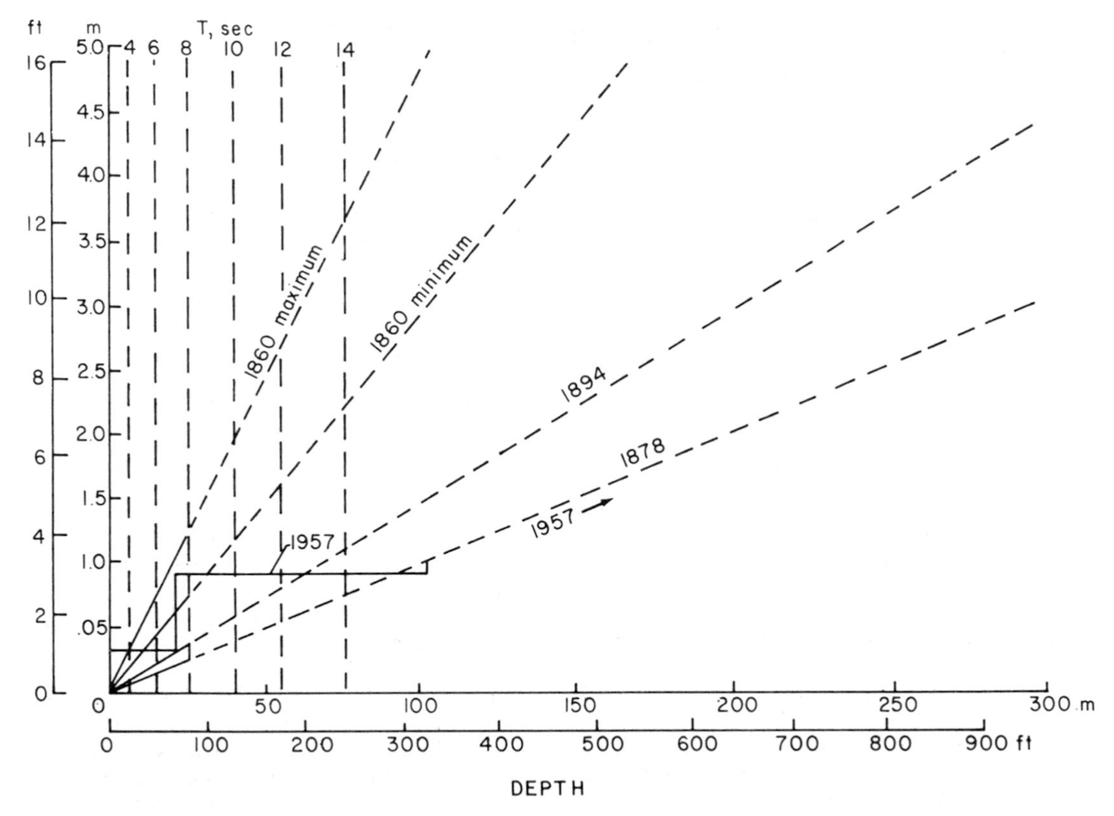

FIG. 4.—Sounding error criteria for the original bathymetric hydrographic sounding sheets used by NOAA (formerly Coast and Geodetic Survey), as compiled in Sallenger and others, 1975. The depths equal to one-fourth wave length of the wave periods used in the VSWCM are superimposed on the diagram (see text).

(1975). Figure 4, taken from Sallenger and others (1975), illustrates the different criteria set by the U.S. Coast Survey and its successor agencies for surveys of different dates. The depths at which waves of the periods used in this study are first significantly refracted by the sea floor irregularities are plotted over the Coast Survey accuracy criteria to give an indication of the depth errors influencing the wave climate model. Only four of the charts which were used in the depth accumulation were surveyed prior to 1915, and only three of these charts were surveyed prior to 1870. These charts (prior to 1915) used for the model were surveyed where the depths did not exceed 27 meters.

Approximately 100,000 of these original uninterpolated depths were transferred from the 61 sounding sheets and other data, using latitude and longitude onto a transverse Mercator map projection 2.4 by 1.2 meters, specially constructed for the present study. Then 84,420 of these depths were read from the map grid of 0.5 nautical mile squares and punched on cards.

Wave conditions.—The second major input to the Wave Climate model is a wide variety of wave conditions. Approximately 200 to 250 wave orthogonals were propagated shoreward from deep water[1] for each of 122 wave conditions. The wide variety of wave conditions (Table 1) was chosen in order to model as many different combinations of wave period, direction, height, and tidal condi-

[1]For the purpose of starting the 14-second waves, deep water was considered to be where the waves were not appreciably affected by the bottom or at depths $\leq 1/4\ L_o$.

FIG. 3.—Bathymetry of the wide, shallow continental shelf of the Virginian Sea (Goldsmith and others, 1973). The major shelf relief elements influencing the waves are indicated. The shelf edge canyons (Washington and Norfolk) head at approximately 70 meter depths.

Table 1.—Initial deep-water wave conditions

Tide		Wave direction,	Wave periods,	Wave height	
m	ft	deg	sec	m	ft
0	0	0	4, 6, 8, 10	0.61, 1.83	2, 6
		22.5	4, 6, 8, 10, 12, 14	.61, 1.83	2, 6
		45 NE	4, 6, 8, 10, 12, 14	.61, 1.83	2, 6
		67.5	4, 6, 8, 10, 12, 14	.61, 1.83	2, 6
		90 E	4, 6, 8, 10, 12, 14	.61, 1.83	2, 6
		112.5	4, 6, 8, 10, 12, 14	.61, 1.83	2, 6
		135 SE	4, 6, 8, 10, 12, 14	.61, 1.83	2, 6
		167.5	4, 6, 8, 10, 12, 14	.61, 1.83	2, 6
		180 S	4, 6, 8, 10, 12, 14	.61, 1.83	2, 6
+1.22	+4.0	45	4, 6, 8, 10, 12, 14	.61	2
		90	4, 6, 8, 10, 12, 14	.61	2
		135	4, 6, 8, 10, 12, 14	.61	2

tions as possible from amongst the infinite variety of conditions that occur in nature.

Thus, a "library file" of a wide variety of wave conditions is accumulated so that it can be used in conjuction with other geological, biological, and chemical studies of the shelf and shoreline and as an aid to resource managers charged with selecting sites for offshore ports and shoreline defense structures.

One might ask the question as to why use this "scattergun" approach with respect to wave input conditions. Why not concentrate on just the most significant waves for calculating the wave parameters? There are two reasons why a wide variety of wave conditions is calculated. First, anyone can testify to the almost infinite variety of wave conditions that may occur over a long span of time. Second, data for determining the precise percentage of time that a given wave condition will occur are presently unavailable in most areas. The large spectrum of conditions is also needed in order to calculate parameters such as mean wave height at a shelf location and total shoreline wave energy along a stretch of coast during an average year. This could be easily calculated by summing up, based on frequency of occurrence, the data for a given location from each of the calculated wave parameters. Also, in order to determine the effects of storm waves along a shoreline, the sequence of weather fronts and resulting storm-generated waves is needed.

Such data come from four sources: (1) wave measurements by instruments in deep water, (2) wave measurements by instruments in shallow water and along the shoreline (i.e., on piers, anchored buoys, etc.), (3) shipboard wave observations compiled by U.S. Naval Oceanographic Office using 10° squares called Marsden squares, and (4) wave hindcast calculations. None of these methods has produced data considered adequate for the Virginian Sea.

Data from source (1) are quite rare, and where available, are generally of insufficient duration to be statistically valid. Summaries of shallow water wave measurements calculated from coastal wave gages were found to bear little relationship to individual shipboard wave height and period observations off the east coast of the United States (Thompson and Harris, 1972). These authors further concluded that if adequate data were available from shipboard observations, wave refraction methods would be useful in determining shallow water and shoreline wave parameters. Furthermore, no procedures presently exist which use wave parameters determined from coastal gages as input for propagating waves seaward incorporating the effects of bottom friction.

Shipboard wave observation data are not accurate enough for determining the percentage frequency of occurrence for a given wave condition, as there are several inherent biases built into the present data collection system. Several of these biases, such as the awkward computer forms, are discussed by Harris (1972). Another bias is suspected from the interpretation of summary graphs of shipboard wave observations from the Marsden squares adjacent to Cape Cod, Massachusetts, (Goldsmith, 1972) and the southern portion of the Virginian Sea (Goldsmith and others, 1974). In these two area summaries, the dominant waves appear on an annual basis to approach from the west, despite the proximity of land to the west and more than 3000 nautical miles of ocean to the east. One possible explanation for this suspected bias is related to the fact that the shipboard wave observations are recorded as part of a volunteer program. Ships tend to avoid extreme wave conditions, and when they do encounter severe conditions, the assigned observer might find that he has more important duties to perform than filling out wave forms. Nevertheless, shipboard observations appear to be the best information available at present for summing up individual wave conditions. These data have been used with some success in making littoral drift calculations along the coast of Florida (Walton, 1973).

The final method used in summarizing wave conditions utilizes wave hindcast calculations. Hindcast calculations using the Bretschneider-revised Sverdrup-Munk significant wave method have been computed for four stations along the U.S. East Coast, including one adjacent to the Chesapeake Bay entrance, by using data from weather stations for the 3-year period, 1948–1950 (Saville, 1954). There are large discrepancies between the shipboard wave observations and these wave hindcast data (Goldsmith and others, 1974). Important considerations in these discrepancies are the two major assumptions used in wave hindcasting: (1) deep water for 360° around the hindcast station and (2) that meteorological conditions in the 3-year period, 1948–1950, are representative of long-term weather conditions.

VERIFICATION OF WAVE RAY DIAGRAMS

A test of the wave ray diagrams in duplicating complex wave conditions was made by Farrell and others (1971) and is illustrated in Figure 5. The input for the diagram was based on the actual wave conditions as closely as they could be determined from the lower right margin of the photograph in "deep" water. Note the excellent qualitative comparison, even in the caustic regions down-wave from the island.

SHELF GEOMORPHOLOGY

Two important aspects of the shelf geomorphology in this area are the great width and relatively shallow nature of the continental shelf (Fig. 3). The abrupt increase in gradient at the shelf edge is between depths of 61 and 91 meters and is located as much as 60 nautical miles from shore. Thus, a great expanse of the continental shelf and superimposed relief elements is available for influencing wave behavior.

A closer examination of the detailed bathymetric map of the sea floor (Fig. 3) reveals that the shelf surface is not a smooth plain but instead consists of numerous irregularities. These irregularities may be divided into two groups: (1) Large-scale morphogeometry consists mainly of erosional forms cut into the shelf such as terraces, channels and valleys, and shelf-edge canyons. (2) Small-scale shelf relief elements consist of low relief features (i.e., less than 9 meters) of probable depositional origin, most notably ridge and swale bathymetry and arcuate (e.g., cape-associated) shoals. Whereas the origin of group (1) features is directly related to a lowered sea level, group (2) features probably formed since the last rise in sea level under the present shelf hydraulic conditions. The most recent eustatic sea level lowering reached its maximum extent approximately 15,000 years ago on the Atlantic Continental Shelf (Milliman and Emery, 1968).

Terraces.—The most pronounced terraces adjacent to Chesapeake Bay are at 24, 30, 40, and 86 meters. The presence of these terraces on the sea floor indicates a step-like bathymetric profile. The effect of the steeper portions of the profiles on the incoming waves will depend primarily on the angle of wave approach to these rises. However, even the steepest rises have relatively gentle slopes. The slope is 0°07′19″ for the rise between depths of 87.8 and 62.2 meters as compared with a slope of 0°01′58″ for the total shelf landward of the depth contour of 62.2 meters.

Subaqueous stream drainage.—The major relief features remaining from the Pleistocene stream drainage are the shelf valleys at the mouths of Delaware and Chesapeake Bays which are generally perpendicular to the strike of the terraces. Swift (1973) has suggested, however, that the Delaware shelf valley is an estuary retreat path and not a drowned river valley. Both these southeastern-oriented valleys have a pronounced influence on the wave refraction patterns, with areas of confused seas forming over the seaward rim of the shelf valleys.

Most of the relict Pleistocene river channel network has been filled in with sediments. However, subtle changes in relief in some areas of the shelf surface of the Virginian Sea are suggestive of former channels. Examples of these transverse shelf valleys are found between the mouth of Chesapeake Bay and Norfolk Canyon (Susquehanna Valley), from the Delaware Bay shelf valley to the shelf edge (Delaware Valley), from the Chesapeake Bay shelf valley southeastward to the shelf edge (Virginia Beach Valley), from the vicinity of Oregon Inlet, North Carolina southeastward to the shelf edge (Albermarle Valley), and from the Metomkin-Assawoman Island, Virginia, vicinity east-southeastward to Washington Canyon. The valley names are adopted from Swift and others (1972) (Fig. 3).

Virginia Beach Massif.—The Virginia Beach Massif, between the Susquehanna Valley and the Virginia Beach Valley, is an extensive shallow, relatively level-topped topographic high lying approximately between the depth contours of 18.3 and 21.9 meters (Fig. 3). This imposing large-scale relict feature, of probable interfluve origin, contains a superimposed irregular ridge and swale bathymetry, which is delineated by the depth contour of 18.3 meters. The Virginia Beach Valley, flanked to the northeast by the Virginia Beach ridges on the topographic high and to the southeast

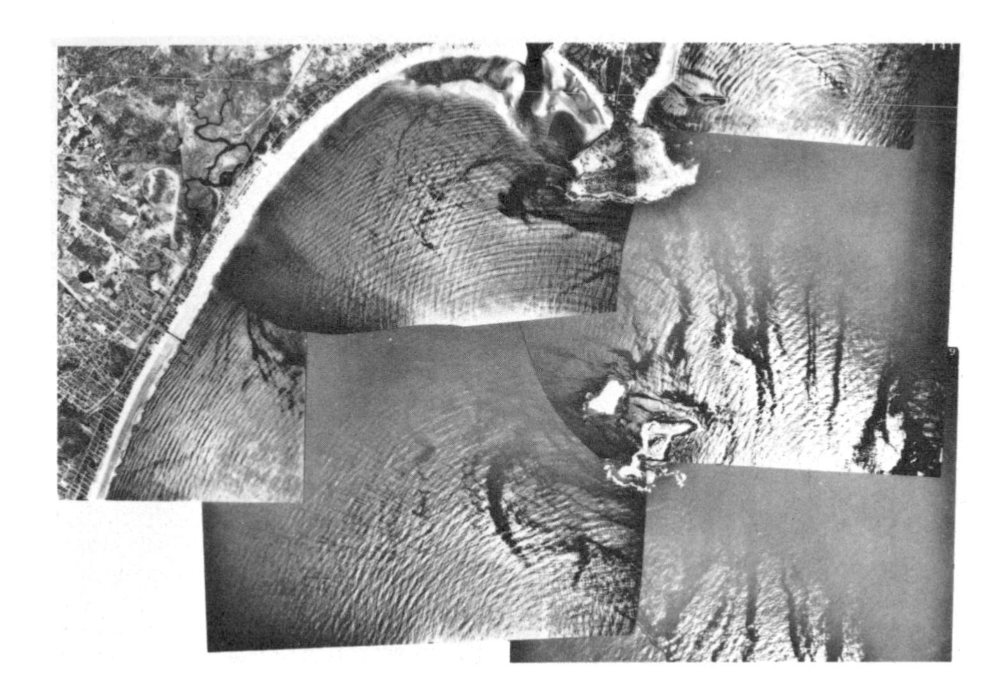

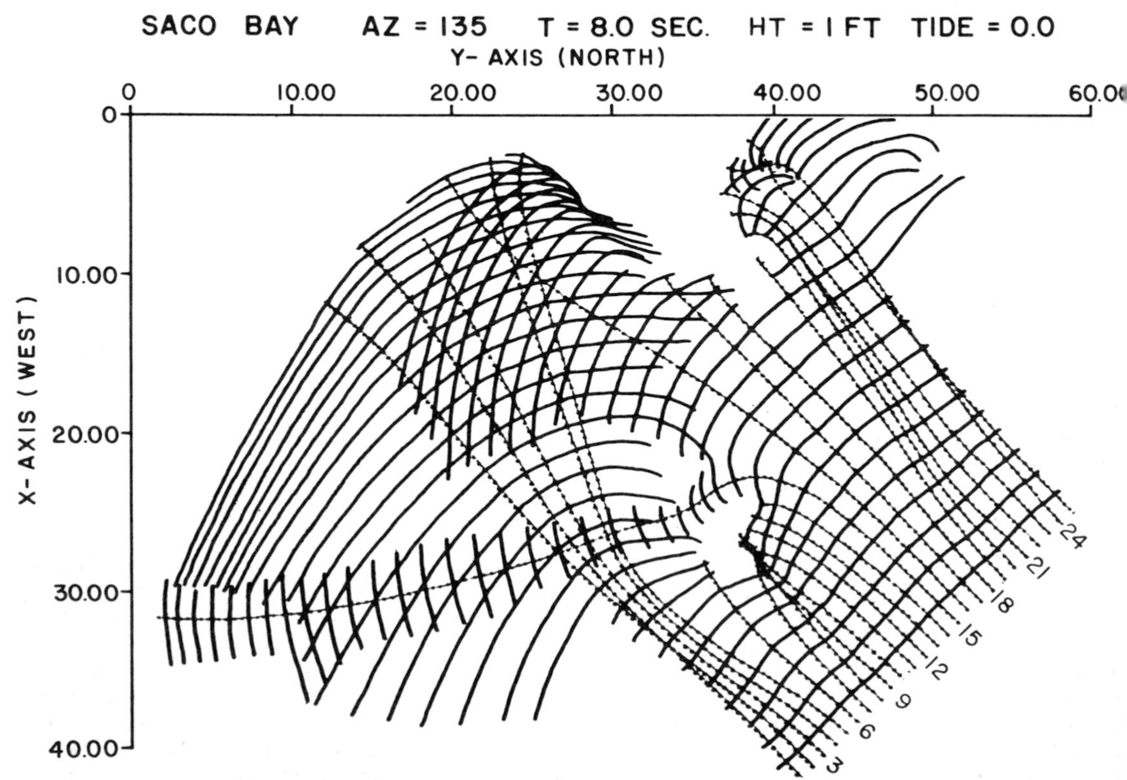

FIG. 5.—Comparison of vertical aerial photograph of Saco Bay, Maine, and wave refraction diagrams computed for waves from the southeast with a period of 8 sec (from Farrell and others, 1971).

by the False Cape ridges, is indeed suggestive of a series of relict ebb-tidal deltas formed as the sea level rose and the estuary mouth retreated as hypothesized by Swift and others (1972).

This complex topographic high, originating as an interfluve feature, with subsequent superimposed tidal-delta-associated ridges that have been modified under the present shelf hydraulic regime, has been named the Virginia Beach shoal retreat massif by Swift, and others (1972).

Linear ridges.—Superimposed on the larger relief elements is an undulating ridge and swale bathymetry composed of shoals with less than 9 meters of relief, with the long axis generally extending from 1 to 10 nautical miles and oriented such that it forms a small angle (<35°) with the present shoreline. Linear ridges, separated by valleys called swales, are most prominent opposite the shorelines of Delaware and Maryland, the southern Delmarva Peninsula, the Virginia-North Carolina state line, and Oregon Inlet to Rodanthe, North Carolina.

Arcuate shoals.—The arcuate shoals are most prominent when associated with capes such as within Chincoteague Shoals opposite the south end of Assateague Island. They are even more extensive immediately south of the study area, within Diamond Shoals opposite Cape Hatteras, North Carolina. Arcuate shoals are also located opposite the mouths of nearly all inlets along the coast of the Virginian Sea. The formation of the inlet shoals (i.e., ebb-tidal deltas) is related to the tidal-current-wave interaction. These sediment bodies have an important effect on the nearshore wave refraction patterns.

Probably the largest arcuate shoal in the study area is one associated with the entrance to Chesapeake Bay. Though highly bisected and cut by tidal channels, the distinct convex-seaward arcuate shape of this intermittent sand body, encompassing the mouth of the Bay, can be delineated from the detailed bathymetry. This huge sand body, suggestive of an ebb-tidal delta, may also be directly related to the origin of linear ridges adjacent to False Cape. Indeed, many of the linear ridges, especially those attached to shore, as well as many of the arcuate shoals, may owe their origin, in part, to the formation of now relict ebb-tidal deltas.

DATA PRESENTATION

The wave refraction calculations for this wide shallow shelf and adjacent shoreline area have been presented within several formats in a continuing series of publications, in order to encourage the widest possible usage.

The wave ray diagrams clearly illustrate the importance of these aforementioned east coast shelf-relief elements and shoreline wave energy distribution (Figs. 6–8) much as Munk and Traylor (1947) found in their manual wave refraction diagrams drawn for the United States west coast.

A specific example of such shelf-shoreline interaction, quite prominent on much of the east coast shelf, is schematically presented in Figure 9. In addition to the development of alternate shoreline zones of wave ray convergence and divergence by the wave refraction over these

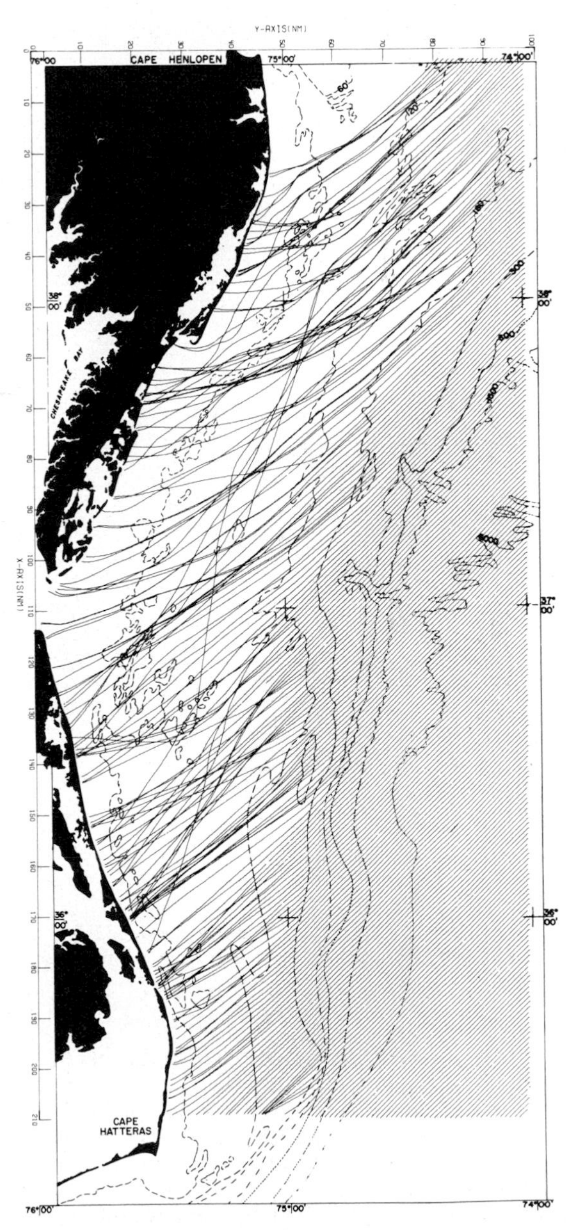

Fig. 6.—Wave refraction diagram computed in the VSWCM (Goldsmith and others, 1974); waves from the northeast with a period of 10 sec.

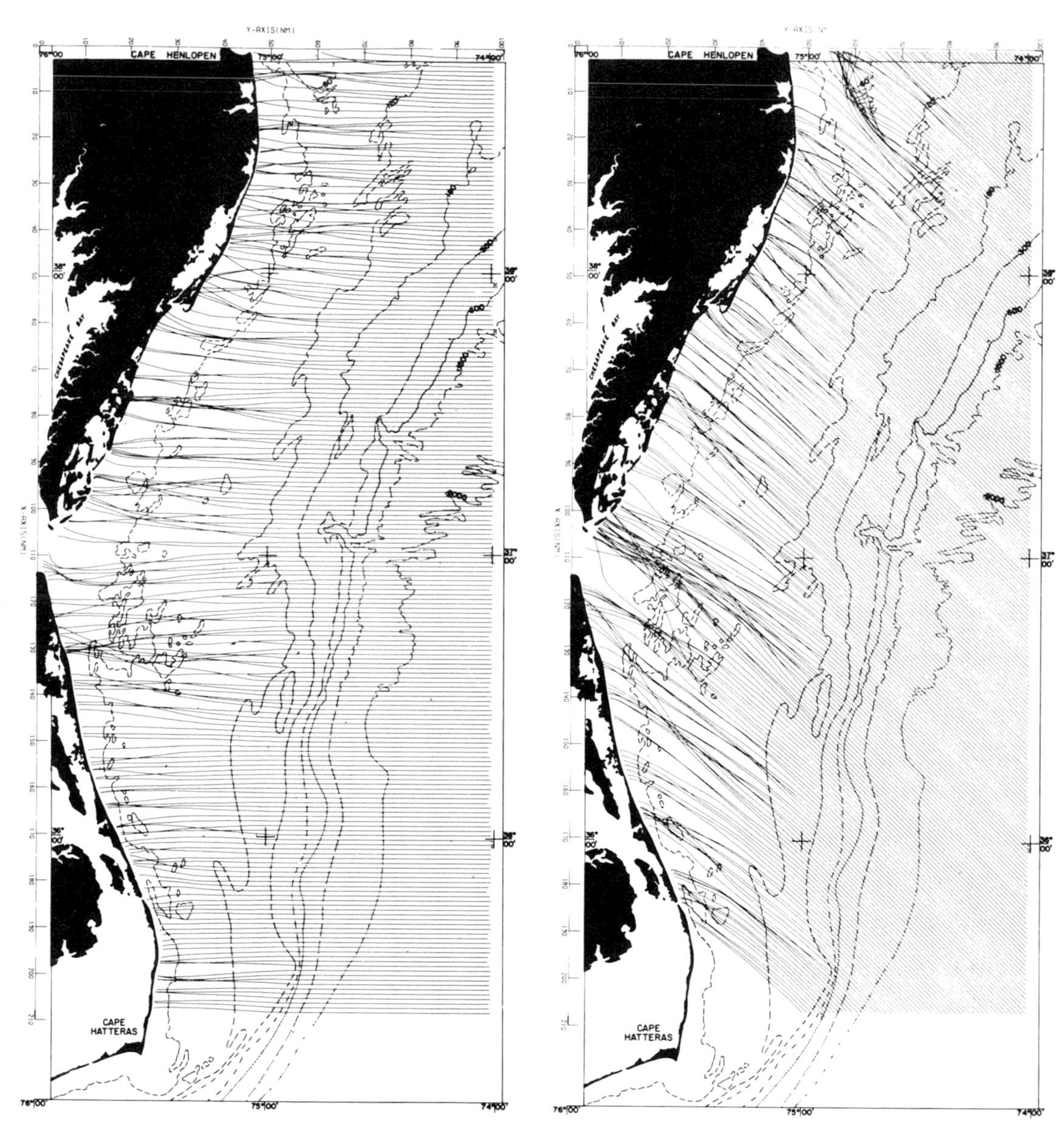

FIG. 7.—Wave refraction diagram computed in the VSWCM; waves from the east with a period of 10 sec.

FIG. 8.—Wave refraction diagram computed in the VSWCM; waves from the southeast with a period of 10 sec.

abundant linear ridges, these studies further suggest that such wave refraction may be an important process whereby these linear ridges are developed and maintained (Goldsmith, 1972). The bending of the wave fronts over ridges tends to encourage sand movement both downwave and toward the long axes of ridges. This process would result in the observed shape and orientation of the linear ridges with their long axes oriented perpendicular to the dominant wave approach directions on much of the shelf (Uchupi, 1968; fig. 14).

Shelf contour diagrams (Goldsmith and others, in preparation) of wave height and maximum bottom wave orbital velocities have been prepared for the Virginian Sea shelf by contouring these values. The parameters were calculated along each of the wave rays for 122 specific wave conditions (Figs. 10–12).

Note that both the wave heights and orbital velocities are higher over the relatively shallow

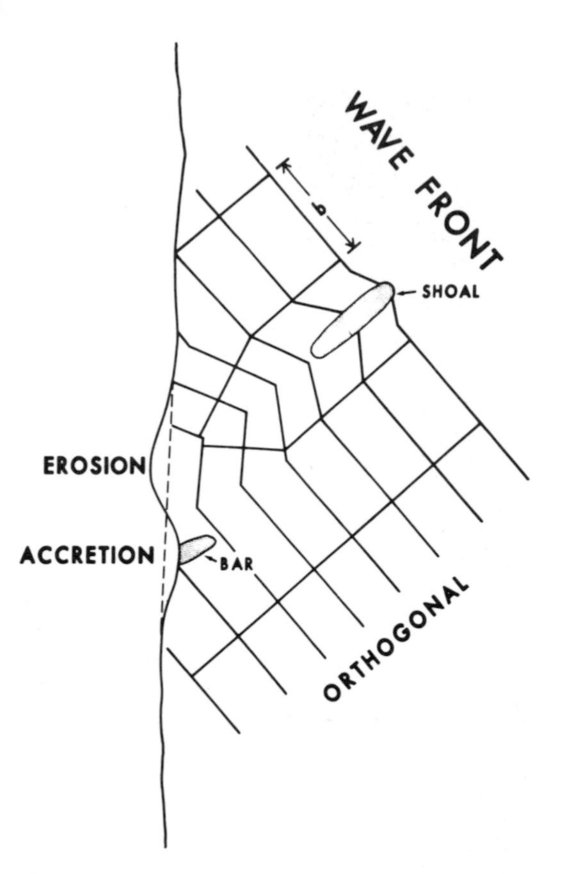

FIG. 9.—Schematic illustrating a mechanism proposed to explain growth and maintenance of linear ridges on shelf and its effect on shoreline wave energy distribution (Goldsmith, 1972).

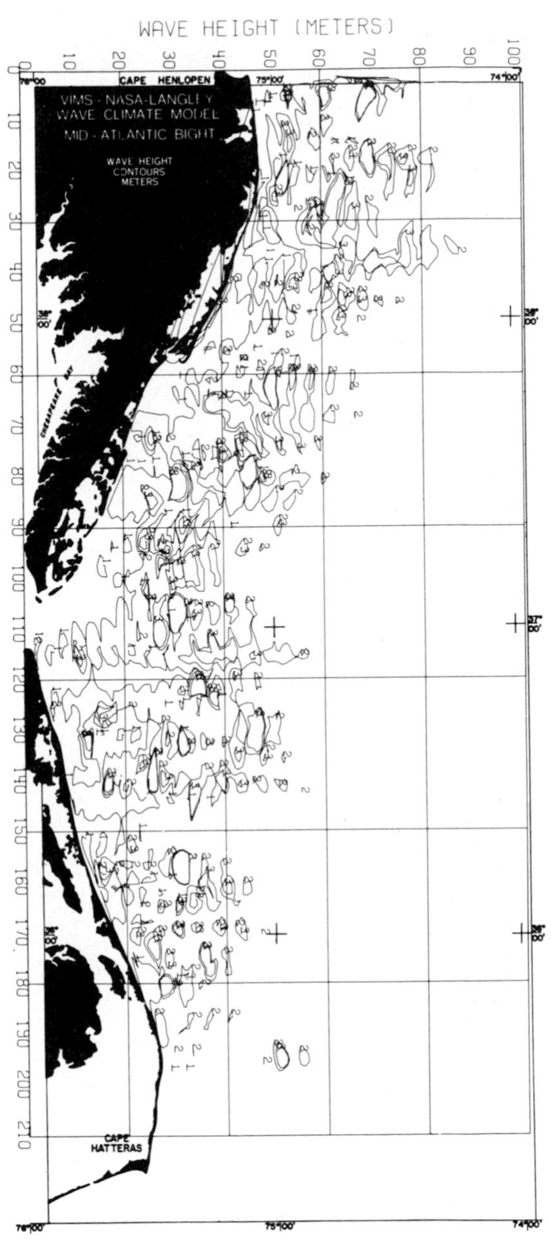

FIG. 10.—Continental shelf contour diagram summarizing computations of wave parameters made in the VSWCM along wave rays (Goldsmith and others, 1975); wave heights for waves from the east with a period of 10 sec.

Virginia Beach Massif (Fig. 3), and as a result are lower downwave from this feature, and seaward of the Virginia Beach area. Also note the tendency for the wave rays to converge, and hence, greater wave heights at the inlets of the Virginia Eastern Shore barrier islands (Fig. 11). This relationship exists for many of the wave ray diagrams in Goldsmith and others (1974).

Shoreline histograms of wave height and wave energy (Goldsmith and others, in preparation) are drawn by computer from the values of height and energy calculated at the ends of each of the wave rays. Where two or more wave rays impinge upon the shore quite close together, the wave heights from these superimposed rays are added. Thus, for this and other reasons important differences do occur although there is generally a qualitative agreement between shoreline histograms of wave ray frequency and wave height and energy (Fig. 13).

Examples of these shoreline histograms at one nautical mile class intervals for 200 nautical miles of coast, are presented in Figures 14–16. The first two diagrams illustrate shoreline wave heights, and the third illustrates shoreline wave energy distribution. Thus, Figures 7, 10–12, 15 and 16 all display wave information for waves from the

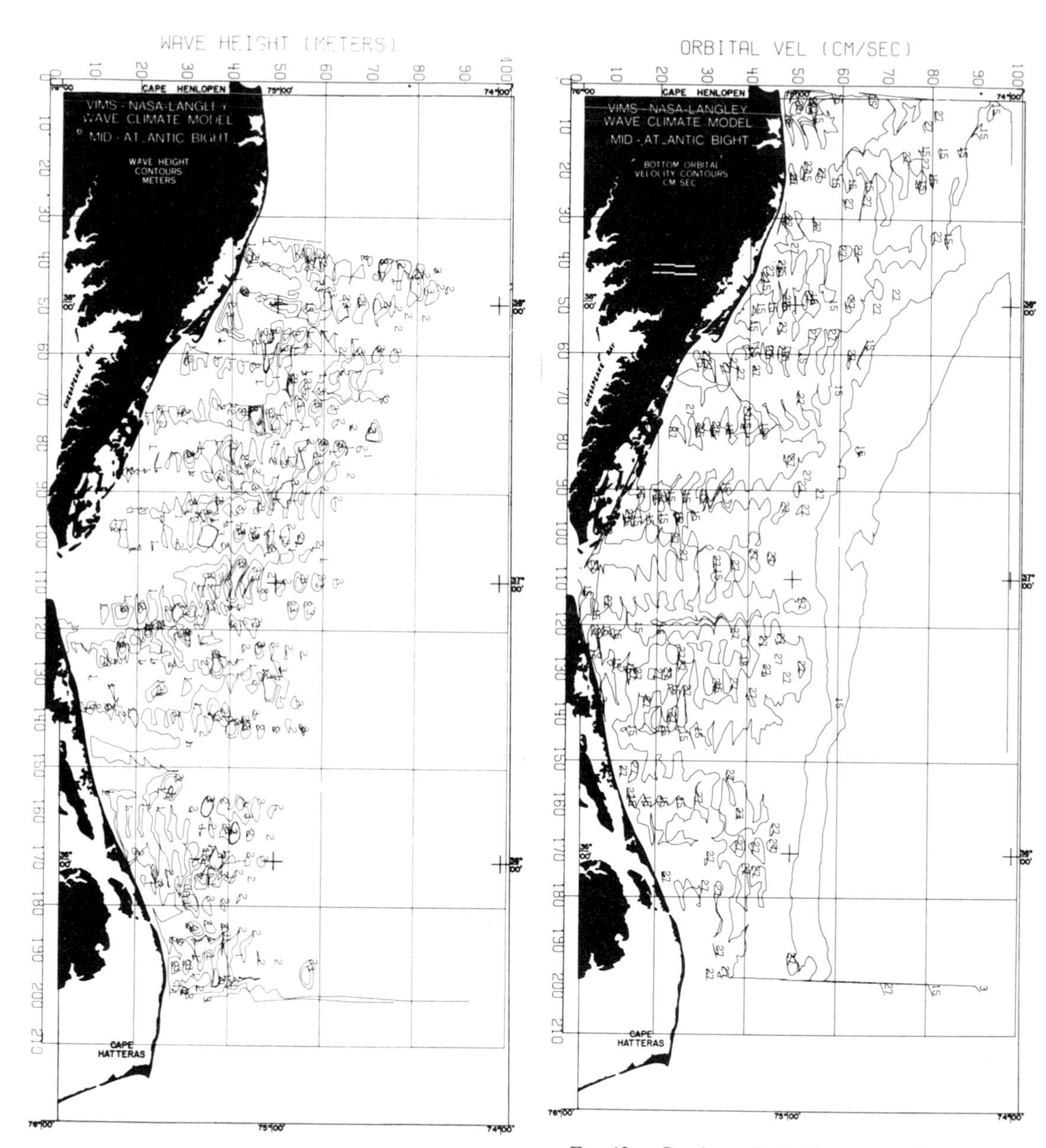

FIG. 11.—Continental shelf contour diagram summarizing computations of wave parameters made in the VSWCM along wave rays; wave heights for waves from the east with a period of 14 sec.

FIG. 12.—Continental shelf contour diagram summarizing computations of wave parameters made in the VSWCM along wave rays; maximum bottom orbital velocities for waves from the east with a period of 10 sec.

east with a period of 10 seconds, but with a variety of formats.

Note the dramatic shoreline variations in wave ray convergences and divergences, in wave heights and wave energy, all for waves from the east. This is caused by extensive wave refraction over many shelf relief elements superimposed on the wide, shallow shelf (discussed in an earlier section and shown in Fig. 3).

SHORELINE RESPONSE

Large variations along the shoreline in the computed wave parameters should be reflected in the shoreline processes, and these processes

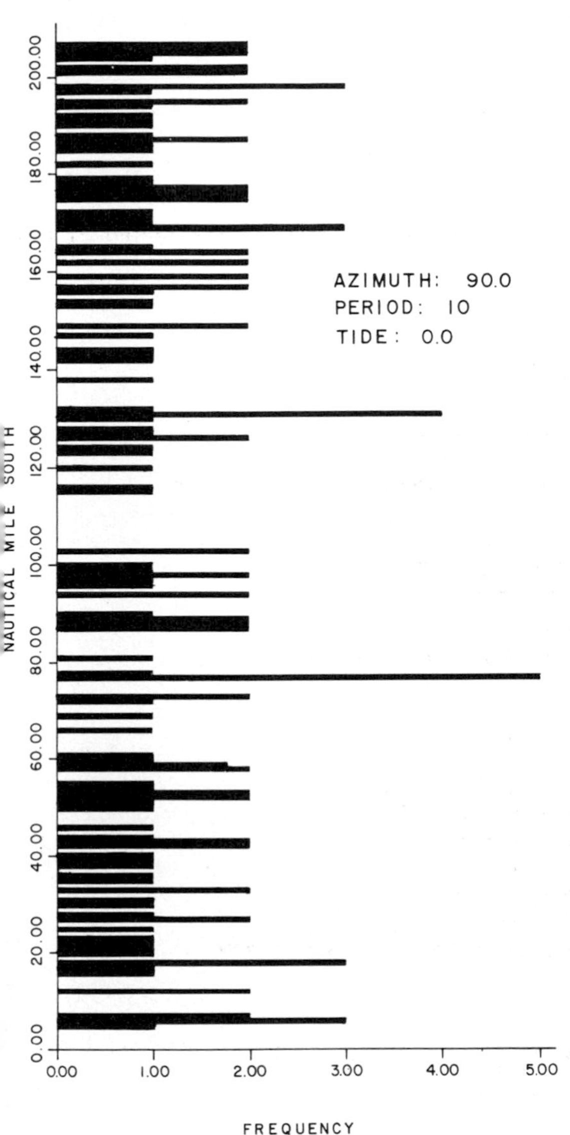

FIG. 13.—Shoreline wave ray histogram for waves from the east with a period of 10 sec; compare with Figures 14 and 16.

should be reflected in the shoreline response. The most obvious parameters that may be used to delineate the shoreline response are long-term historical shoreline changes and perhaps grain size of beach sediments.

Historical shoreline changes.—Shoreline changes for the area between Cape Hatteras and the Delaware-Maryland line are shown for approximately 48 to 105 year intervals (as indicated), depending on when the oldest and most recent surveys were made (Fig. 17). Several aspects of these data are of interest:

1. The largest amount of erosion (i.e., shoreline recession) occurs in the vicinity of inlets.
2. Even in areas away from inlets (e.g., south of Chesapeake Bay entrance) there is a large variability in shoreline erosion.
3. Other than at the inlets, the major erosion in this area occurs at Cape Hatteras and south of the Virginia-North Carolina state line.
4. The major accretion in this area occurs at the Maryland-Virginia state line and approximately 15 miles north of Cape Hatteras. There is little shoreline net change at about 36°N latitude.

Thus, the large variability in shoreline wave heights and wave energy appears to be reflected in the large variability in historical shoreline changes. With respect to wave ray diagrams, the tendency for wave ray convergence at these inlets has already been mentioned. Of course, tidal processes are also involved. However, the relationship between inlets and the computed wave energy concentrations may be an indication of the origin of the inlets at these particular locations, with the size of the tidal prisms being a major factor in maintaining the inlets.

With respect to shoreline histograms there appears to be a direct correlation between the shoreline recession at the Virginia-North Carolina state line (up to 5 meters/year) and the larger calculated wave heights (e.g., waves from the east with 10 and 14 second periods; Figs. 14, 15). Similarly, areas of accretion or small shoreline changes, in (4) above, compare well with low calculated waves heights (Figs. 14, 15).

It needs to be pointed out, however, that these are qualitative correlations, and that the correlations may be better for some wave conditions than for others. Present work at VIMS is being directed at quantifying and statistically representing these relationships for the 122 computed wave conditions. Spectral analysis techniques are being applied to shoreline spacing of wave heights. Preliminary analyses suggest strong spectral peaks at spacings of 5.3 and 12.0 nautical miles for the 200 nautical miles of coast for waves from the east (Goldsmith and Colonell, 1974).

Grain size distribution.—Mean grain size and standard deviation have been determined for the beach berm crests in four different investigations, at different times (Fig. 18). Despite the expected variability, some trends are apparent in this summary:

1. Whereas the northern area (Cape Henlopen to Chesapeake Bay entrance) shows a decrease in mean grain size to the south, the data for the southern area (i.e., the Outer Banks), indicates a general increase in mean grain size toward the south.
2. The coarsest beach sand is located north

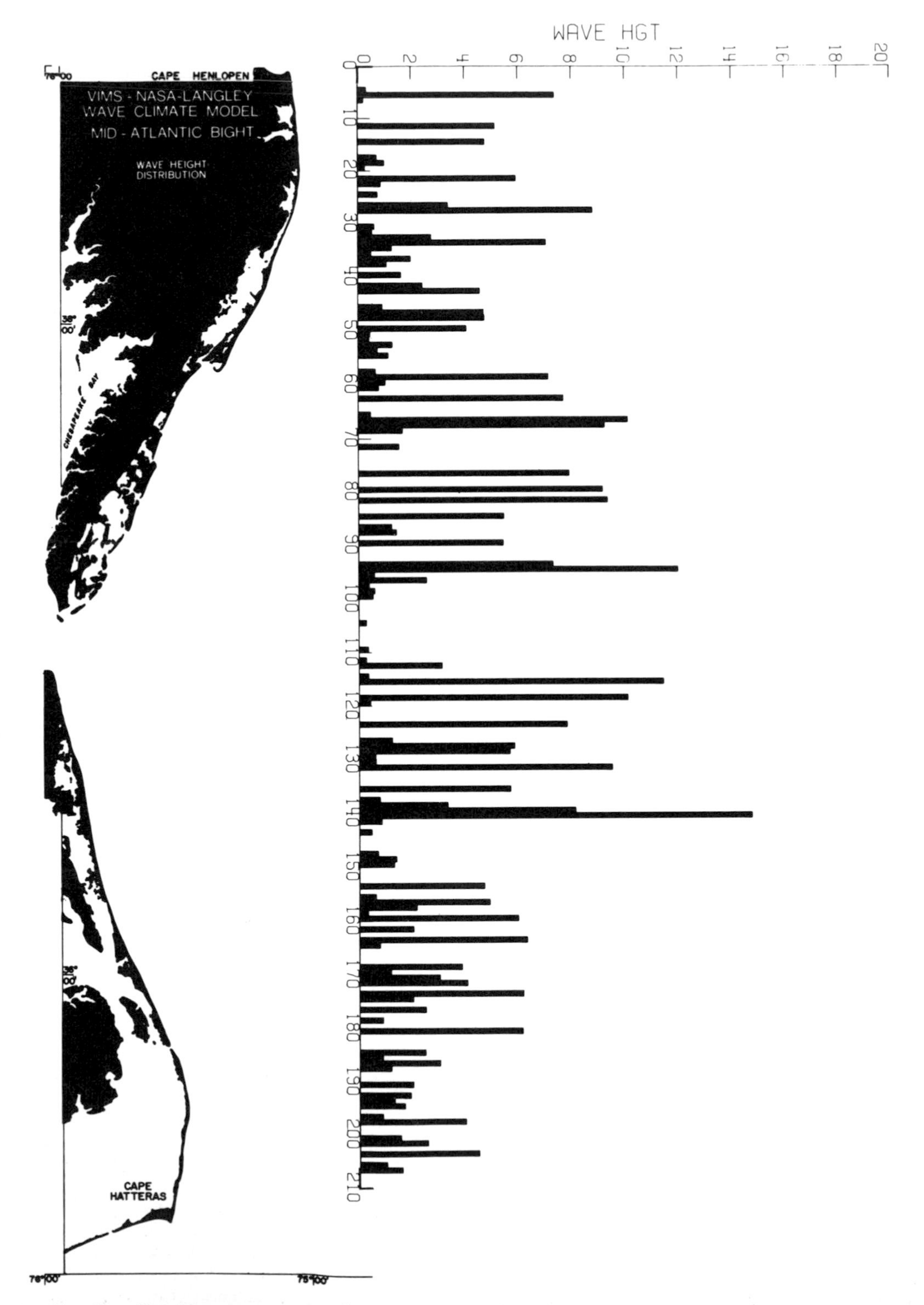

FIG. 14.—Shoreline histogram summarizing computations of wave parameters made in the VSWCM at the shoreline ends of the wave rays (Goldsmith and others, 1975); wave heights for waves from the east with a period of 10 sec.

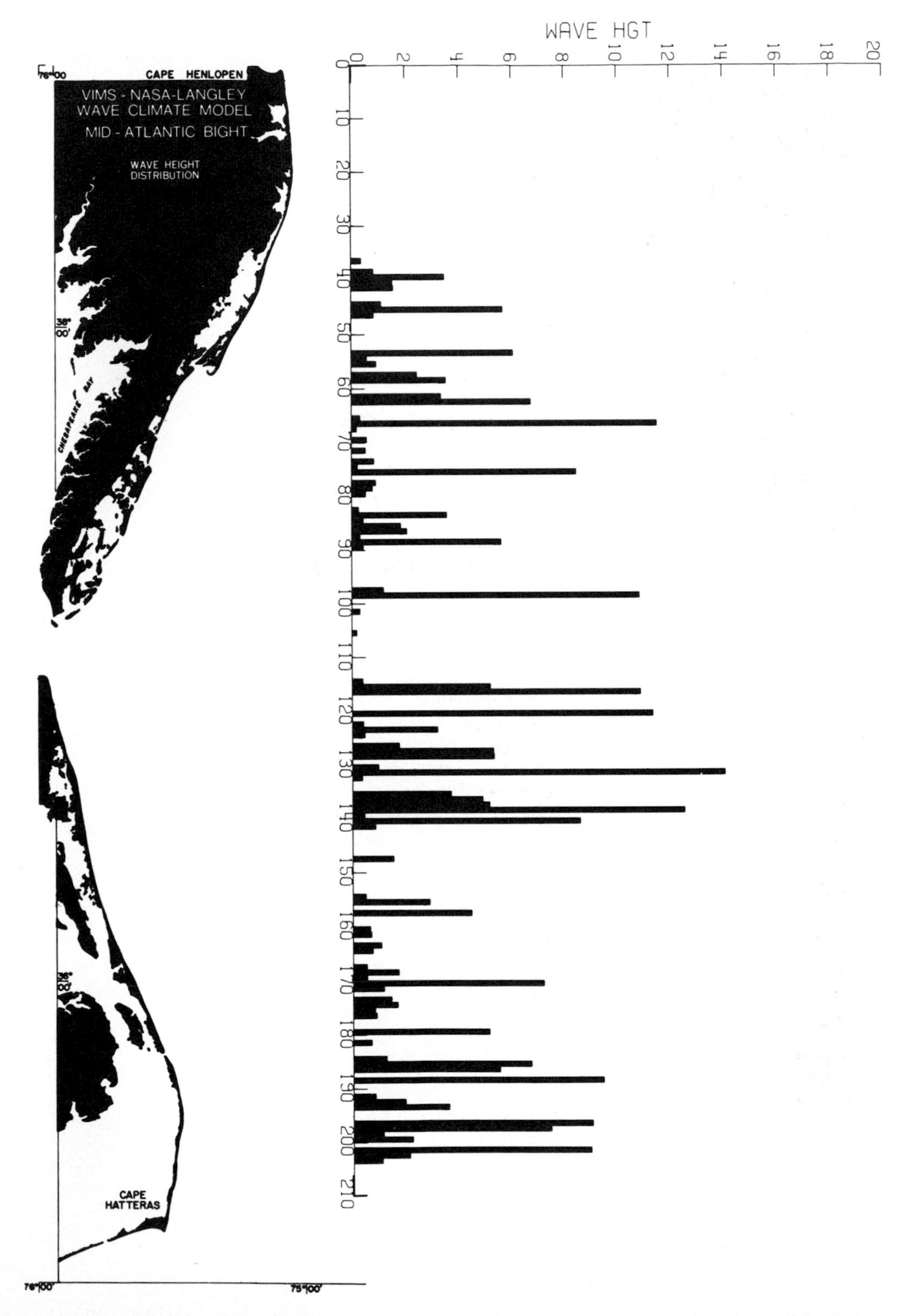

FIG. 15.—Shoreline histogram summarizing computations of wave parameters made in the VSWCM at the shoreline ends of the wave rays; wave heights for waves from the east with a period of 14 sec.

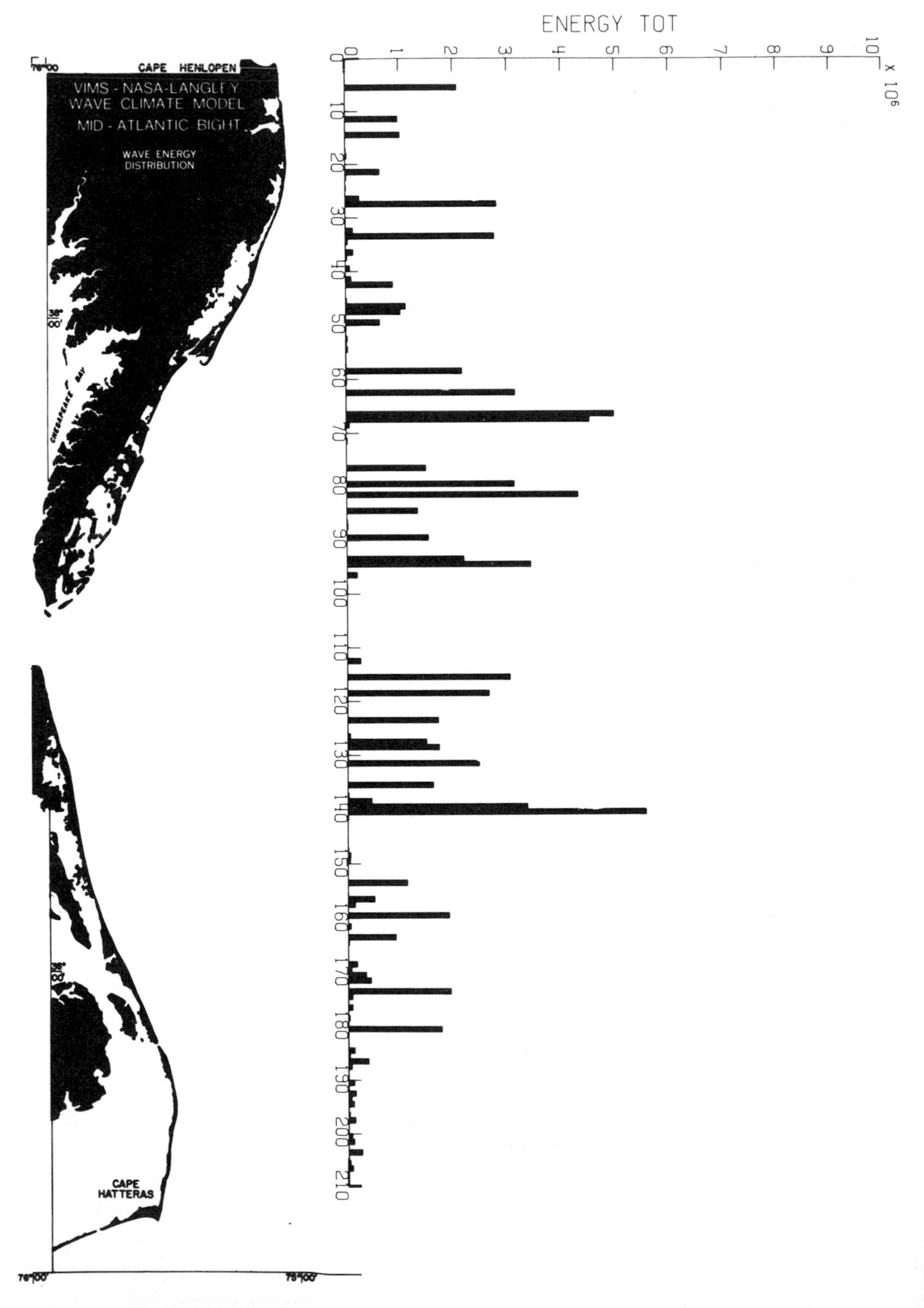

FIG. 16.—Shoreline histogram summarizing computations of wave parameters made in the VSWCM at the shoreline ends of the wave rays; wave energy for waves from the east with a period of 10 sec.

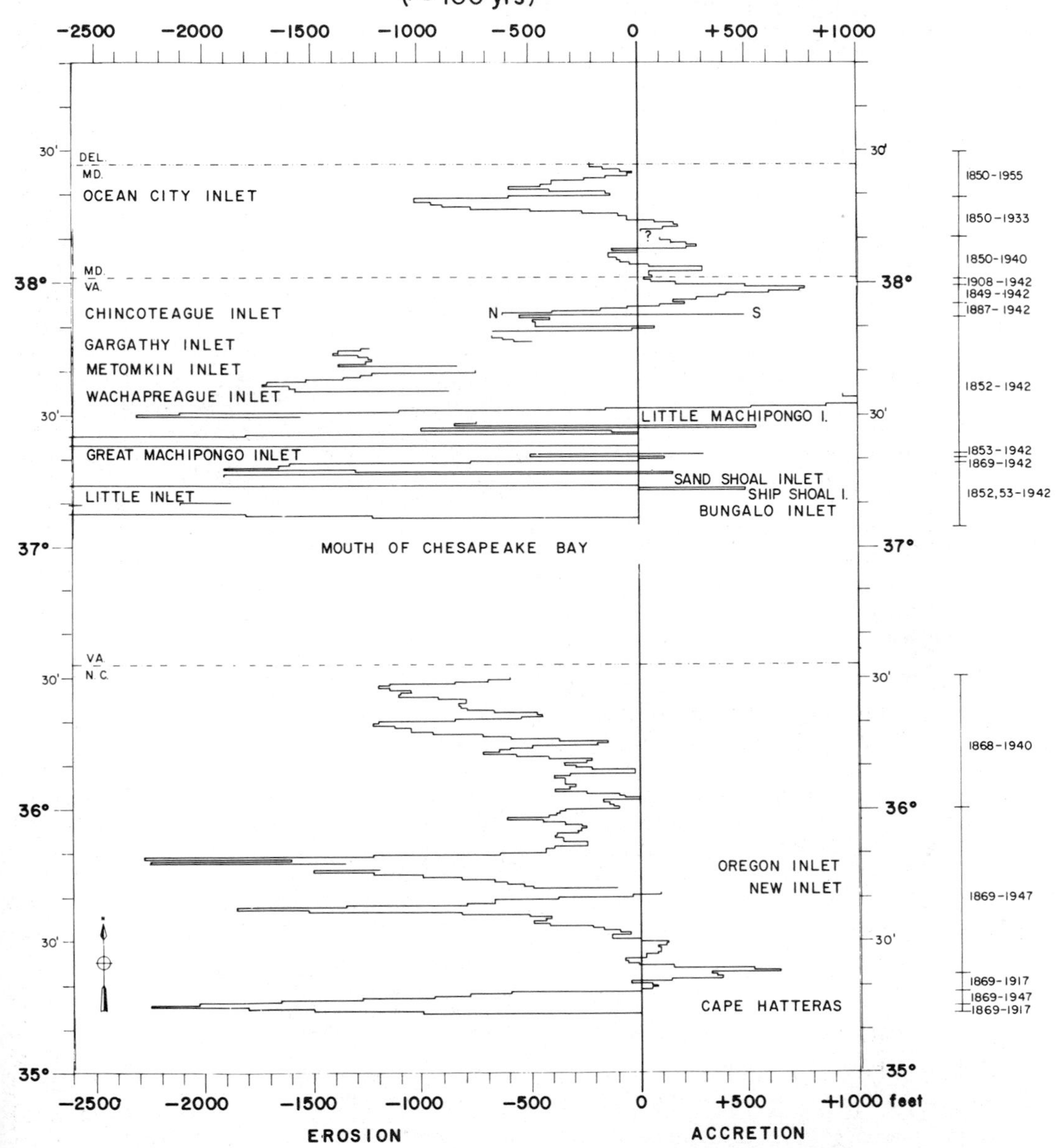

FIG. 17.—Historical shoreline changes for intervals of 48 to 105 years (as indicated) for the shoreline of the Virginian Sea between the Delaware-Maryland state line and Cape Hatteras, North Carolina. Compare with Figures 10–12 and 14–16.

of Duck, North Carolina, and at the south end of Assateague Island.

With respect to the former, it has been suggested that the increase in grain size to the south along the southerly Outer Banks is due to an increase in shoreline wave energy due to the abrupt narrowing of the shelf to the south (Shideler, 1973). This hypothesis seems to be substantiated by the computed shoreline wave heights for easterly 14-second waves (Fig. 15), but not easterly 10-second waves (Fig. 14). Does this mean that the 14-second wave is more important along the Outer Banks? Again, we need to expand our efforts along these lines by quantifying and further exa-

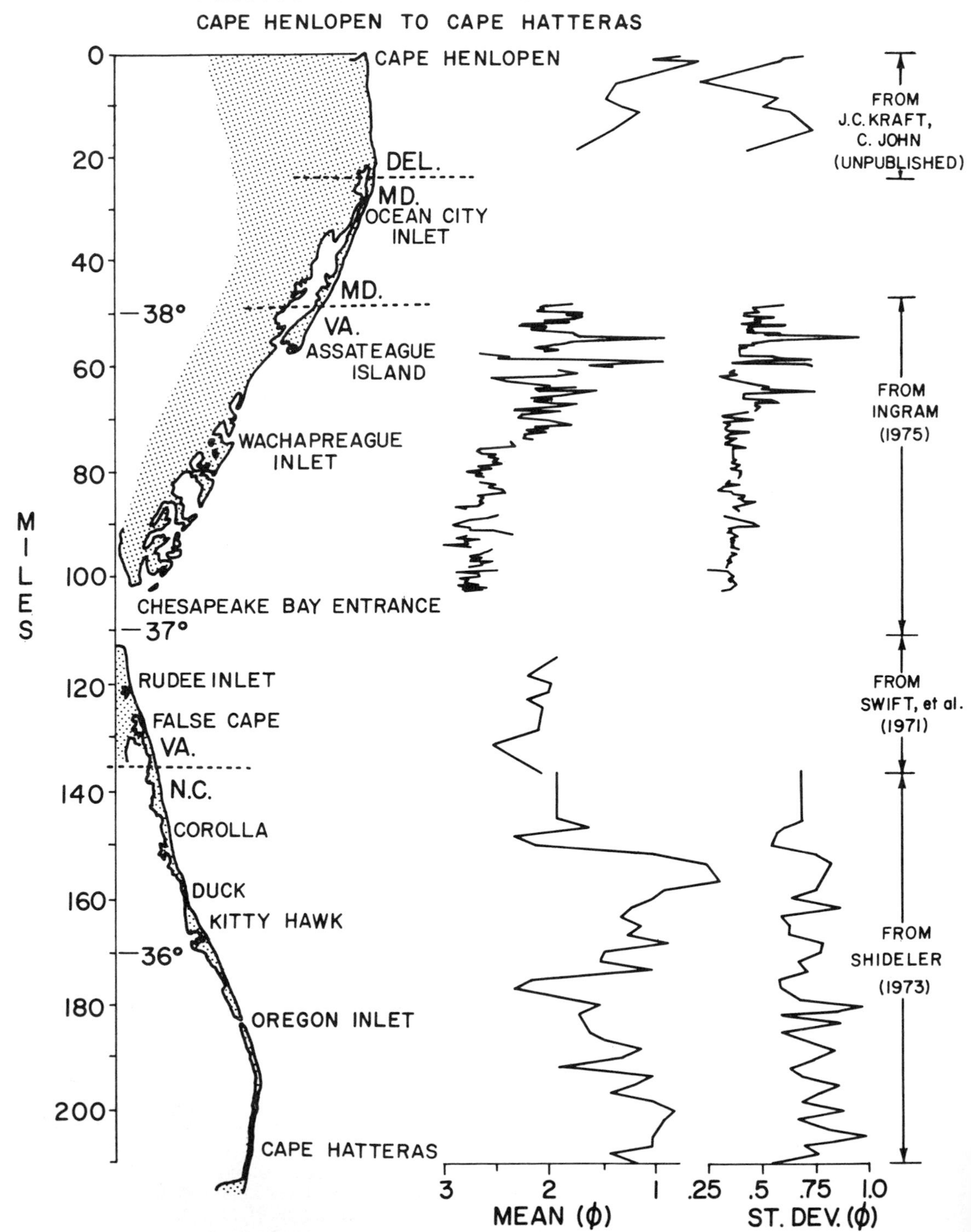
MEAN GRAIN SIZE (Ø) AND ST. DEV. (Ø)
ALONG BERM CREST
CAPE HENLOPEN TO CAPE HATTERAS
CAPE HENLOPEN
DEL.
MD.
OCEAN CITY
INLET
MD.
VA.
ASSATEAGUE
ISLAND
WACHAPREAGUE
INLET
CHESAPEAKE BAY ENTRANCE
RUDEE INLET
FALSE CAPE
VA.
N.C.
COROLLA
DUCK
KITTY HAWK
OREGON INLET
CAPE HATTERAS
38°
37°
36°
MILES
0
20
40
60
80
100
120
140
160
180
200
FROM
J.C. KRAFT,
C. JOHN
(UNPUBLISHED)
FROM
INGRAM
(1975)
FROM
SWIFT, et al.
(1971)
FROM
SHIDELER
(1973)
3
2
1
.25
.5
.75
1.0
MEAN (Ø)
ST. DEV. (Ø)

mining these relations. With respect to the very coarse zone north of Duck, North Carolina, reconnaissance surveys by the author suggest that this zone is due to an additional local source of sediment. The local sediment source, which is most probably a relict deposit, could be on the shelf and is being reworked and moved landward by waves, or could have originated when Currituck Sound had a direct opening through the Outer Banks barrier into the ocean.

STATE OF THE ART

Despite the wide usage of the wave refraction, and application to many coastal problems, theoretical advances in the approach since the work of Munk and Traylor has been surprisingly limited. In general, great confidence exists in the ability of the computed wave ray diagrams to reproduce wave behavior, even in areas of complex conditions (Fig. 5). Somewhat less confidence exists in the results of the wave computations involving wave height, such as wave energy and bottom orbital velocities. These data can be, and should continue to be, used for understanding coastal processes, computation of design wave conditions, and other uses. This cautionary note is merely meant to heighten the awareness that there are still unsolved problems, some of which are discussed below.

Depth fitting procedures.—All of these methods involve some application of a technique to smooth a surface fitted to the depth at the "local" grid points. Dobson (1967) fitted a quadratic surface to a grid of 12 adjacent depths in the form of a cross in order to determine the "local depth" (i.e., depth and locations between grid points). This is the scheme used in the VSWCM.

There is probably not another aspect of wave refraction computation, except for crossed-wave patterns, that produces so much controversy as that of local depth determination. Most of the current schemes employ a grid of measured depths for this purpose. There seem to be as many schemes for handling this problem as there are programs. While each scheme has both advantages and disadvantages, none is clearly superior to all others. This state of affairs is largely due to the fact that there does not seem to be a general agreement on what criteria should be used to evaluate the various schemes. Until such agreement exists, those schemes having the loudest and most persistent advocates will enjoy the highest regard.

Thus, in view of the limitations in the available depth information (discussed previously), it seems of little consequence to incorporate "better" depth-finding subroutines until we make vast improvements in the quality of our basic depth information (Fig. 4).

Crossed-wave patterns.—The effects of shoaling and refraction can be estimated by linear wave theory. For example, the propagation of surface waves into shallow water is analyzed by consideration of the wave energy between two vertical planes which are orthogonal to the wave crests and which intersect with the surface to produce wave rays. Energy is assumed not to be transmitted along the wave crest; thus, it is not transmitted across wave rays. If it is also assumed that the wave period is constant and that there is no loss or gain of energy from reflection or percolation, then linear wave theory provides the well-known result,

$$\frac{H}{H_o} = K_r K_s K_f \tag{1}$$

where H is modified wave height, H_o is initial deep water wave height, K_r is coefficient of refraction, K_s is coefficient of shoaling, and K_f is the wave height reduction factor due to friction and is proportional to $1/f$ where f is the friction coefficient.

The coefficient of refraction is given by

$$K_r = \frac{b_o^{1/2}}{b} \tag{2}$$

where b_o is initial distance between adjacent rays and b is distance between adjacent rays.

It can be seen that in the calculation of the wave refraction coefficient (K_r) that as b goes to 0.0 when the wave rays cross, the resulting wave height will approach infinity; that is, according to linear wave theory. Wave observations and subsequent theoretical work prove that this is certainly not the case; i.e., wave heights do not become infinitely high. The proper interpretation of crossed-wave rays (or fronts) in the refraction diagrams (i.e., caustics), such as those in many of the wave refraction diagrams, does not appear to be the problem it was once thought to be. Chao, in a thorough series of theoretical (Chao, 1970, 1971), wave tank (Chao and Pierson, 1970, 1972), and continental shelf (Chao, 1972,

←

FIG. 18.—Mean grain size and standard deviation along the berm crests between Cape Henelopen, Delaware, and Cape Hatteras, North Carolina. Compare with Figures 10–12 and 14–16.

1974) refraction studies of the caustic phenomena, has reached the following conclusion for such wave refraction studies (Chao, 1974, p. 32): "The rays, after escaping from the caustic regions, eventually follow the continued ray path and the wave conditions are determined by the b factor just as if no caustic had occurred except that there has been a phase shift, which is unobservable because of the randomness of waves in nature. These conditions eliminate the necessity of the evaluations of the waves near a caustic . . ." Although some wave height changes may occur in the waves that pass through a caustic region, theoretical and wave tank studies (Chao and Pierson, 1972) suggest that such changes seaward of the zone of breaking waves may be minimal and well within the bounds set by other limiting factors, such as depth information.

The qualitative correlation between the Saco Bay photograph and diagram (Fig. 5) suggests that the computational procedure is reasonably valid for this situation which is characterized by a complex shoreline and irregular bathymetry.

Wave energy dissipation due to bottom friction.—The VSWCM incorporates the effects of bottom friction, which, in the wide shallow shelf of the Virginian Sea, acts to reduce wave energy and consequently, wave heights approximately 50 to 75 percent for the longer wave periods. Thus, these computations of bottom friction are quite important. The calculations of frictional loss were adopted from computer routines developed by Coleman and Wright (1971) based upon equations for calculating bottom friction developed by Putnam and Johnson (1949) and modified by Bretschneider and Reid (1954) and Bretschneider (1954).

Bretschneider and Reid (1954) presented a general theoretical solution for Putnam's equations for deriving wave energy loss due to bottom friction. In these equations a friction coefficient of 0.01 was used, but only with a carefully phrased preface alluding to the numerous assumptions that were made. These assumptions involve steady wave conditions, and hence, a stable bottom, because studies by Savage (1953) and others had demonstrated large variations in K_f with changes in bottom sand ripples. This results in up to 25 percent loss in wave height due to the presence of bottom sand ripples. Changes in bottom material along wave paths was also mentioned by Bretschneider and Reid (1954) as critical with respect to choosing a value for f.

In a succeeding report, Bretschneider (1954) reported on the results of a field investigation which was conducted, as a companion study to the theoretical work, in order to refine the equations for wave energy loss from bottom friction. The average of 10 values of f was 0.053 (Table 3, p. 9), and was derived from field measurements of wave height changes between two offshore oil platforms in the Gulf of Mexico. Because of the surprisingly high value of f, it was decided to carry the investigation further (Bretschneider, 1954, p. 10). A solution was theoretically derived to account for some of this energy loss through nonfrictional processes; i.e., having a non-rigid, impermeable bottom, participate in the wave motion.

A thorough review of the methodologies for directly or indirectly measuring bottom friction under oscillatory flow was made by Jonsson (1966). He concluded (p. 140) that, "In nature the boundary layer is always rough and turbulent (thus, there is more friction than with laminar flow). The friction factor here will often exceed the value of 0.02 adopted by Bretschneider (1965). This is also confirmed by the observations of Iwagaki and Kakinuma (1965)." More recent studies have further supported this contention.

In summary then, both Bretschneider's theoretical and field studies fall far short of predicting a particular value for the frictional coefficient. Surprisingly, the value of 0.01 for f, which resulted largely from theoretical rather than field considerations (e.g., Bretschneider, 1954; Iwagaki and Kakinuma, 1965) has been generally adopted (Coastal Engineering Research Center, 1973, p. 3-46) despite much subsequent work which indicates that a much higher value should be used for f.

In the VSWCM a value of $f = 0.02$ was used in order to adopt a conservative approach, since the higher the value of f the greater the frictional reduction in wave height during wave progression through shallow water. However, the potential wave analyst should be alert to this new body of literature and evaluate the wave results accordingly.

Zone of breaking waves.—Because of the complexity of the wave processes within, and landward of the zone of breaking waves, few attempts have been made to continue the wave refraction computations into this zone. Thus, nearly all of these refraction programs, including the VSWCM, end at the point of wave breaking. In reality, however, in areas of low nearshore gradients the waves generally break, reform, and break again, adding to the complexity of interpretation.

The major significance of this limitation is in the use of wave climate models in the computations of longshore drift, which is certainly a critical parameter for understanding the coastal processes of a particular area. There are, at present, several promising efforts in longshore drift computations that are applicable to adoption within presently existing wave refraction schemes, such as Komar (1975), Tanner (1974), Walton (1973), Galvin (1973), and Fox and Davis (1973).

CONCLUSIONS

Wave climate models have become an important tool for assisting in the understanding of coastal processes. This is because of the general recognition that the most important wave process on the continental shelf is the interaction between the ocean waves and the various shelf relief elements, resulting in the observed nonuniform wave energy distribution over the shelf and along the nearshore zone.

A regional approach to the study of this nonuniform wave energy distribution by the VSWCM has shown that the large variations in computed shoreline wave heights and wave energy are reflected in the large variations in the observed historical shoreline changes and, to a lesser extent, in the beach grain size variations.

Basic wave refraction patterns result in a wave energy concentration downwave from a topographic high (e.g., linear ridges) and a wave energy diminution downwave from a topographic low (e.g., shelf canyons). Also, of importance are areas of "confused seas", or crossed wave patterns (i.e., straight caustics), that occur over the downwave side of canyon rims for particular wave periods and approach directions, and other shelf areas.

The ability of wave refraction diagrams to accurately duplicate wave behavior, even under complex crossed wave patterns, has been largely verified. The results of computations of wave parameters based on wave heights should be interpreted in terms of the present state of the art with respect to the problems of bottom friction, crossed-wave patterns, and wave behavior beyond the zone of breaking waves.

ACKNOWLEDGEMENTS

This research was supported by Sea Grant Program grant No. 04-5-158-49 to VIMS from the Office of Sea Grant Programs, National Oceanic and Atmospheric Administration, Department of Commerce, under Public Law 39-688, and the Commonwealth of Virginia. The Model computations were provided by NASA-Langley under the direction of W. D. Morris.

An undertaking of this vast scope required the assistance of many able individuals. Special appreciation goes to C. H. Sutton for assisting in much of the data preparation (including the historical shoreline changes), and to R. J. Byrne and J. M. Colonell for their competent advice on all phases of the project. R. Bradley, K. Thornberry and C. Otey provided expert assistance in drafting, photography and typing, respectively.

J. C. Kraft and C. John kindly furnished unpublished grain size data for the Delaware coast, and R. J. Byrne furnished original data of shoreline changes along the Eastern Shore of Virginia.

REFERENCES

Anonymous, 1950, Wave refraction at Long Beach and Santa Barbara, California: U.S. Army Corps of Engineers, Beach Erosion Board Bull., v. 4, p. 1-14.

Arthur, R. S., 1951,The effect of islands on surface waves: Scripps Inst. Oceanography Bull., v. 6, p. 1-26.

Bates, C. C., 1949,Utilization of wave forecasting in the invasions of Normandy, Burma and Japan: New York Acad. Sci. Annals, v. 51, p. 545-569.

Bascom, W. J., 1974, The control of stream outlets by wave refraction: Jour. Geology, v. 62, p. 600-605.

Bretschneider, C. L., 1954, Field investigation of wave energy loss in shallow water ocean waves: U.S. Army Corps of Engineers, Beach Erosion Board Tech. Memo., No. 46, 21 p.

——, 1965, Generation of waves by wind—State of the art: Nat. Eng. Sci. Co., Washington, D. C., 96 p.

——, and Reid, R. O., 1954, Modification of wave height due to bottom friction, percolation and refraction: U.S. Army Corps of Engineers, Beach Erosion Board Tech. Memo 45, 36 p.

Chao, Y. Y., 1970, The theory of wave refraction in shoaling water, including the effects of caustics and the spherical earth: New York Univ. Sch. Eng. Dep. Meterology and Oceanography, Geophys. Sci. Lab., TR-70-7, 70 p.

——, 1971, An asymptotic evaluation of the wave field near a smooth caustic: Jour. Geophys. Res., v. 76, p. 7401-7408.

——, 1972, Refraction of ocean surface waves on the continental shelf: New York Univ. Sch. Eng., Dep. Meteorology and Oceanography, Geophys. Sci. Lab., Contr. 124, 10 p.

——, 1974, Wave refraction phenomena over the continental shelf near the Chesapeake Bay entrance: U.S. Army Corps of Engineers, Coastal Eng. Res. Center Tech. Memo. 47, 53 p.

——, Jones, A. M., and Roney, J. R., 1975, Wave environment prediction in the coastal zone [abs.]: Am. Soc. Civil Engineers, Proc. Civil Eng. in the Oceans/III, June 9-12, Newark, Delaware, Abs. Vol., p. 135-136.

——, and Pierson, W. J., 1970, An experimental study of gravity wave behavior near a straight caustic: New York Univ. Sch. Eng., Dep. Meteorology and Oceanography, Geophys. Sci. Lab., TR-70-17, 40 p.

—— and ——, 1972, Experimental studies of the refraction of uniform wave trains and transient wave groups near a straight caustic: Jour. Geophys. Res., v. 77, p. 4545-4554.

Coastal Engineering Research Center, 1973, Shore protection manual: U.S. Army Corps of Engineers, Coastal Eng. Res. Center, 3 vols.

Coleman, J. M., and Wright, L. D., 1971, Analysis of major river systems and their deltas: Procedures and rationale with two examples: Louisiana State Univ. Coastal Studies Inst., Tech. Rep. 95 (Contract N00014-69-A-0211-0003), 125 p.

DOBSON, R. S., 1967, Some applications of a digital computer to hydraulic engineering problems: Stanford Univ. Dep. Civil Eng., Tech. Rep. 80 [Contract Nonr 225 (85)], p. 7-35, Appendix B.

FARRELL, S., RHODES, E., AND COLONELL, J., 1971, Saco Bay, Maine, a test of computer models for wave forecasting and refraction [abs.]: Geol. Soc. America Abstracts with Programs, v. 3, p. 562-563.

FOX, W. T., AND DAVIS, R. A., JR., 1973, Simulation model for storm cycles and beach erosion on Lake Michigan: Geol. Soc. America Bull., v. 84, p. 1769-1790.

GALVIN, C. J., JR., 1973, Longshore energy flux, longshore force, and longshore sediment transport [abs.]: Am. Geophys. Union Trans., EOS, v. 54, p. 334.

GOLDSMITH, V., 1972, Coastal processes of a barrier island complex and adjacent ocean floor: Monomoy Island-Nauset Spit, Cape Cod, Massachusetts: Ph.D. Dissertation, Univ. Massachusetts, 469 p.

——, BYRNE, R. J., SALLENGER, A. H., JR., AND DRUCKER, D. M., 1973, The influence of waves on the origin and development of the offset coastal inlets of the Delmarva Peninsula: Second Internat. Estuarine Conf.: Geol. Sec.—Coarse-grained sediment transport and accumulation in estuaries, Myrtle Beach, South Carolina, October 15-18, 1973, p. 183-200.

——, AND COLONELL, J., 1970, Effects of nonuniform wave energy in the littoral zone: Am. Soc. Civil Engineers, Proc. 12th Conf. on Coastal Eng., p. 767-785.

—— AND ——, 1974, Results of ocean wave-continental shelf interaction: Am. Soc. Civil Engineers, Proc. 14th Conf. on Coastal Eng., p. 586-600.

——, MORRIS, W. D., BYRNE, R. J., AND WHITLOCK, C. H., 1974, Wave climate model of the mid-Atlantic shelf and shoreline (Virginian Sea): Model development, shelf geomorphology, and preliminary results: Virginia Inst. Marine Sci., Spec. Rep. Appl. Marine Sci. and Ocean Engr. 38, 146 p.

——, SUTTON, C. H., AND DAVIS, J. S., 1973, Bathymetry of the Virginian Sea-continental shelf and upper slope, Cape Henelopen to Cape Hatteras: Bathymetric map: Virginia Inst. Marine Sci., Spec. Rep. Appl. Marine Sci. and Ocean Engr. 39, map, 1 sheet.

HARRIS, D. L. 1972, Characteristics of wave records in the coastal zone: *In* R. E. Meyer (ed.), Waves on beaches and resulting sediment transport: Academic Press, Inc., New York, p. 1-51.

HARRISON, W., AND WILSON, W. S., 1964, Development of a method for numerical calculation of wave refraction: U.S. Army Corps of Engineers, Coastal Eng. Res. Center Tech. Memo. 6, 64 p.

HORRER, P. L., 1950, Southern hemisphere swell and waves from a tropical storm at Long Beach, California: U.S. Army Corps of Engineers, Beach Erosion Board Bull. 4(3), 18 p.

INGRAM, CAREY, 1975, Beach sands of the southern Delmarva Peninsula, patterns and causes: Unpub. M.S. Thesis, The College of William and Mary, 90 p.

IWAGAKI, Y., AND KAKINUMA, T., 1967, On the bottom friction factors off five Japanese coasts: Coastal Eng. in Japan, v. 10, p. 13-22.

JOHNSON, J. R., 1948, Recent contributions of wave research to harbor engineering: Univ. California, Fluid Mechanics Lab. Tech. Rep. HE-116-267, 10 p.

——, O'BRIEN, M. P., AND ISAACS, J. D., 1948, Graphical construction of wave refraction diagrams: U.S. Navy Hydrographic Office Pub. 605, 45 p.

JONSSON, I. G., 1966, Wave boundary layers and friction factors: Am. Soc. Civil Engineers, Proc. 10th Conf. on Coastal Eng., p. 127-148.

KEULEGAN, G. H., AND HARRISON, J., 1970, Tsunami refraction diagrams by digital computer: Am. Soc. Civil Engineers, Jour. Waterways, Harbors and Coastal Eng. Div., v. 96, p. 219-233.

KOMAR, P. D., 1975, Summary of longshore drift computations: Am. Soc. Civil Engineers, Proc. Civil Eng. in the Oceans/III, June 9-12, Newark, Delaware, in press.

LEPETIT, J. P., 1964, Etude de la réfraction de la house monochromatique par le calcul numérique: Bull. Cent. Recher. & Essais de Chatou No. 9, p. 3-25.

LEWIS, D., 1974, Wind, wave, star, and bird: Natl. Geog. Mag., v. 146, p. 747-781.

MAY, J. P., AND TANNER, W. F., 1973, The littoral power gradient and shoreline changes: *In* D. R. Coates (ed.), Coastal geomorphology: State Univ. New York at Binghamton, p. 43-60.

MILLIMAN, J. D., AND EMERY, K. O., 1968, Sea levels during the past 35,000 years: Science, v. 162, p. 1121-1123.

MUNK, W. H., AND ARTHUR, R. S., 1951, Wave intensity along a refracted ray: Natl. Bur. Standards Circ. 521, p. 95-108.

—— AND TRAYLOR, M. A., 1947, Refraction of ocean waves; a process linking underwater topography to beach erosion: Jour. Geology, v. 55, p. 1-26.

O'BRIEN, M. P., 1942, A summary of the theory of oscillatory waves: U.S. Army Corps of Engineers, Beach Erosion Board Tech. Rep. 2, 43 p.

——, 1947, Wave refraction at Long Beach and Santa Barbara, California: Univ. California Fluid Mechanics Lab., Tech. Rep. HE-116-246, 14 p.

PIERSON, W. J., JR., 1951, The interpretation of crossed orthogonals in wave refraction phenomena: U.S. Army Corps of Engineers, Coastal Eng. Res. Center Training Memo. 21, 83 p.

——, 1972, The loss of two British trawlers—A study in wave refraction: Jour. Navigation, v. 25, p. 291-304.

PUTNAM, J. A., AND JOHNSON, J. W., 1949, The dissipation of wave energy by bottom friction: Am. Geophys. Union Trans., v. 30, p. 69-74.

ROBERTS, J. A., AND KAUPER, E. K., 1964, The effects of wind and precipitation on the modification of south beach, Crescent City, California: Atmos. Res. Group, Altadena, California, 32 p.

SALLENGER, A. H., GOLDSMITH, V., AND SUTTON, C. H., 1975, Bathymetric comparisons: A manual of methodology, error criteria and applications: Virginia Inst. Marine Sci., Spec. Rep. Appl. Marine Sci. and Ocean Engr. 66, 34 p.

SAVAGE, R. P., 1957, Laboratory study of wave energy losses by bottom friction and percolation: U.S. Army Corps of Engineers, Beach Erosion Board Tech. Memo. 31, 25 p.

SAVILLE, THORNDIKE, JR., 1954, North Atlantic coast wave statistics hindcast by Bretschneider-revised Sverdrup-Munk method: U.S. Army Corps of Engineers, Beach Erosion Board Tech. Memo. 55, 18 p.

SHEPARD, F. P., AND INMAN, D. L., 1950, Nearshore circulation related to bottom topography and wave refraction: Am. Geophys Union Trans., v. 31, p. 196–212.

SHIDELER, G. L., 1973, Textural trend analysis of coastal barrier sediments along the middle Atlantic bight: Jour. Sed. Petrology, v. 43, p. 748–764.

SWIFT, D. J. P., 1973, Delaware Shelf Valley: Estuary retreat path, not drowned river valley: Geol. Soc. America Bull, v. 84, p. 2743–2748.

——, SANFORD, R. B., DILL, C. E., JR., AND AVIGNONE, N. F., 1971, Textural differentiation of the shore face during erosional retreat of an unconsolidated coast, Cape Henry to Cape Hatteras, western north Atlantic shelf: Sedimentology, v. 16, p. 221–250.

——, KOFOED, J. W., SAULSBURY, F. P., AND SEARS, PHILLIP, 1972, Holocene evolution of the shelf surface, central and southern Atlantic shelf of North America: *In* D. J. P. Swift, D. B. Duane and O. H. Pilkey (eds.), Shelf sediment transport: Dowden, Hutchinson & Ross, Inc., Stroudsburg, Pennsylvania, p. 499–574.

TANNER, W. F. (ED.), 1974, Sediment transport in the nearshore zone; Proceedings of a symposium at the Florida State University, January 26, 1974: Florida State Univ. Dep. Geol., Coastal Res. Notes, 147 p.

THOMPSON, E. F., AND HARRIS, D. L., 1972, A wave climatology for U.S. coastal waters: Fourth Ann. Offshore Technology Conf. Preprints, v. II, p. II-675–II-688.

UCHUPI, ELAZAR, 1968, Atlantic continental shelf and slope of the United States—Physiography: U.S. Geol. Survey Prof. Paper 529-C, 30 p.

WALKER, J. R., PALMER, R. Q., AND KUKEA, J. K., 1972, Recreational surfing on Hawaiian reefs: Am. Soc. Civil Engineers, Proc. 13th Conf. on Coastal Eng., 1972, p.

WALTON, T. L., JR., 1973, Littoral drift computations along the coast of Florida by means of ship wave observations: Univ. Florida, Coastal and Oceanogr. Eng. Lab. Tech Rep. 15 (NOAA Grant NG-3-73), 97 p.

WILSON, B. W., AND TORUM, A., 1968, The tsunami of the Alaska earthquake, 1964: Engineering evaluation: U.S. Army Corps of Engineers, Coastal Eng. Res. Center Tech. Memo. 25, 401 p.

EVALUATION OF WAVE-GENERATED LONGSHORE CURRENT VELOCITIES AND SAND TRANSPORT RATES ON BEACHES

PAUL D. KOMAR
Oregon State University, Corvallis, 97331

ABSTRACT

Currents associated with the nearshore cell circulation, including rip currents, redistribute beach sands into a variety of rhythmic topographies, but do not produce longshore sand transport continuously along the shoreline. Waves breaking at an angle to the beach generate longshore currents flowing parallel to the shoreline. These currents in turn interact with the wave surf to produce a longshore transport of sand. A simple equation has been found with which this longshore current velocity can be evaluated for the mid-surf position where data are available. Theoretical relationships have been formulated for the complete longshore current distribution across the beach width, but data are lacking. However, the distribution can be made to agree with the available data at mid-surf by the proper selection of the drag coefficient. Equations have also been obtained for the evaluation of the sand transport rate, caused either by waves breaking at an angle to the shoreline or by longshore currents generated in other ways. This gives the total sand transport rate. Theoretical relationships have been determined for the distribution of longshore sand transport across the beach width, but again data are lacking to test the equations. The distribution can be calibrated such that, when summed across the beach, it gives the correct total sand transport rate.

INTRODUCTION

When waves break at an angle to the shoreline they generate a longshore current flowing parallel to the shoreline, confined almost entirely to the nearshore zone between the breakers and the shoreline. Interacting with the incoming waves, this longshore current in turn produces sand transport along the beach. The purpose of this present report is to review briefly our state of knowledge concerning the generation and evaluation of these longshore currents and the resulting sand transport. This paper is a capsule summary of several papers that have already appeared in the literature [principally Komar and Inman (1970), Komar (1975; in press, a; in press, b)] and therefore represents my own view.

In order to remain brief, this summary will not include a review of the currents associated with the nearshore cell circulation of which rip currents are the principal feature. There is presently some debate as to the origin of such currents; reference can be made to Bowen (1969b), Bowen and Inman (1969), and Hino (1975) as to opinions on the origin of rip currents. Rip currents and associated longshore currents redistribute beach sands into a variety of rhythmic topography which in turn has a strong feedback in controlling the nearshore currents. Again this is beyond the scope of the present review. Sonu (1972), Fox and Davis (1973), and Bowen and Inman (1971) are important recent papers on the origin of rhythmic topography and the effect of the topography on the nearshore currents. Guza and Inman (1975) examine the related problem of the origin of beach cusps.

A more complete understanding of nearshore currents and the sand transport must include consideration of rip currents and irregular beach topographies. However, under such conditions the only possible approach is a numerical computer model of the wave field, the generated currents, and the sand transport distribution. This will differ for each beach in question so that general predictions and analytical equations are not possible. Before that stage is reached, it is necessary to better understand the less complex case of longshore current and sand transport distributions on beaches where topographic effects are minimal.

Geologists perhaps tend to overemphasize beaches with pronounced topography in that they provide more interesting sedimentary structures. It should also be recognized that the processes and features we see at low tide or within the inner surf zone may not necessarily be most important at high tide or within the breaker zone and outer surf zone. For example, such features are observed on Oregon beaches (Fox and Davis, 1974) but my feeling is that they probably have little effect on the sediment transport at high tide or within the outer surf and breaker zones.

LONGSHORE CURRENTS DUE TO OBLIQUE WAVES

A number of equations has been proposed relating the longshore current velocity to the parameters of the incoming waves. Galvin (1967) reviewed the early theories and concluded that none of them were successful in prediction. Subsequent to that time considerable progress has been attained, principally through the studies of Bowen (1969a), Longuet-Higgins (1970a, b), and Thornton (1971). Longshore currents under an oblique wave approach were attributed to the longshore component of the radiation stress which is the momentum flux associated with the wave propagation (Longuet-Higgins and Stewart, 1964). Excluding any horizontal mixing in the surf zone, Longuet-Higgins (1970a) derived the simple relationship

$$v_1 = \frac{5\pi}{16}\gamma\frac{\tan\beta}{c}\sqrt{g h_b}\,\sin\alpha_b \qquad (1)$$

for the maximum longshore current v_1 at the breaker zone, where h_b is the water depth at breaking, α_b is the breaker angle, β is the angle of the beach slope, c is a frictional drag coefficient, and γ is the ratio of the wave height to water depth (γ is on the order of 0.8 to 1.2 for most beach slopes). Under this derivation the longshore current decreases linearly from a maximum v_1 value at the breaker zone to zero at the shoreline.

Equation (1) can be rewritten as

$$v_1 = \frac{5\pi}{8}\frac{\tan\beta}{c}u_m \sin\alpha_b \qquad (2)$$

where u_m is the maximum value of the horizontal orbital velocity evaluated at the breaker zone with

$$u_m = \left[\frac{2E_b}{\rho h_b}\right]^{1/2}. \qquad (3)$$

E_b is the energy density of the breaking waves and ρ is the density of water.

The sand transport studies of Komar and Inman (1970) had earlier suggested that

$$v_\ell = 2.7\,u_m \sin\alpha_b \qquad (4)$$

for the longshore current, v_ℓ, at the mid-surf position. This was prompted by the agreement between two seemingly independent estimates of the littoral sand transport, the agreement being possible only if the longshore current is given by equation (4).

Komar (1975, in press, a) has undertaken a complete review of the longshore current relationship of equation (4), testing it against the available field data (Fig. 1) and laboratory measurements. In the laboratory tests much greater angles of wave breaking are achieved, and the results indicate that the relationship should be modified to

$$v_\ell = 2.7\,u_m \sin\alpha_b \cos\alpha_b. \qquad (5)$$

Breaker angles in the field are generally small so that $\cos\alpha_b \simeq 1$. In Komar (in press, b) I review many of the past theories of longshore current generation and conclude that equation (5) offers by far the best prediction of the longshore current velocity at mid-surf as generated by an oblique wave approach. It also has the firmest theoretical foundation as it is basically the same as equation (2) derived by Longuet-Higgins (1970a).

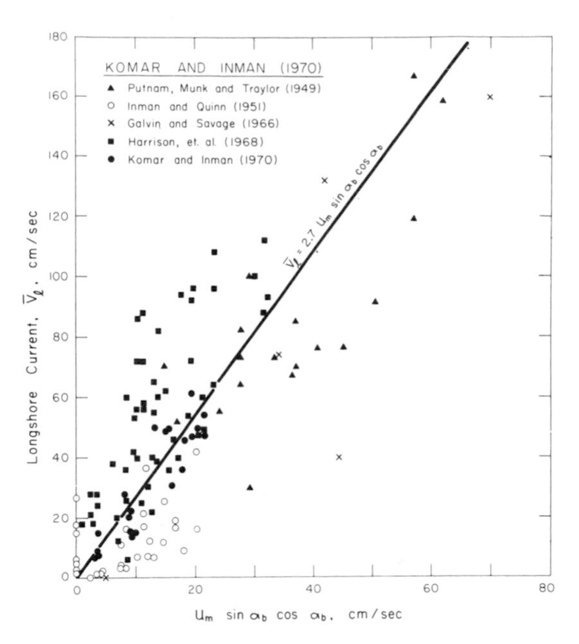

FIG. 1—Test of equations (4) and (5) for the longshore current velocity at mid-surf, utilizing the available field data. References to data sources can be found in Komar (1975, and in press, a).

LONGSHORE CURRENT DISTRIBUTION

Solutions have been obtained for the complete distribution of longshore current velocities across the nearshore, extending to beyond the breaker zone (Bowen, 1969a; Longuet-Higgins, 1970b; Thornton, 1971; Komar, 1975). These include a horizontal eddy mixing effect which produces an onshore-offshore transfer of momentum. This causes the position of maximum velocity to shift shoreward from the breaker zone, and allows the driving forces within the surf zone to produce a longshore current beyond the breaker zone, decreasing in velocity with distance outside the breakers. The resulting distributions appear reasonable.

Although the mathematical solution for the complete distribution at first looks formidable, its application is actually fairly straight forward. Komar (1975, in press, a) provides a solution that is a slight modification of that given by Longuet-Higgins (1970b). Insufficient data exist on the actual distributions of longshore currents, but Komar (1975, in press, a) provides a graph for the proper selection of a drag coefficient such that the velocity distribution has the same velocity at mid-surf as given by equation (5) which we have seen agrees with the available data.

LONGSHORE SAND TRANSPORT

Attempts at evaluating the sand transport rate have relied mainly on the relationship

$$I_\ell = 0.77\,(ECn)_b \sin\alpha_b \cos\alpha_b = 0.77\,P_\ell \quad (6)$$

where $(ECn)_b$ is the wave energy flux evaluated at the breaker zone. I_ℓ is the immersed weight sediment transport rate which is related to the more familiar volume transport rate, S_ℓ, by

$$I_\ell = (\rho_s - \rho)\,g\,a'\,S_\ell \quad (7)$$

where ρ_s is the sand density and a' is the correction factor for the pore-space of the beach sand (approximately 0.6 for most beach sands). In the *cgs* system of units, S_ℓ has units cm^3/sec and I_ℓ has units dynes/sec. I_ℓ has the same units as wave energy flux and therefore as P_ℓ, so that equation (6) is dimensionally correct (the 0.77 factor is dimensionless). Therefore a correlation between I_ℓ and P_ℓ is good for beaches composed of coral sand, coal, and so on, as well as for quartz sand beaches; a correlation between S_ℓ and P_ℓ depends on the grain density. The use of the immersed weight transport rate, I_ℓ, is based on considerations by Bagnold (1963) of the problem of sediment transport in general. The data comparison of equation (6) is given in Figure 2 which establishes the 0.77 coefficient for the field data. It is seen in this figure that the laboratory data show poor agreement with the relationship. Agreement should not be expected and the straight line based on the field data should form an upper

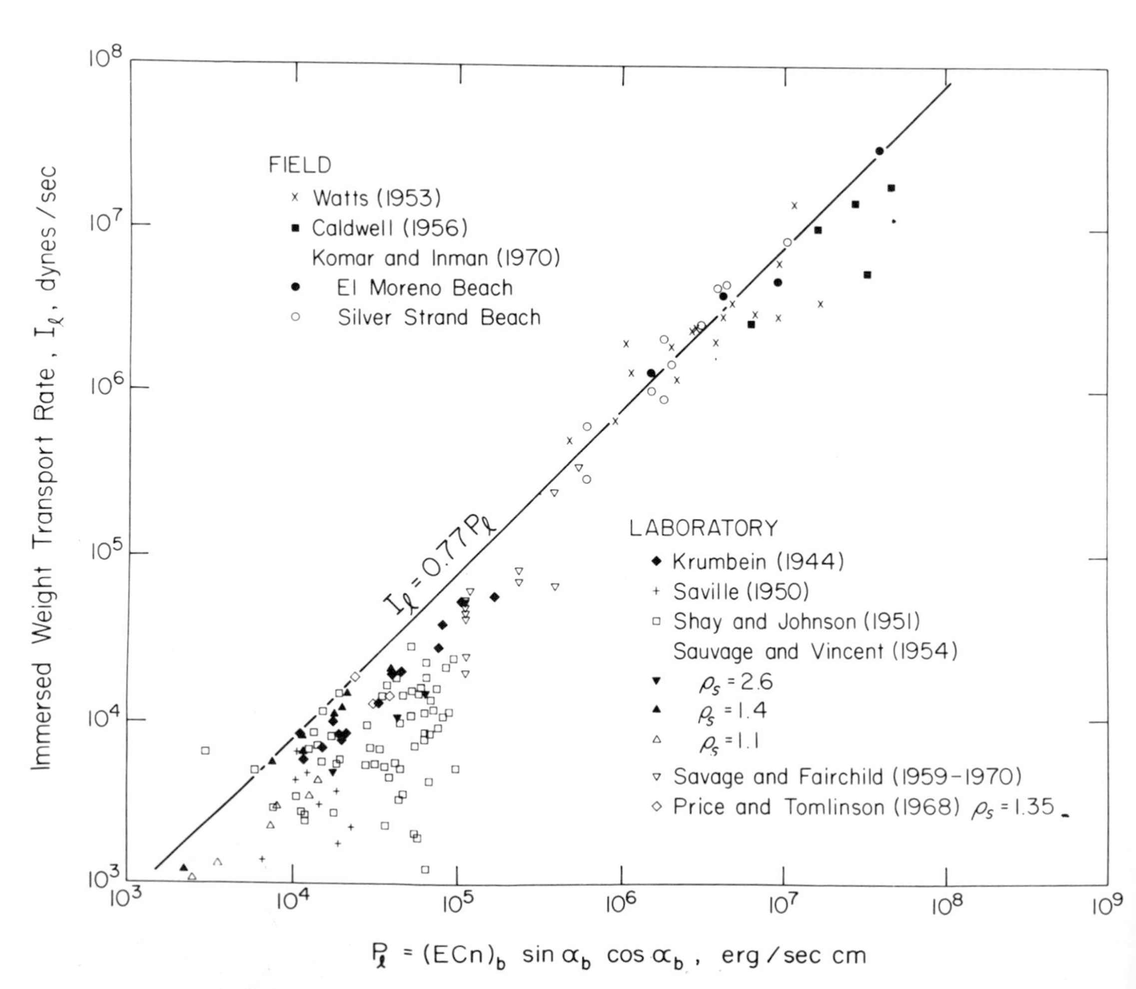

FIG. 2—Data test of the relationship of equation (6) for the total sand transport rate. References to the data sources can be found in Komar and Inman (1970). The straight-line is fitted to the field data alone. [From Komar and Inman (1970)]

limit to the laboratory data as it is seen to do. This is discussed more in Komar and Inman (1970) and in Komar (1975).

The relationship of equation (6) strictly applies only to sand transport produced by an oblique wave approach. A more fundamental examination of the processes of sand transport by combined waves and currents was undertaken by Bagnold (1963) and applied to sand transport on beaches by Inman and Bagnold (1963). Conceptually the model involves the waves placing the sand in motion and a superimposed current causing a net sand transport. The model yields the relationship

$$I_\ell = 0.28\,(ECn)_b \cos\alpha_b \frac{v_\ell}{u_m} \tag{8}$$

where u_m is given by equation (3). The 0.28 coefficient is based on the field data check of Komar and Inman (1970). Komar and Inman also showed that equation (6) results from the more basic equation (8) under an oblique wave approach if the longshore current is given by equation (4). Komar and Inman actually worked backward, obtaining equation (4) for the longshore current by a simultaneous solution of equations (6) and (8). It is apparent that equations (6) and (8) for the sand transport rate, and equation (4) or (5) for the longshore current, are mutually supported by the available data for both currents and sand transport. In light of the $\cos\alpha_b$ factor of equation (5) being needed for large breaker angles, the inclusion of the $\cos\alpha_b$ in equation (8) may not be correct; however, this has little importance in field conditions because, as we have already seen, $\cos\alpha_b \simeq 1$.

The importance of the introduction of equation (8) is that it is more general than equation (6), and applies to sand transport situations where v_ℓ is not due entirely to an oblique wave approach and thus given by equation (5). For example, if v_ℓ were due to the local wind stress, tidal currents, longshore currents of the cell circulation, or any combinations thereof, it would no longer be given by equation (5). Under these circumstances the resulting sand transport must be evaluated with

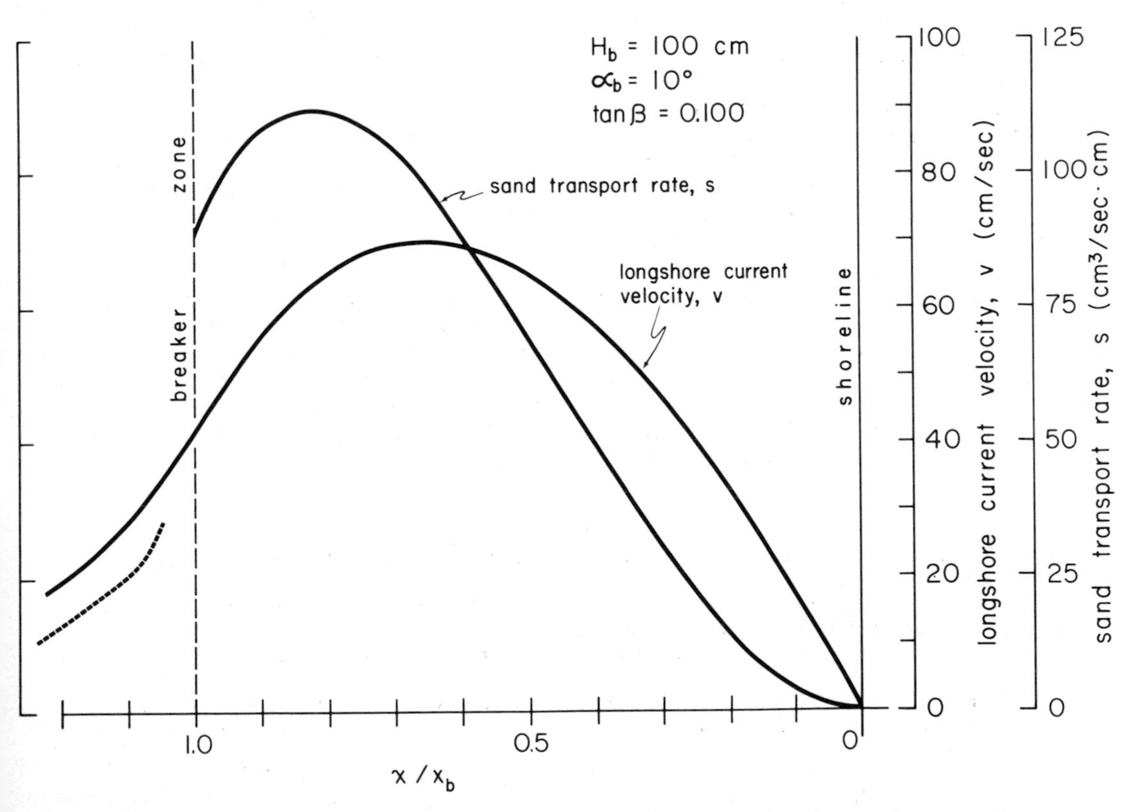

FIG. 3—Distributions of the longshore current and sand transport rate across the nearshore zone for a wave breaker height of 100 cm and breaker angle of 10° on a beach of slope 0.100, calculated by the approach presented in Komar (in press, a). The longshore current distribution agrees with the available data at the mid-surf position, and the sand transport distribution gives the correct total sand transport when summed across the nearshore. [From Komar (in press, a)]

equation (8), not equation (6). Either direct measurements of v_ℓ must be made or some appropriate prediction, and these values employed in equation (8) to estimate the sand transport. In future considerations of sand transport on beaches where topographic effects and cell circulation are important, equation (8) will experience increased application and equation (6) should encounter decreased use.

THE SAND TRANSPORT DISTRIBUTION

Equations (6) and (8) give the total sand transport along the beach. As with the longshore currents, we would also like to know its distribution across the nearshore. This has been investigated by Bijker (1971), Thornton (1973) and Komar (1971; in press, a) Both Thornton and Komar apply the sediment transport model of Bagnold (1963) for combined currents and waves. This is the same model that led to equation (8), but here it is applied to the local sand transport rather than to the total sand transport. In Komar (in press, a) an unknown proportionality coefficient in the solution is evaluated by integrating this transport distribution across the surf zone and placing it equal to the total transport. Theoretical distributions are thus predicted whose summation yields the correct total sand transport along the beach for a given set of wave conditions. One example is given in Figure 3. The longshore current distribution is that obtained with the solution given by Komar (1975, in press, a) such that it agrees with equation (5) and thus with the available data at the mid-surf position. The theoretical sand transport distribution of Figure 3 is based on Komar (in press, a) and yields the correct total sand transport when summed across the nearshore. It is seen that the sand transport rate reaches a maximum situated between the maximum in the longshore current velocity and the breaker line position. This is because the sand transport rate is dependent upon the magnitude of the longshore current velocity and the stress exerted by the waves which is a maximum at the breaker zone, decreasing shoreward.

DISCUSSION

Although the above theoretical examinations of longshore currents and sand transport assume an evenly sloping beach, the data to which they are compared come from real beaches with modest amounts of natural variability. Thus the results should have reasonably wide application. As discussed earlier, when the beach topography effects are appreciable, then the above relationships are probably not applicable. In addition, if breaker angles are very small then the nearshore currents can be expected to be controlled more by the cell circulation and its rip currents. Again the above considerations will be oversimplified. Complete numerical models would be necessary in such circumstances.

More work is required even on beaches of small topography. More measurements are required on the longshore current velocities, especially on the entire distribution of velocities across the nearshore. Nearly all the existing data were obtained prior to the above theoretical developments and so are not entirely satisfactory. More measurements are needed of littoral drift, especially of coarse materials such as on gravel or shingle beaches. Obtaining meaningful data on the distribution of sand transport across the nearshore is still a difficult problem.

ACKNOWLEDGEMENTS

This work is a result of research sponsored in part by the Oregon State University Sea Grant Program, supported by NOAA Office of Sea Grant, Department of Commerce, under Grant #04-5-158-2. I would like to thank R. A. Davis, Jr., for his useful suggestions in preparing this paper.

REFERENCES

BAGNOLD, R. A., 1963, Mechanics of marine sedimentation: *In* M. N. Hill (ed.), The Sea: Interscience Pub., New York, p. 507-582.

BIJKER, E. W., 1971, Longshore transport computations: Am. Soc. Civil Engineers Proc., Jour. Waterways, Harbors and Coastal Eng. Div., WW4, p. 687-701.

BOWEN, A. J., 1969a, The generation of longshore currents on a plane beach: Jour. Marine Res., v. 37, p. 206-215.

——, 1969b, Rip currents, 1. Theoretical investigations: Jour. Geophys. Res., v. 74, p. 5467-5478.

—— AND INMAN, D. L., 1969, Rip currents, 2. Laboratory and field observations: Jour. Geophys. Res., v. 74, p. 5479-5490.

—— AND ——, 1971, Edge waves and crescentic bars: Jour. Geophys. Res., v. 76, p. 8662-8671.

FOX, W. T., AND DAVIS, R. A., JR., 1973, Simulation model for storm cycles and beach erosion on Lake Michigan: Geol. Soc. America Bull., v. 84, p. 1769-1790.

—— AND ——, 1974, Beach processes on the Oregon coast: Williams College, Tech. Rep. 12, ONR Contract 388-092, 86 p.

GALVIN, C. J., JR., 1967, Longshore current velocity: A review of theory and data: Rev. Geophys., v. 5, p. 287-304.

Guza, R. T., and Inman, D. L., 1975, Edge waves and beach cusps: Jour. Geophys. Res., v. 80, p. 2997-3012.
Hino, Mikio, 1975, Theory on formation of rip current and cuspidal coast: Am. Soc. Civil Engineers, Proc. 14th Conf. on Coastal Eng., p. 901-919.
Inman, D. L., and Bagnold, R. A., 1963, Littoral processes: *In* M. N. Hill (ed.), The Sea: Interscience Pub., New York, p. 529-553.
Komar, P. D., 1971, The mechanics of sand transport on beaches: Jour. Geophys. Res., v. 76, p. 713-721.
——, 1975, Nearshore currents: Generation by obliquely incident waves and longshore variations in breaker height: *In* J. Hails and A. Carr (eds.), Nearshore sediment dynamics and sedimentation: John Wiley & Sons, Ltd., New York, p. 17-45.
——, In press, a, Distributions of longshore current velocity and sand transport across beaches: Am. Soc. Civil Engineers, Proc. Civil Eng. in the Oceans/III, June 9-12, Newark, Delaware.
——, In press, b, Beach processes and sedimentation: Prentice-Hall, Inc., Englewood Cliffs, New Jersey.
—— and Inman, D. L., 1970, Longshore sand transport on beaches: Jour. Geophys. Res., v. 75, p. 5914-5927.
Longuet-Higgins, M. S., 1970a, Longshore currents generated by obliquely incident waves, 1: Jour. Geophys. Res., v. 75, p. 6778-6789.
——, 1970b, Longshore currents generated by obliquely incident waves, 2: Jour. Geophys. Res., v. 75, p. 6790-6801.
—— and Steward, R. W., 1964, Radiation stress in water waves, a physical discussion with applications: Deep-Sea Res., v. 11, p. 529-563.
Sonu, C. J., 1972, Field observations of nearshore circulation and meandering currents: Jour. Geophys. Res., v. 77, p. 3232-3247.
Thornton, E. B., 1971, Variations of longshore current across the surf zone: Am. Soc. Civil Engineers, Proc. 12th Conf. on Coastal Eng., p. 291-308.
——, 1973, Distribution of sediment transport across the surf zone: Am. Soc. Civil Engineers, Proc. 13th Conf. on Coastal Eng., p. 1049-1068.

WAVE MODELING AND HYDRODYNAMICS

J. IAN COLLINS
Tetra Tech, Inc., Pasadena, California 91107

ABSTRACT

The first order approximate wave theory is summarized to give the basic mathematical properties of water waves. Following this, a brief summary is presented of the factors affecting wind wave generation in deep water and subsequent wave modifications as shallow water is approached. As the shoreline is approached by incoming waves and wave breaking starts to occur, momentum forces are released as the wave energy is dissipated. These forces are then available to drive longshore currents, rip currents and produce wave set-up. The mathematical formulation of these effects is briefly summarized and numerous further references indicated.

A singular wave model is assumed in the three first parts of the paper, i.e., a model sea is assumed to be represented by one wave height, one wave period and one wave direction. For many geological and engineering purposes this assumption may be sufficient or even necessary; however, a more realistic model which attempts to consider the statistical randomness of a sea state may be required. The fourth section of the paper presents the basic definition of the directional energy density spectrum and its modification by wind, bottom topography, bottom friction and nonlinear interactions between wave components in the spectrum.

INTRODUCTION

The intent of this paper is to provide a background to the hydrodynamics of water waves. By their very nature the hydrodynamic aspects require the expression of their concepts in mathematical terms. It is recognized that much of the mathematics may be beyond the general background of many sedimentologists; however, the movement of sediments by water has traditionally been studied by two schools of investigators: the "empiricists" and the "mathematicians." More recently these two schools appear to be approaching each other and it is hoped that this paper may help to contribute toward that merging by providing a summary of some useful algebraic expressions for the determination of desired wave parameters. Where practical the interested reader is referred to more detailed works.

As waves approach the shoreline they bring energy which is used to rework the nearshore sediments and modify the topography. Waves in the ocean actually serve as a mechanism which can abstract energy from wind systems, store it in the form of potential and kinetic energies, and transmit it toward shorelines. The dissipation of wave energy occurs near shore in a relatively narrow zone. Sedimentary processes associated with wave action are of interest to geologists because of their effects on generating sedimentary structures preserved in the rock record and because of the need to understand coastal environments in order to utilize and manage them properly. This paper attempts to present an overview of the hydrodynamics of wave action. The following sections present summaries of the basic hydrodynamic properties, the mechanics of wave generation and propagation into shoaling water, and a brief summary of the hydrodynamic aspects of the surf zone. It is hoped that these discussions can serve as a background for some of the succeeding papers.

In the first section the first order approximate wave theory is summarized to give the basic mathematical properties of water waves. The velocity field, wave profile and pressure response are presented in a useful summary table. Section two summarizes the factors of fetch, duration and wind speed as they affect wave height and period. A nomograph, based on the so-called Sverdrup-Munk-Bretschneider wave forecasting method, is given which should enable the user to obtain an estimate of wave characteristics from a knowledge of the governing parameters of fetch, duration and wind speed. The subsequent parts of section two present a review of the factors which lead to substantial modification of waves as a shoreline is approached, leading up to the occurrence of wave breaking.

As the shoreline is approached by incoming waves and wave breaking starts to occur, momentum forces are released as the wave energy is dissipated. These forces are then available to drive longshore currents, rip currents and produce wave set-up. The mathematical formulation of these effects is briefly summarized and numerous further references indicated. The momentum stress tensor is defined as producing the driving forces. Simple cases involving a plane beach are solved relatively easily but irregular offshore topography and rip current systems produce complex algebraic formulations which in general can only be solved numerically.

In the three first parts of the paper a singular wave model is assumed, i.e., a model sea is assumed to be represented by one wave height, one wave period and one wave direction. For many geological and engineering purposes this assumption may be sufficient or even necessary, however, a more realistic model which attempts to consider the statistical randomness of a sea state may be required. The fourth section of the paper presents the basic definition of the directional energy density spectrum and the last section indicates the procedures (currently still being refined) to handle the modification of the entire wave spectrum by wind, bottom friction and nonlinear interactions between wave components in the spectrum.

HYDRODYNAMICS OF WAVES

Linear theory.—Mathematically, the problem of linear wave theory and its solution in Eulerian coordinates, OX, coinciding with the level of the still-water and OZ, positive upwards (Fig. 1) is as follows:

a) continuity: $\nabla^2 \phi = o$ (1)

b) boundary condition: on the bottom at $z = -d$

$$w = \frac{\partial \phi}{\partial z} = 0 \tag{2}$$

c) The free surface condition: on the free surface of the water the pressure is constant. Thus

$$\frac{\partial^2 \phi}{\partial t^2} + g \frac{\partial \phi}{\partial z} = 0 \tag{3}$$

The progressive wave solution to this first-order approximate theory with these boundary conditions is:

$$\phi = \frac{gH}{\omega} \frac{\cosh k(d+z)}{\cosh kd} \sin(kx - \omega t) \tag{4}$$

where $K = \frac{2\pi}{L}$ and $\omega = \frac{2\pi}{T}$.

From the free surface condition, k and ω are derived and hence the wave length, L, as a function of depth and period, namely

$$L = \frac{gT^2}{2\pi} \tanh \frac{2\pi d}{L} \tag{5}$$

Thus C, the wave speed $\left(= \frac{L}{T}\right)$, is given by

$$C = \sqrt{\frac{gL}{2\pi} \tanh \frac{2\pi d}{L}} \tag{6}$$

Figure 2 shows the variation of wave length, L, with water depth, d, for various wave periods, T.

Using the value of the potential function, ϕ, which satisfies the given conditions the velocity components of a water particle are given by

$$u = \frac{\partial \phi}{\partial x} = \frac{kH}{2} \frac{\cosh k(d+z)}{\sinh kd} \cos(kx - \omega t) \tag{7}$$

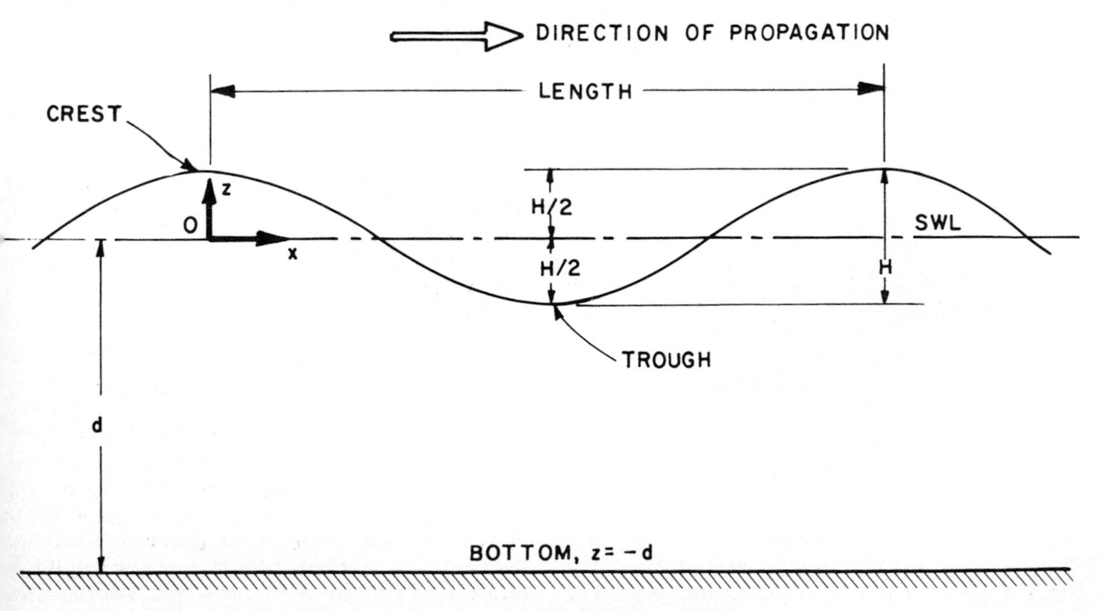

FIG. 1.—Definition of terms and coordinate system.

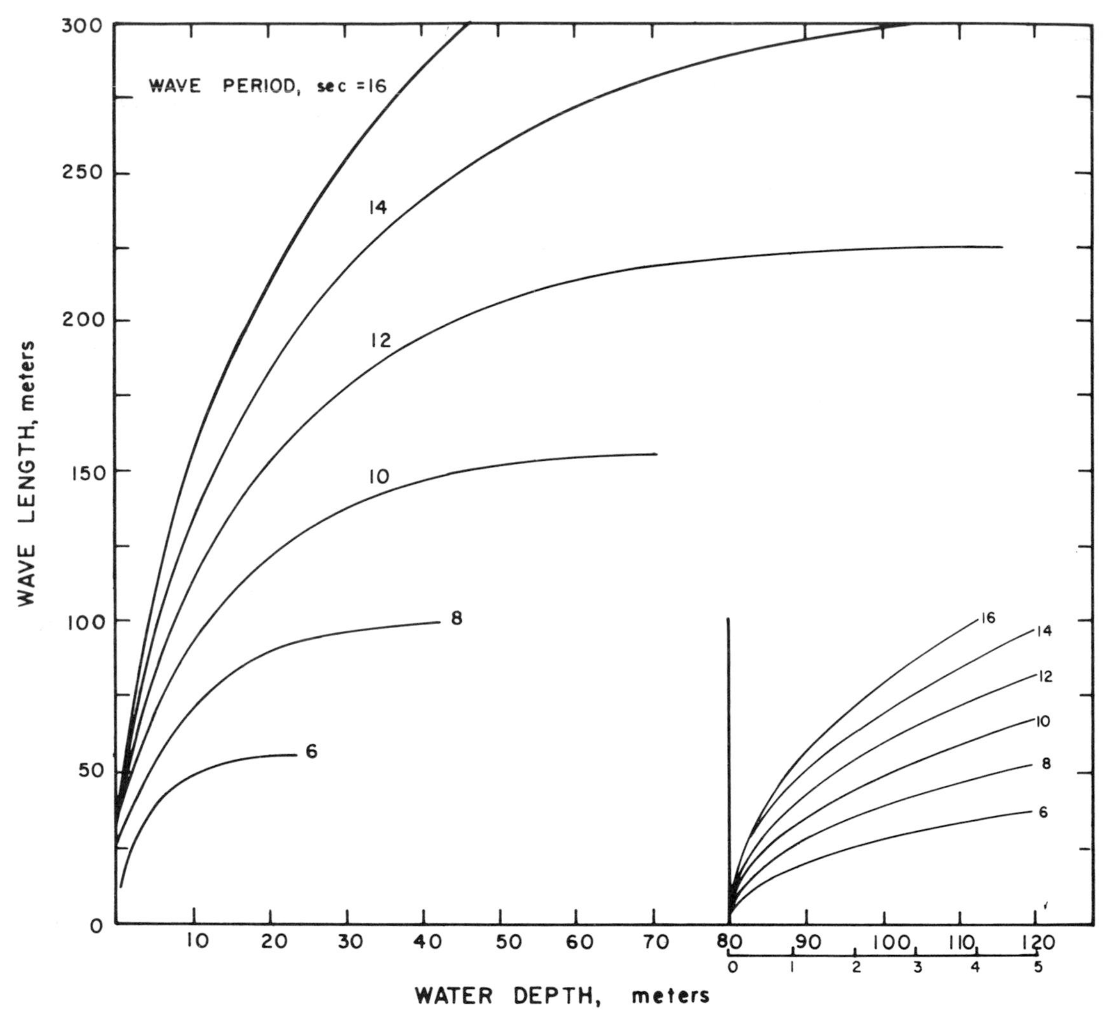

FIG. 2.—Wave length versus period and depth.

$$w = \frac{\partial \phi}{\partial z} = \frac{kH}{2} \frac{\sinh k(d+z)}{\sinh kd} \sin(kx - \omega t) \quad (8)$$

The position of any particle at time t, is given in Lagrangian coordinates, by

$$x = x_o - \frac{H}{2} \frac{\cosh k(d+z_o)}{\sinh kd} \sin(kx_o - \omega t) \quad (9)$$

$$z = z_o + \frac{H}{2} \frac{\sinh k(d+z_o)}{\sinh kd} \cos(kx_o - \omega t) \quad (10)$$

The particles in motion describe ellipses, the ratio of the major and minor axes depending on the depth and the depth to wave-length ratio. For deep water, $d/L > 0.5$, the orbits are circles of radii $H/2\ e^{kz}$. On the bottom in shallow water the motion is straight with a horizontal semi-amplitude of $H/2 \sinh kd$.

The pressure at any point is given in Eulerian coordinates by

$$p = -\rho g z + \frac{gH}{2} \cdot \frac{\cosh k(d+z)}{\cosh kd} \cdot \cos(kx - \omega t) + \text{constant} \quad (11)$$

If a pressure fluctuation recorder for measuring wave amplitude is placed at z in water of depth d, the pressure response factor, k, (defined as the pressure amplitude to surface wave amplitude ratio) is $\cosh k(d+z)/\cosh kd$. At the free surface, $z = \zeta$, and $d + \zeta \approx d$, giving,

TABLE 1.—SUMMARY OF FIRST ORDER WAVE FORMULAE

	Infinite Depth: $\frac{d}{L} > 0.5$	Any Depth
Potential Function ϕ	$\frac{H}{2} \cdot \frac{L}{T} \cdot e^{2\pi z/L} \sin 2\pi \left(\frac{x}{L} - \frac{t}{T}\right)$	$\frac{H}{2} \cdot \frac{L}{T} \cdot \frac{\cosh \frac{2\pi}{L}(d+z)}{\sinh \frac{2\pi d}{L}} \cdot \sin 2\pi \left(\frac{x}{L} - \frac{t}{T}\right)$
Velocity Components $u = \frac{\partial \phi}{\partial z}$	$u = \frac{\pi H}{T} \cdot e^{2\pi z/L} \cdot \cos 2\pi \left(\frac{x}{L} - \frac{t}{L}\right)$	$u = \frac{\pi H}{T} \cdot \frac{\cosh \frac{2\pi}{L}(d+z)}{\sinh \frac{2\pi d}{L}} \cdot \cos 2\pi \left(\frac{x}{L} - \frac{t}{T}\right)$
$w = \frac{\partial \phi}{\partial z}$	At the free surface $u_{max} = \frac{\pi H}{T}$	At the bed, $z = -d$, $w = 0$ $u_{max} = \frac{\pi H}{T} \cdot \frac{1}{\sinh \frac{2\pi d}{L}}$
Particle Orbits	Circles of Radius $= \frac{H}{2} \cdot e^{2\pi z/L}$ At the free surface, Radius $= \frac{H}{2}$ At the bed, Radius $= 0$	Ellipses: Horiz. Radius $= \frac{H}{2} \cdot \frac{\cosh \frac{2\pi}{L}(d+z)}{\sinh \frac{2\pi d}{L}}$ Vert. Radius $= \frac{H}{2} \cdot \frac{\sinh \frac{2\pi}{L}(d+z)}{\sinh \frac{2\pi d}{L}}$ At the free surface, Horiz. Rad. $= \frac{H}{2} \cdot \coth \frac{2\pi d}{L}$ Vert. Rad. $= \frac{H}{2}$ At the bed, Horiz. Rad. $= \frac{H}{2} \cdot \frac{1}{\sinh \frac{2\pi d}{L}}$ Vert. Rad. $= 0$
Free Surface	$\zeta = \frac{H}{2} \cdot \cos 2\pi \left(\frac{x}{L} - \frac{t}{T}\right)$	$\zeta = \frac{H}{2} \cdot \cos 2\pi \left(\frac{x}{L} - \frac{t}{T}\right)$
Gauge Pressure	$\frac{p}{\rho g} = -z + \frac{H}{2} \cdot e^{2\pi z/L} \cdot \cos 2\pi \left(\frac{x}{L} - \frac{t}{T}\right)$	$\frac{p}{\rho g} = -z + \frac{H}{2} \cdot \frac{\cosh \frac{2\pi}{L}(d+z)}{\cosh \frac{2\pi d}{L}} \cdot \cos 2\pi \left(\frac{x}{L} - \frac{t}{T}\right)$

TABLE 1.—(*continued*)

Wave Speed, Length	$C = \sqrt{\frac{gL}{2\pi}} = \frac{gT}{2\pi} = \frac{L}{T}$	$C = \sqrt{\frac{gL}{2\pi} \cdot \tanh \frac{2\pi d}{L}} = \frac{gT}{2\pi} \cdot \tanh \frac{2\pi d}{L} = \frac{L}{T}$ for $d/L < 0.05$, $C \simeq \sqrt{gd}$: $L \simeq T\sqrt{gd}$
Group Velocity	$V = \frac{C}{2} = \frac{1}{2} \cdot \frac{gT}{2}$	$V = \frac{C}{2} \left[1 + \frac{\frac{4\pi d}{L}}{\sinh \frac{4\pi d}{L}} \right]$ for $d/L < 0.05$, $V \simeq C$

$$\zeta = \frac{H}{2} \cos(kx - \omega t) \tag{12}$$

Table 1 presents a useful summary of the pertinent properties of the first order (linear) solution for gravity waves.

Nonlinear theory.—The mathematics of nonlinear wave theories is the subject of specialized studies and is considered to be beyond the scope of this particular summary of wave mechanics. The interested reader is referred to the works of Dean (1965), LeMehaute, Divoky and Lin (1968), Skjelbreia and Hendrickson (1961), Laitone (1960), Biésel (1952), Longuet-Higgins (1953) and others.

The effects of nonlinearities are far from just mathematical curiosities. Several significant physical effects arise from nonlinear effects: (a) wave particle orbits are no longer closed but exhibit a drift with time, referred to as mass transport; (b) wave crests and troughs are not symmetrical; the crests are higher and more peaked whereas the troughs are flatter and longer; (c) wave length of the waves is a function of the wave height; and (d) high particle velocities reduce the pressure readings obtained by pressure transducers.

WAVE GENERATION AND PROPAGATION INTO SHALLOW WATER

In the past, it has proved useful to perform wave generation, propagation and dissipation computations following the approximation of a single wave height, period and direction, chosen to represent a more realistic random sea characterized by a directional spectrum. Clearly, this must be a very crude approximation but it does permit the solution of a number of geological and engineering problems which could not be attempted otherwise. This section presents methods which can be used for computations.

Wave generation.—The wave generation relationships are stated in dimensionless form as:

$$\frac{C}{W} = \psi_1 \left[\frac{gF}{W^2}, \frac{gt}{W} \right] \tag{13}$$

$$\frac{gH}{W^2} = \psi_2 \left[\frac{gF}{W^2}, \frac{gt}{W} \right] \tag{14}$$

where C = wave speed, W = wind speed, F = fetch length, t = duration of wind, H = wave height.

These equations represent the wave generation parameters for deep water, based on dimensional considerations. ψ_1 and ψ_2 are functional relations that must be determined by use of observed data. gH/W^2, C/W, gF/W^2 and gt/W are defined respectively as the wave height parameter, the wave speed parameter, the fetch parameter, and the wind duration parameter. The wave speed parameter can also be written $\frac{gT}{2\pi W}$, which is preferred because the wave period is more easily measured than the wave speed.

Using the above parameters, Bretschneider (1951) revised the original forecasting relations of Sverdrup and Munk (1947). Further revisions of these relationships have been made again by Bretschneider (1958). These forecasting relationships have acquired the name *S-M-B method* for Sverdrup, Munk and Bretschneider. Wave forecasting relationships given in Figure 3 are based

FIG. 3.—Deep water wave forecasting curves as a function of wind speed, fetch length, and wind duration.

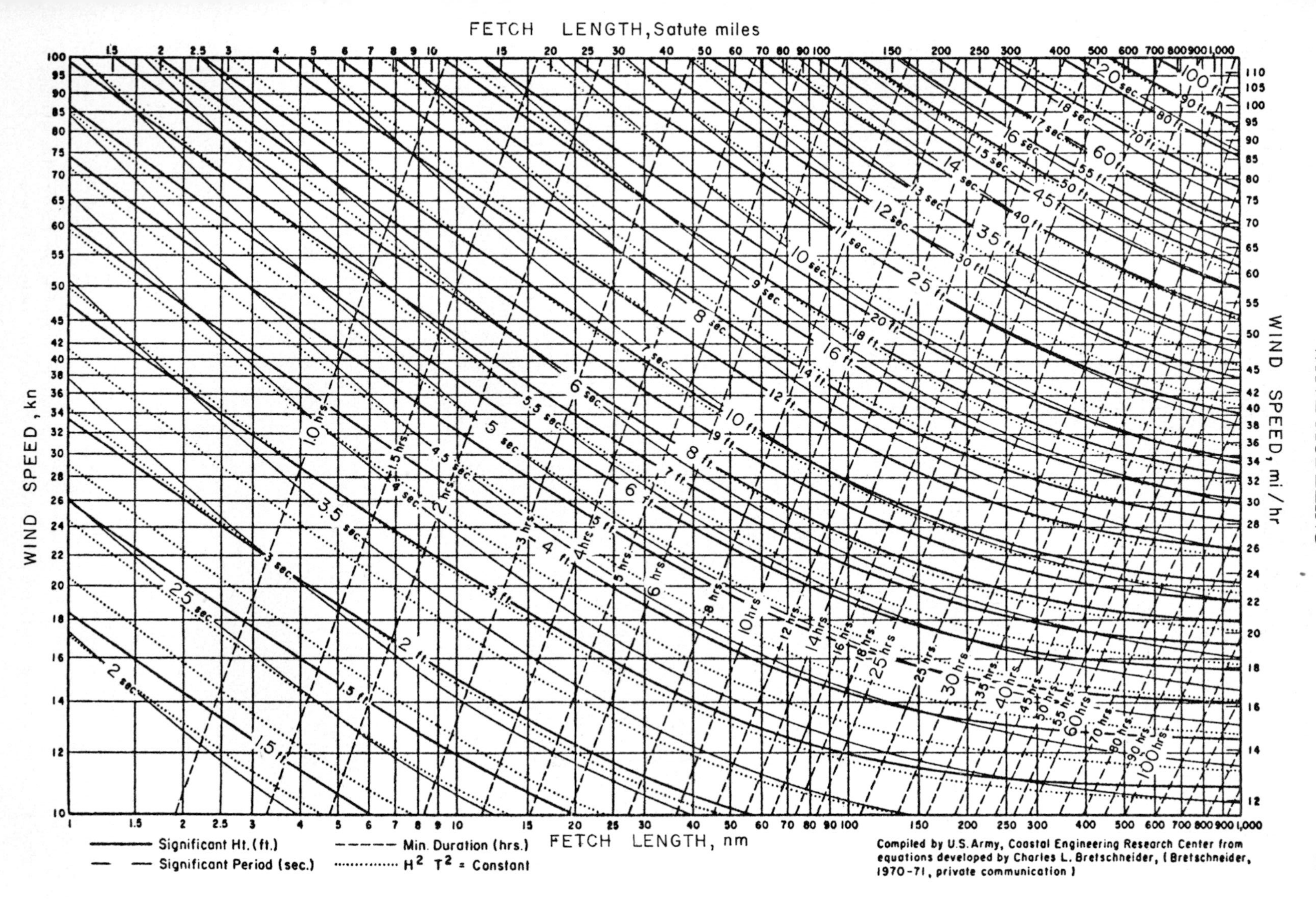
FETCH LENGTH, Satute miles
WIND SPEED, kn
WIND SPEED, mi / hr
FETCH LENGTH, nm
Significant Ht. (ft.)
Significant Period (sec.)
Min. Duration (hrs.)
H^2 T^2 = Constant
Compiled by U.S. Army, Coastal Engineering Research Center from equations developed by Charles L. Bretschneider, (Bretschneider, 1970-71, private communication)

on the dimensionless parameters of equations 13 and 14.

Wave shoaling.—The average energy flux per unit of wave crest through a fixed vertical plane parallel to the wave crest is:

$$F_{av} = \frac{1}{T}\int_{t}^{t+T}\int_{-d}^{\eta} \left(\rho\frac{V^2}{2} + p + \rho g z\right) u \, dz \, dt \quad (15)$$

i.e., by application of the Bernoulli equation where $f(t)$ is assumed to be taken into account in $\frac{\partial \phi}{\partial t}$:

$$F_{av} = -\rho\frac{1}{T}\int_{t}^{t+T}\int_{-d}^{t} \frac{\partial \phi}{\partial t}\frac{\partial \phi}{\partial x} dz \, dt. \quad (16)$$

This formula is general and can be applied by inserting the value of the potential function, ϕ, for any kind of irrotational wave, linear or nonlinear. In the case of a linear periodic progressive wave

$$F_{av} = \frac{1}{16}\rho g H^2 C\left[1 + \frac{2\,kd}{\sinh 2\,kd}\right]. \quad (17)$$

It is assumed in the case of a wave traveling over a very gentle slope that the flux of transmitted energy is a constant. The wave height at depth, d, is determined as a function of the deep water wave height from the formula

$$F_{av}\Big|_{d} = F_{av}\Big|_{d\infty}. \quad (18)$$

Thus,

$$\frac{1}{16}\rho g H^2 \frac{g}{\omega}\tanh kd\left[1 + \frac{2\,kd}{\sinh 2\,kd}\right] = \frac{1}{16}\rho g H_o^2 C_o \quad (19)$$

i.e.,

$$\left(\frac{H}{H_o}\right)_s = \left\{\frac{1}{\tanh kd\left[1 + \frac{2\,kd}{\sinh 2\,kd}\right]}\right\}^{1/2} = K_s \quad (20)$$

where K_s is the shoaling coefficient. The value of H/H_o versus d/L_o is given in Figure 4. Note that $(H/H_o)_s$ denotes ratio of wave heights due to shoaling effects only.

Wave refraction.—A wave orthogonal is a line defining the minimum travel time for a wave as it propagates. When a wave arrives at an angle with the bottom contours it is refracted; the orthogonals are bent toward the normal if the depth is decreasing. The distance between two wave orthogonals which cross a contour at an angle, α (see Figure 5), varies by a factor of:

$$\frac{b_2}{b_1} = \frac{\cos\theta_2}{\cos\theta_1} \quad (21)$$

with θ the angle made by the bottom contour and the wave crest. Subscripts 1 and 2 refer to the two referenced water depths on both sides of the bottom contour. It is generally assumed that the wave energy flux between wave orthogonals is constant, i.e., $F_{av}\,b$ = constant, or

$$\frac{1}{16}\rho g H^2 c\left[1 + \frac{2\,kd}{\sinh 2\,kd}\right] = \text{constant} \quad (22)$$

i.e.,

$$H_1^2 K_{s1}^2 b_1 = H_2^2 K_{s2}^2 b_2 \quad (23)$$

Hence, the wave height varies with a refraction coefficient equal to the square root of the inverse ratio of the distance between wave orthogonals.

The refraction coefficient, K_r, is $(b_o/b)^{1/2}$, where b_o is the distance between wave orthogonals in the deep water. Hence, the combined effect of refraction and shoaling give the ratio of wave height at depth, d, to the wave height in deep water as:

$$\left(\frac{H}{H_o}\right)_{s,r} = K_s K_r \quad (24)$$

Bottom friction and wave damping.—The velocity at the bed is

$$u_b = \frac{\partial \phi}{\partial x}\Big|_{z=-d} = \frac{\pi H}{T\sinh kd}\cos\omega t \quad (25)$$

The friction stress, τ, for a turbulent boundary layer is given by

$$\tau = \rho c_f u|u|, \quad (26)$$

where c_f is the friction factor. The average work done by this friction stress is:

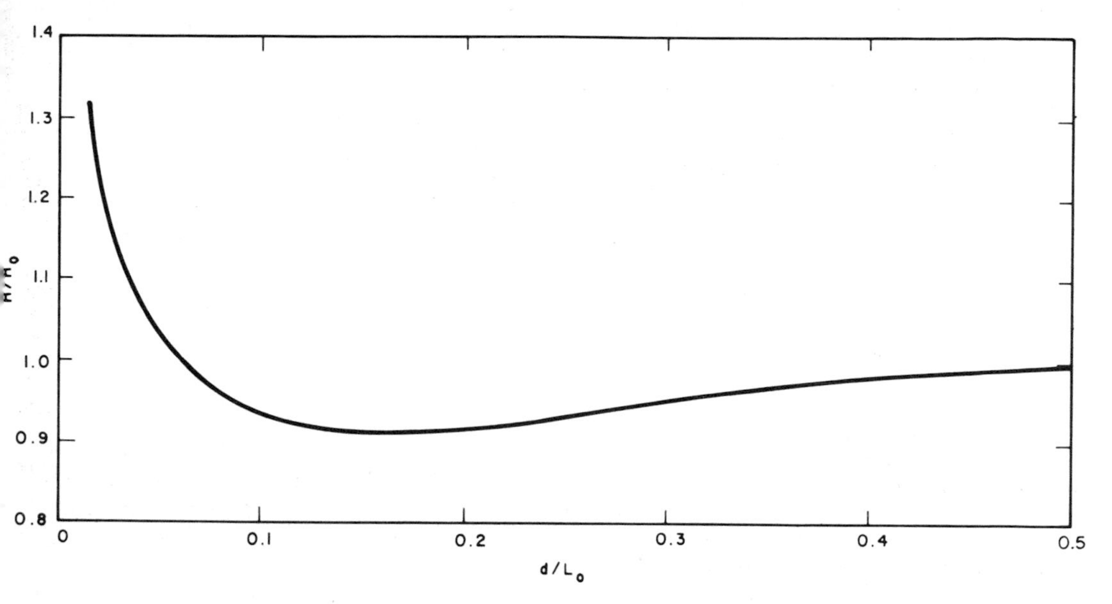

FIG. 4.—Relative wave height variation with depth.

$$\overline{\tau_b u_b} = \frac{1}{T}\int_o^T \rho c_f u_b^2 |u_b| \, dt. \quad (27)$$

$$= \frac{4}{3}\rho c_f \pi^2 \frac{H^3}{T^3 \sinh^3 kd} \quad (28)$$

The variation of energy flux equals the rate of energy dissipation, i.e.,

$$\frac{d}{dx}(F_{av}) = -\overline{\tau_b u_b} \quad (29)$$

or

$$\left(\frac{H}{H_o}\right)_{s,f} = K_s K_f \quad (30)$$

The damping factor, K_f, is

$$K_f = \left[1 - \frac{8}{\rho g}\int_{-\infty}^{d(x)} \overline{\tau_b u_b}\right] dx, \quad (31)$$

which has to be solved numerically.

The combined effects of shoaling, damping and refraction are obtained by multiplying the deep water wave height by the shoaling coefficient, K_s, the damping coefficient, K_f, and the refraction coefficient, K_r, so that

$$\left(\frac{H}{H_o}\right)_{s,f,r} = K_s K_f K_r. \quad (32)$$

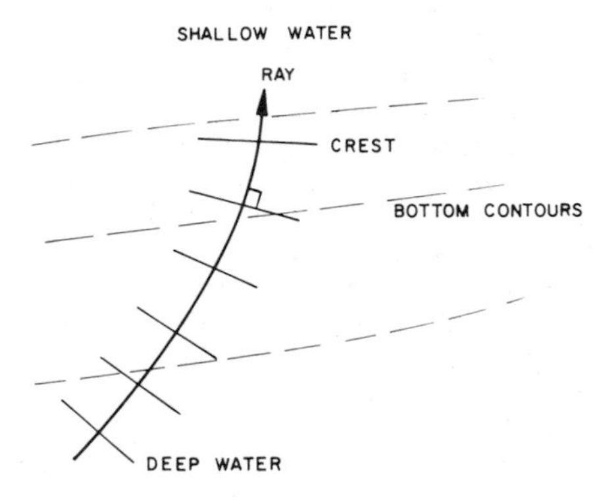

(a) DEFINITION OF RAYS, CRESTS, AND BOTTOM CONTOURS

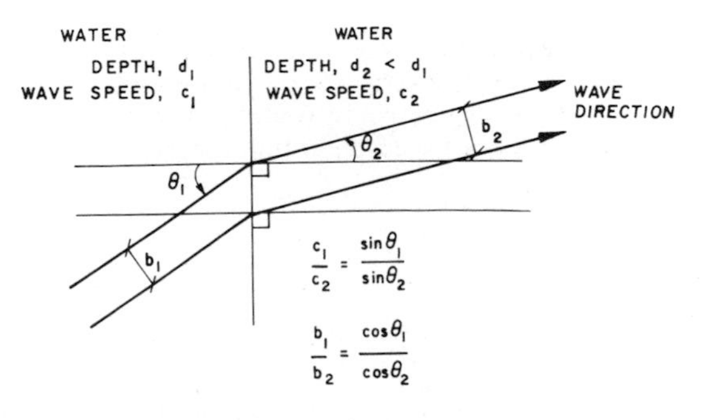

(b) SCHEMATIC REFRACTION EFFECTS

FIG. 5.—Illustration of refraction effects.

Wave breaking.—Three conditions for wave breaking are recognized:

1) The vertical acceleration of a surface particle cannot exceed $1/2\ g$ for a progressive wave (it may approach g in the case of a standing wave).

2) The horizontal velocity of any water particle cannot exceed the speed of propagation of the wave.

3) The free surface pressure condition has to match the atmospheric pressure.

Each of these conditions has a mathematical expression but generally, breaking criteria are considered to be empirical. For progressive waves these empirical criteria are: (1) wave steepness limited breaking in deep water, $H/L \not> 0.142$; (2) wave steepness limited breaking in shoaling water, $H/L \not> 0.14 \tanh kd$; and (3) water depth limited breaking in shallow water, $H/d \not> 0.78$.

It is generally acknowledged that any one of these conditions will produce breaking. In practice, breaking usually occurs before these limits are reached over a horizontal bed but the limits can easily be exceeded on a sloping bed.

SURF ZONE INTERACTION

Longshore currents and wave set-up.—In the nearshore area, arriving waves continuously bring in momentum, energy and mass (mass transport). Theories for longshore currents and wave set-up generally consider one or more of these phenomena. This paper will utilize momentum. It is extremely difficult to develop a longshore current theory based on energy because it is known that only a very small percentage of the wave energy is needed to drive the longshore current. The derivation of a longshore current based on an energy budget is practically hopeless. The occurrence of mass transport in progressive waves is known but the precise volumes of water which propagate toward the shoreline due to this phenomenon are practically impossible to predict. In any case, the momentum of the waves must drive the mass transport to give longshore currents and wave set-up. The conservation of momentum is used.

The integrated (over a vertical) momentum and continuity equations for an incompressible fluid having a free surface can be written, (acceleration) + (surface slope) = (momentum forces) − (friction forces), i.e.

Momentum:

$$\frac{\partial u}{\partial t} + u\frac{\partial u}{\partial x} + v\frac{\partial u}{\partial y} + g\frac{\partial \zeta}{\partial x} = M_{fx} - F_x \tag{33}$$

$$\frac{\partial v}{\partial t} + u\frac{\partial v}{\partial y} + v\frac{\partial v}{\partial x} + g\frac{\partial \zeta}{\partial y} = M_{fy} - F_y \tag{34}$$

Continuity:

$$\frac{\partial \zeta}{\partial t} + \frac{(uD)}{\partial x} + \frac{(vD)}{\partial y} = 0 \tag{35}$$

where, $\zeta(x,y)$ is the surface disturbance from the still water level (positive upward); F_x is the bottom friction in the x-direction, F_y is the bottom friction in the y-direction; and M_{fx}, M_{fy} are the external driving force components. In fact,

$$M_{fx} = -\frac{1}{\rho D}\left[\frac{\partial \sigma_{xx}}{\partial x} + \frac{\partial \tau_{xy}}{\partial y}\right] \tag{36}$$

$$M_{fy} = -\frac{1}{\rho D}\left[\frac{\partial \sigma_{yy}}{\partial x} + \frac{\partial \tau_{xy}}{\partial y}\right] \tag{37}$$

Where D is the total water depth, $d + \zeta$, σ_{xx} σ_{yy} and τ_{xy} are given by the momentum flux tensor (Longuet-Higgins and Stewart, 1964),

$$S = \begin{pmatrix} \sigma_{11} , & \tau_{12} \\ \tau_{21} & \sigma_{22} \end{pmatrix} \tag{38}$$

where,

$$\sigma_{11} = \frac{1}{8}\rho g H^2\left[\frac{1}{2} + \frac{2\ kd}{\sinh 2\ kd}\right] \tag{39}$$

$$\sigma_{22} = \frac{1}{8}\rho g H^2\left[\frac{kd}{\sinh 2\ kd}\right] \tag{40}$$

$$\tau_{12} = \tau_{21} = 0 \tag{41}$$

The components of the tensor given by the above equations are relative to the direction of travel of the wave. In a coordinate system which is fixed to the beach at an angle, θ, with the wave front, the momentum stress components in the x (normal) and y (parallel) directions to the beach are,

$$\sigma_{xx} = \sigma_{11}\cos^2\theta + \sigma_{22}\sin^2\theta \tag{42}$$

$$= \frac{1}{8}\rho g H^2\left(2n - \frac{1}{2}\right)\cos^2\theta + \frac{1}{8}\rho g H^2\left(n - \frac{1}{2}\right)\sin^2\theta \tag{43}$$

$$\sigma_{yy} = \sigma_{11}\sin^2\theta + \sigma_{22}\cos^2\theta \tag{44}$$

$$= \frac{1}{8}\rho g H^2\left(2n - \frac{1}{2}\right)\sin^2\theta + \frac{1}{8}\rho g H^2\left(n - \frac{1}{2}\right)\cos^2\theta \tag{45}$$

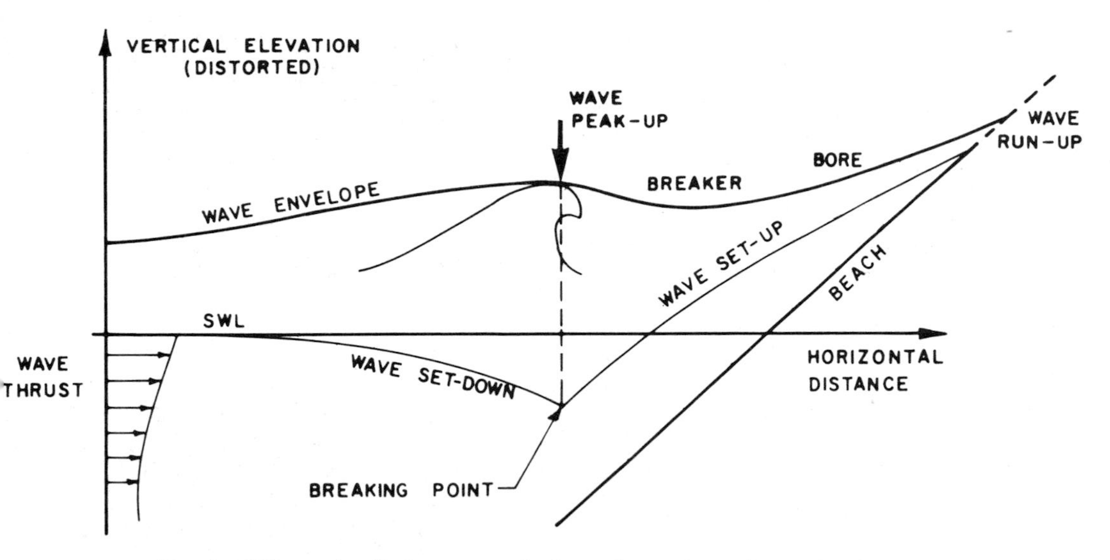

FIG. 6.—Effects of radiation stresses in the surf zone (normal wave incidence).

$$\tau_{xy} = \frac{1}{2}(\sigma_{11} - \sigma_{22}) \sin 2\theta \qquad (46)$$

$$= \frac{1}{16}\rho g H^2 n \sin 2\theta \qquad (47)$$

where, n is the ratio of group velocity to phase speed, i.e.,

$$n = \frac{1}{2} + \frac{kd}{\sinh 2kd} \qquad (48)$$

D is the total water depth, $d + \zeta$. The bottom friction terms, F_x and F_y, for fully turbulent flow are usually taken in quadratic form.

The general solution of the momentum and continuity equations are extremely complex for an irregular shoreline. As a first approximation, consider the case of a nearshore area having straight, parallel bottom contours and further consider a steady state condition. Then, the momentum equations are reduced to,

$$g\frac{\partial \zeta}{\partial x} = M_{fx} \qquad (49)$$

$$0 = M_{fy} - c_f V^2 \qquad (50)$$

where,

$$M_{fx} = \frac{1}{\rho D}\frac{\partial \sigma_{xx}}{\partial x} \qquad (51)$$

$$M_{fy} = \frac{1}{\rho D}\frac{\partial \tau_{xy}}{\partial x} \qquad (52)$$

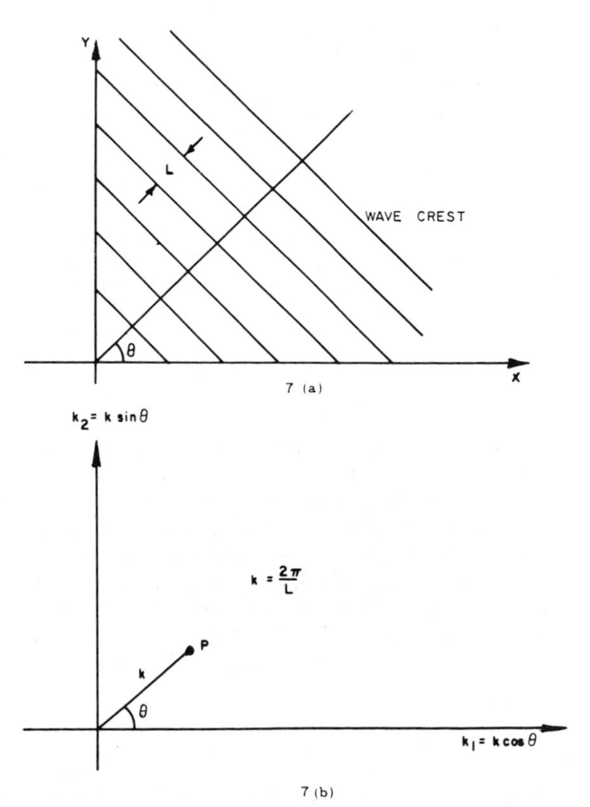

FIG. 7.—Real space and wave number representation of a periodic wave component.

because $\frac{\partial}{\partial t} = \frac{\partial}{\partial y} = 0$ and $u = 0$, the continuity equation is automatically satisfied. Equations (49) and (50) are solved for $\zeta(x)$, the wave set-up and $V(x)$, the longshore current because σ_{xx} and τ_{xy} are functions of the incident wave height, wave period and water depth which are known.

In the special case of normal incidence and parallel bottom contours only the first equation is needed and the solution for wave set-up is illustrated in Figure 6. More involved cases can be found in Bowen (1969a, b), Bowen and Inman (1969), Longuet-Higgins (1970), Thornton (1969), Collins (1972) and Noda (1972, 1973). These referenced works include cases for normal incidence with a longshore variation in wave height, oblique wave incidence (which produces longshore currents), and rip currents arising from an inherent instability in the longshore flow.

ENERGY SPECTRUM

Definition.—Consider a sinusoidal wave having a wave length, L, and wave height, H, traveling over an ocean in an arbitrary direction, θ, relative to some chosen Cartesian coordinate axes, Ox, Oy (see Fig. 7a). If the phase of the wave is arbitrary it can be represented by a point in the wave number plane, (k_1, k_2), at a distance, k, from the origin and an angle, θ, from the $k_1 = k \cos \theta$ axis (Fig. 7b). Furthermore, if the point, P, in the k_1, k_2 plane is considered as a two-dimensional delta function having the value $\frac{1}{8} H^2$ at ($k_1 = k \cos \theta$, $k_2 = k \sin \theta$) then the magnitude of this spike is proportional (by the factor $1/\rho g$) to the energy contained in the periodic wave per unit surface area in the $x-y$ plane.

In reality, a sea state can be considered to consist of the sum of an infinite number of periodic waves each of infinitesimal amplitude having an ensemble of wave lengths and directions of travel. Figure 8 shows a schematic representation of sea surface components. In the wave number plane (k_1, k_2) the sea surface components will be represented by a large (infinite) number of dots such as P in Figure 7b, completely filling the plane. If over an element of area δk_1, δk_2, an average value of the mean square surface fluctuation is defined by the function $E(k_1, k_2)$ then $E(k_1, k_2)$ is the spectral density. That is,

$$\int_{\delta k_1} \int_{\delta k_2} E(k_1, k_2)\, dk_1\, dk_2 = \left\langle \frac{1}{2} a_n' \right\rangle = \left\langle \frac{1}{8} H_n^2 \right\rangle \qquad (53)$$

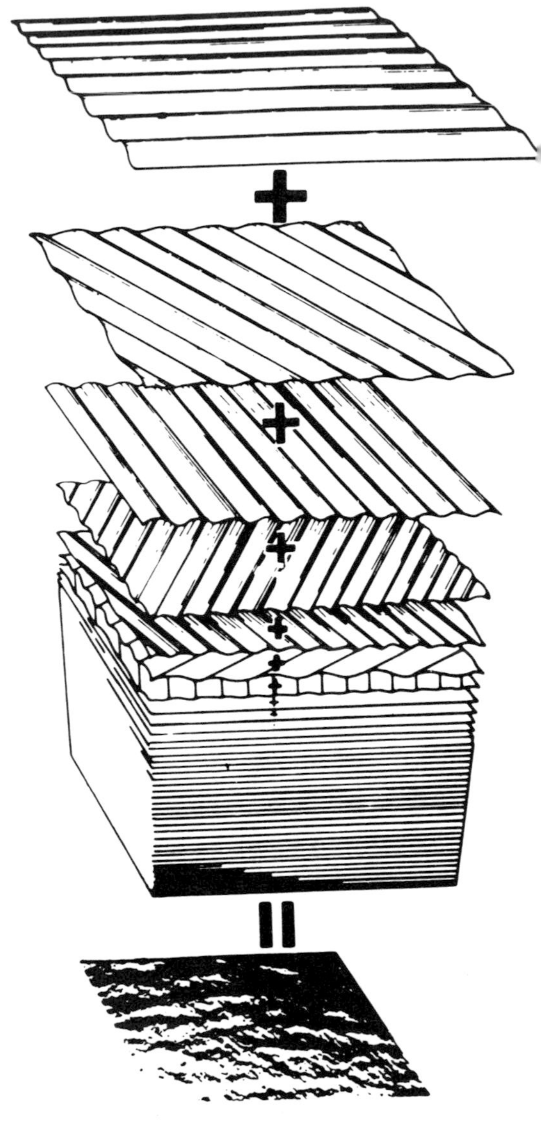

FIG. 8.—A sum of many simple sine waves makes a sea (from U.S. Navy Oceanographic Office, H. O 603).

where the a_n' s are the amplitudes of all the wave elements represented by the dots contained in the element of the area δk_1, δk_2 and the operation $\langle\ \rangle$ indicates mean value. In order that $E(k_1, k_2)$ represents the energy spectral density then a factor, ρg, is required. In common practice this factor is ignored and $E(k_1, k_2)$ is called the energy spectral density or more frequently the energy spectrum (or even just "the spectrum"). In summary then, $E(k_1, k_2)$ represents the energy *per unit area per unit wave number space.*

The spectral density function, $E(k_1, k_2)$ can be considered as a description of the sea in the neighborhood of a point in space, (x,y). The wave spectrum prediction problem consists in defining $E(k_1, k_2)$ [or $E(\underline{k})$] as a function of x and y and also, in general, as a function of time, t. Hence, $E(\underline{k})$ is really a five-dimensional variable, $E(x, y, t, k_1, k_2)$. Alternate forms of the energy density function, E, are often considered. It is emphasized that, E, is a spectral *density* function per unit *area* per unit *wave number space*. If the definition of wave space is changed then the values E will be different.

a) E can be expressed in polar coordinates in wave number space:

$$E(k, \theta) \equiv kE(k_1, k_2). \tag{54}$$

b) E can be defined in polar frequency-direction space:

$$E(\omega, \theta) \equiv \frac{1}{V} E(k, \theta), \tag{55}$$

where $V = \dfrac{\partial \omega}{\partial k}$ (the group velocity with ω and k related by the dispersion relationship $\omega^2 = gk \tanh kd$), d is water depth.

c) $$E(f, \theta) \equiv 2\pi\, E(\omega, \theta) \tag{56}$$

d) $$E(f) \equiv \int_{-\pi}^{\pi} E(f, \theta)\, d\theta, \tag{57}$$

It is noted that

$$\int_{k_1}\int_{k_2} E(k_1, k_2)\, dk_1\, dk_2 = \int_f \int_\theta E(f, \theta)\, df\, d\theta = \sigma^2 \tag{58}$$

where σ^2 is the value of the mean square sea surface fluctuation.

WAVE GENERATION AND MODIFICATION (SPECTRUM APPROACH)

Formulation of spectrum prediction equation.—Consider an arbitrary area, A, (four dimensional) bounded by a closed contour, S, (Fig. 9) containing the total energy $\int E\, dA$ (E is the spectral density which is proportional to the energy per unit area).

Consider an element of S, such as dS, having a unit normal, $\vec{n}$, and a velocity of energy propa-

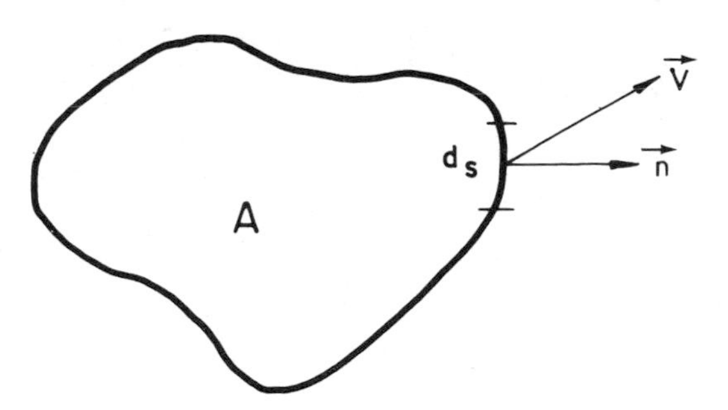

FIG. 9.—Schematic representation of energy conservation.

gation through dS of $\vec{V}$. The area, A, has within it sources (and/or sinks) of strength, Q, per unit area. The expression of conservation of energy for the area, A, is:

$$\frac{\partial}{\partial t}\int E dA + \int E\, \vec{V} \cdot \vec{n}\, dS + \int Q dA = 0. \tag{59}$$

The second term, which is evaluated along the boundary, S, can be transformed into an integral over the area, A, by means of Green's theorem yielding:

$$\frac{\partial}{\partial t}\int E dA + \int \nabla \cdot (E\, \vec{V})\, dA + \int Q dA = Q. \tag{60}$$

Now reversing the order of integration and differentiation in the first term and combining terms

$$\iint \left(\frac{\partial E}{\partial t} + \nabla \cdot (E\, \vec{V}) + Q \right) dA = 0. \tag{61}$$

Since the area, A, in equation (61) is arbitrary then the integrand must everywhere be zero. Therefore:

$$\frac{\partial E}{\partial t} + \nabla \cdot (E\, \vec{V}) + Q - 0. \tag{62}$$

(See Phillips, 1966, p. 147; Karlsson, 1969 and others).

Equation (62) is always valid whatever the measure of E but in the particular case when E is expressed as $E(x, y, t, k_1, k_2)$, equation (62) can be further simplified as follows. Using the Hamiltonian property of the wave field,

$$\dot{x} = \frac{dx}{dt} = \frac{\partial\omega}{\partial k}, \dot{y} = \frac{dy}{dt} = \frac{\partial\omega}{\partial k_2}, \dot{k}_1 = \frac{dk_1}{dt}$$

$$= -\frac{\partial\omega}{\partial x}, \text{ and } \dot{k}_2 = \frac{dk_2}{dt} = -\frac{\partial\omega}{\partial y} \quad (63)$$

then

$$\nabla \cdot E\vec{V} = \vec{V} \cdot \nabla E \quad (64)$$

and equation (62) reduces to

$$\frac{dE}{dt} + Q = 0. \quad (65)$$

(See Phillips, 1966, p. 1947). Equation (65) was first demonstrated by Longuet-Higgins (1957) for the case when $Q = 0$ (no dissipation or generation in the form, $E(\underline{k})$ = constant). Both forms of the energy conservation equation (equations 62 and 65) can be found in the literature but it must be emphasized that equation 65 is only valid when E is expressed as the energy density in wave number space.

The spectrum prediction problem is now reduced to a numerical evaluation of $\frac{dE(x,y,t,k_1,k_2)}{dt}$ and the choice of suitable functions for the generation and dissipation of energy, Q. Q is composed of the three parts: wind generation, G, frictional dissipation, Θ, and non-linear transfer, T.

Wind generation.—Two mechanisms are involved in the generation of waves by wind. The turbulent pressure spectrum in the wind is convected over the water surface, thus generating waves. The wave components whose phase speeds match the wind speed tend to grow by a "resonant" interaction. This is the mechanism first proposed by Phillips (1957). It governs wave growth in the early phases of generation. Once the water surface is disturbed it, in turn, disturbs the air flow which tends to follow the waves and induce even more water and air disturbance. This is the "instability" phase first proposed by Miles (1957). The wind generation mechanism can be expressed in the form,

$$G = (\alpha + \beta E) \quad (66)$$

where α represents the Phillips mechanism, β, represents the Miles mechanism, and $E(\underline{k})$ is the directional energy density spectrum in wave number space. α includes the atmospheric turbulent pressure spectrum over the ocean surface. Several forms of α have been proposed by Phillips (1966), Hasselmann (1960), Barnett (1968), Snyder and Cox (1966), Inoue (1966), and others.

Phillips (1957) will be taken as the basic definition of α,

$$\alpha = \frac{\pi P(k,\omega)}{\rho^2 C^2}, \quad (67)$$

where $C = \omega/k$, ρ = density of water and P is the atmospheric pressure spectrum (three dimensional). Barnett (1968) has proposed a function for $P(\underline{k}, \omega)$ based on the work of Priestley (1966).

Snyder and Cox recommended $\beta = s(\underline{k} \cdot \underline{W} - \omega)$ where s is the ratio of the density of air to water. Barnett (1968) recommended a slight change in this to yield

$$\beta = 5\, sf(W \cos\delta/C - 0.90) \quad (68)$$

In the steady state, for instance, under an offshore wind which has been blowing for some period of time, the spectral growth is given by $dE(f,\theta)/dx = GV \cos\theta$, where V is the group velocity and θ is the direction of propagation. The solution for the spectral component $E(f,0)$ (i.e., directly downwind) when no friction is present over a deep, uniform ocean is (Fig. 10)

$$E(f,0) = [\exp(\beta x/V) - 1]\,\alpha/\beta. \quad (69)$$

Whitecaps and breaking.—The wind generation model shows that selected wave components are amplified exponentially under wind action. It is a well observed fact that under sustained wind action waves grow to some maximum amplitude before the onset of breaking or whitecaps will limit their energy. Clearly any realistic model of wave generation must consider wave breaking and whitecap formation. This phenomenon can be handled in practice by multiplying the wind generation term, G, by a weighting factor which varies from unity at low wave amplitudes and decreases to zero at breaking. It can be further argued that a particular wave component is either breaking or not breaking. An intuitive model of breaking would consist of a coefficient which is unity or zero depending on the choice of a suitable breaking criterion. That is, the wind generation term, G, is multipled by $1 - \mu$ where μ is zero during generation and is unity at some specified maximum value of each spectra component, $E(f,\theta)$. This condition is stated:

$$G = (\alpha + \beta E)(1 - \mu) \quad (70)$$

where, $\mu = 0$ for E less than $E_{saturated}$; $\mu = 1$ for E equal or greater than $E_{saturated}$; and (after Phillips, 1958)

$$E(f,\theta)_{saturated} = A g^2 \omega^{-5} S(\theta) \quad (71)$$

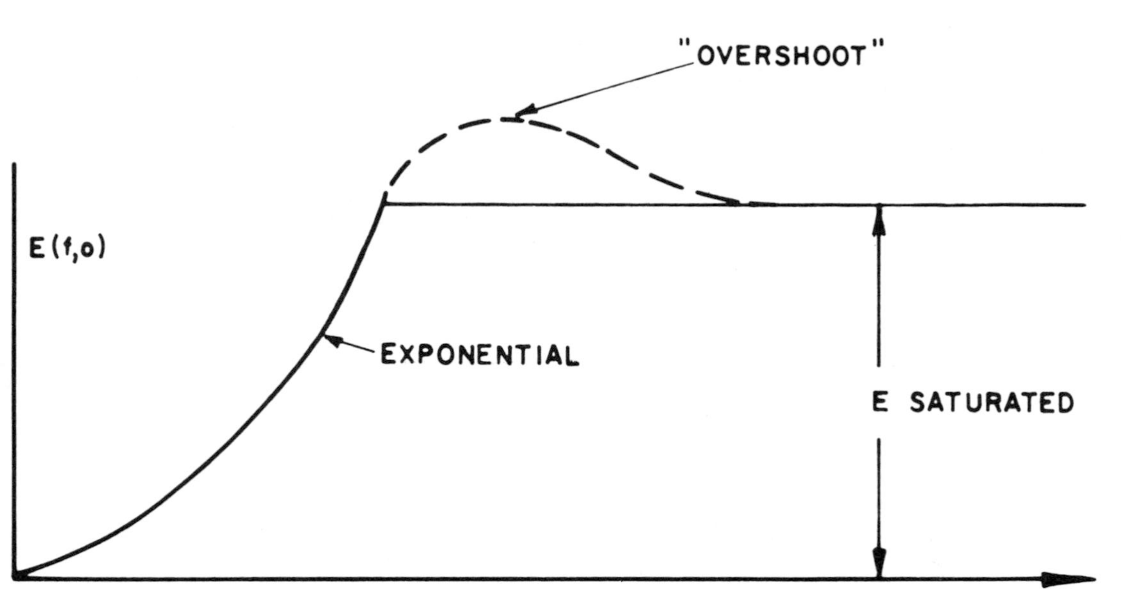

FIG. 10.—Schematic illustration of growth of a downwind spectral component.

and, after Barnett (1968):

$$S(\theta) = \frac{8}{3\pi} \cos^4\theta \tag{72}$$

The coefficient, A, has been determined experimentally by Pierson and others (1962), Burling (1959), Hicks (1960), Kinsman (1960), and others, as a value varying between 0.051 (Longuet-Higgins and others) to 0.093 (Burling) with an average value of about 0.073. The value 0.073 is recommended with an uncertainty of about 10% (see also Phillips, 1966, p. 114).

The function, $S(\theta)$, is really unknown. Barnett (1968) suggested the above form based on private communications with *D.* Cartwright of the National Institute of Oceanography.

Bottom friction dissipation.—Bottom friction dissipation can be considered to include the work done against shear stresses induced at the bed by the water particle motions (turbulent bottom friction) and also the work done against the viscous forces induced by a permeable bottom which permits water particle motions between grains of material (percolation). In general, the former dissipation rate is considerably larger than the latter and becomes significant for large waves propagating over a wide continental shelf such as exists along most of the east coast of the United States.

The bottom friction can be represented by the law

$$\tau_i = \rho c_f u_i |\underline{u}|. \tag{73}$$

The total energy dissipation of the wave field (directional spectrum) is given by

$$\Theta = -\langle \tau_i u_i \rangle = \rho c_f \langle |\underline{u}|^3 \rangle \tag{74}$$

where $\langle\ \rangle$ denotes average, c_f is the friction coefficient and $\underline{u}$ is the véctorial velocity. The functional form of the friction dissipation term, Θ, has been the subject of a number of papers (Hasselmann and Collins, 1968; Collins, 1972).

Nonlinear interactions.—Several types of nonlinear interaction are possible in a wave field. These can be classified into two broad types: strong and weak. Also, as might be expected, the types of nonlinear interaction will be different in deep, shallow, and shoaling water.

Obviously, when waves break there is a strong nonlinear interaction. The wave energy at the frequency of the wave which breaks is transformed into high frequency energy (even turbulence) and possibly low frequency energy due to mass displacements. It has been demonstrated that nonlinear interactions are extremely important. They appear to account for the "overshoot" phenomenon sketched in Figure 10 and also lead to the implication that there is no "steady state" equilibrium spectrum which can be associated with a steady wind. The overall effect of the nonlinear interactions is to tend to flatten the wind generated spectrum by "spilling" energy from the spectral peak into high and low frequencies.

REFERENCES

Barnett, T. P., 1968, On the generation, dissipation and prediction of ocean wind waves: Jour. Geophys. Res., v. 73, p. 513-530.

Biesel, F., 1952, Equations générales au seconde ordre de la houle irreguliere: La Houille Blanche, No. 3, p. 372-376.

Bowen, A. J., 1969a, Rip currents, 1. Theoretical investigation: Jour. Geophys. Res., v. 74, p. 5467-5478.

——, 1969b, The generation of longshore currents on a plane beach: Jour. Marine Res., v. 27, p. 206-215.

—— and Inman, D. L., 1969, Rip currents, 2. Laboratory and field observations: Jour. Geophys. Res., v. 74, p. 5479-5490.

Bretschneider, C. L., 1951, Revised wave forecasting curves and procedures: Univ. California-Berkeley, Inst. Eng. Res., 28 p.

——, 1958, Revisions in wave forecasting: Deep and shallow water: Am. Soc. Civil Engineers, Proc. 6th Conf. on Coastal Eng., p. 30-67.

Burling, R. W., 1959, The spectrum of waves at short fetches: Deutsche Hydrogr. Zeitschr., v. 12, p. 45-64.

Collins, J. I., 1972, Prediction of shallow water spectra: Jour. Geophys. Res., v. 77, p. 2693-2707.

Dean, R. G., 1965, Stream function representation of non-linear ocean waves: Jour. Geophys. Res., v. 70, p. 4561-4572.

Hasselmann, K., 1960, Grundgleichen der Seegangsvoraussage: Schiffstechnik, v. 7, p. 191-195.

—— and Collins, J. I., 1968, Spectral dissipation of finite-depth gravity waves due to bottom friction: Jour. Marine Res., v. 26, p. 1-12.

Hicks, B. L., 1960, The energy spectum of small wind waves: Univ. Illinois, C. S. L. Rep. M-92, 131 p.

Inoue, T., 1966, On the growth of the spectrum of a wind generated sea according to a modified Miles-Phillips mechanism: New York Univ., Geophys. Sci. Rep. TC 66-6, 74 p.

Karlsson, T., 1969, Refraction of continuous ocean wave spectra: Am. Soc. Civil Engineers Proc., Jour. Waterways, Harbors and Coastal Eng. Div., v. 95, p. 437-448.

Kinsman, B., 1960, Surface waves at short fetches and low wind speed—A field study: Johns Hopkins Univ., Chesapeake Bay Inst., Tech. Rep. 19, 581 p.

Laitone, E. V., 1960, The second approximation to cnoidal and solitary waves: Jour. Fluid Mechanics, v. 9, p. 430-464.

Lemehaute, Bernard, Divoky, D. M., and Lin, A. C., 1968, Shallow water waves: A comparison of theory and experiment: Am. Soc. Civil Engineers, Proc. 11th Conf. on Coastal Eng., p. 86-107.

Longuet-Higgins, M. S., 1953, Mass transport in water waves: Royal Soc. London Philos. Trans., Ser. A, v. 245, p. 535-581

——, 1957, On the transformation of a continuous spectrum by refraction: Cambridge Philos. Soc. Proc., v. 53, p. 226-229.

——, 1970, Longshore currents generated by obliquely incident sea waves: Jour. Geophys. Res., v. 75, p. 6778-6801.

Miles, J. W., 1957, On the generation of surface waves by shear flows: Jour. Fluid Mechanics, v. 3, p. 185-204.

Noda, E. K., 1972, Wave-induced circulation and longshore current patterns in the coastal zone: Tetra Tech Inc., Pasadena, California, Rep. TC-149-3, 120 p.

——, 1973, Rip currents: Am. Soc. Civil Engineers, Proc. 13th Conf. on Coastal Eng., p. 653-668.

Phillips, O. M., 1957, On the generation of waves by turbulent wind: Jour. Fluid Mechanics, v. 2, p. 417-445.

——, 1966, The dynamics of the upper ocean: Cambridge Univ. Press, New York, 261 p.

Pierson, W. J., Jr. (Ed.), 1962, The directional spectrum of a wind generated sea as determined from data obtained by the stereo wave observation project: New York Univ. Coll. Eng., Meteorological Paper 2, No. 6, 88 p.

——, Neumann, Gerhard, and James, R. W., 1955, Observing and forecasting ocean waves: U.S. Navy Hydrographic Office Pub. 603, 284 p.

Priestley, J. T., 1966, Correlation studies of pressure fluctuations on the ground beneath a turbulent boundary layer: Natl. Bur. Standards Rep. 8942 (unpublished).

Skjelbreaia, L. V., and Hendrickson, J. C., 1961, Fifth order gravity wave theory: Am. Soc. Civil Engineers, Proc. 7th Conf. on Coastal Eng., p. 184-196.

Snyder, R. L., and Cox, C. S., 1966, A field study of the wind generation of ocean waves: Jour. Marine Res., v. 24, p. 141-167.

Sverdrup, H. V., and Munk, W. H., 1947, Wind, sea and swell: Theory of relations for forecasting: U.S. Navy Hydrographic Office Pub. 601, 44 p.

Thornton, E. B., 1969, Longshore current and sediment transport: Univ. Florida Coll. Eng., Tech. Rep. 5, 171 p.

SAND FOUNTAINS IN THE SURF ZONE

BENNO M. BRENNINKMEYER, S. J.
Boston College, Chestnut Hill, Massachusetts 02167

ABSTRACT

Measurements with photo-electric cells show that suspension of sediment within the nearshore zone occurs in pulses of short duration. These explosions of the bottom—sand fountains—occur primarily at the still water level. In this transition zone, where the swash meets the backwash, a hydraulic jump may suspend much sediment. Also the air entrapped by the breaking of an incoming bore has sufficient suctional effect, when the bubbles penetrate the bottom, to lift sediment above the substrate.

Twice as many suspension clouds occur on the rising tide as on the falling tide. This may be due to the lag of several hours of the ground water fluctuation with respect to the tidal level. The sand fountains range in duration from 0.5 to 53.0 seconds, averaging 10.2 seconds. The average height they attain is 23.6 cm. The height of the water above the bottom is not significant in effecting the heights attained by the sand fountains.

Except at the still water level, suspension of sediment is infrequent. Even within the breaker zone, suspension of sand 15 cm above the bottom is not common. In the outer surf zone sand fountains are almost nonexistent. Therefore, it is only infrequently that a suspension cloud can be followed across the entire width of the surf zone.

At Point Mugu, California, 150-300 m^3 of sand per tidal cycle is thrown into suspension, 15 cm above the bottom for a strip 30 cm wide. This amounts to 105-210 $\times 10^3$ m^3 per year or 2-5 times the quantity moved by longshore currents.

INTRODUCTION

The development of effective remedial programs for the severe erosional problems encountered on many beaches requires a thorough understanding of short and long terms patterns of sand migration in the foreshore and nearshore zones. Elegant models of sediment transport in these zones have been constructed. They are supported by extensive laboratory tests in wave tanks and solutions based on wave mechanics. For long time periods and large scale changes, data is generally gathered by means of onshore and offshore profiles and measurements from topographic maps and air photos. For shorter intervals and smaller scale changes, sand movement is studied by tracers and comparisons of the sediment level to stakes implanted in grids in the sand and by use of various sediment traps. There is a need for in situ and inexpensive, continous measurement techniques of short-term sand movement patterns taken over extended periods of time so that the volume of sediment transported may be determined accurately. This is especially true of the high energy breaker and surf zones where sand movement is pronounced and rapid. These zones, however, are decidedly hostile for instruments and sampling, especially during storms when most of the erosion takes place.

PREVIOUS FIELD STUDIES OF SUSPENDED SAND SAMPLING

Accurate measurements of the amount of sediment in suspension have been difficult. Until recently, few direct methods existed whereby the instantaneous suspended sediment load of the water column could be determined. Most of the laboratory techniques, such as syphon samples (Shinohara and others, 1958; Fairchild, 1959), photography (Bijker, 1971), and various electro-optical meters (Homma, Horikawa and Kajima, 1965; Bhattacharya and others, 1969; Horikawa and Watanabe, 1970) or particle counters (Hattori, 1969) are too delicate to take in the field. Most of the numerous types of field samplers are for use only in unidirectional stream flow (U.S. Interagency Committee on Water Resources, 1963; University of Iowa, 1970). Some samplers average the concentration over a number of wave cycles or units of time (Inman, 1949a; Watts, 1953; Nagata, 1961; Homma and Horikawa, 1963; Cook, 1970) or integrate the volume over depth (Federal Interagency Sedimentation Project, 1970). Basinski and Lewandowski (1975) approached the problem by using radioactive Co^{60} and a detector lowered and raised in two vertical cylinders 40 cm apart. Wenzel (1975) is developing a sand concentration device based on ultrasonics. Shideler and McGrath, (1974) used pressure transducers.

Data comparisons of the many field studies are difficult, due to diverse sampling methods and different wave conditions, bottom configurations, sediment sizes and variance in reporting results. Tests by the Beach Erosion Board (U.S. Army Corps of Engineers, 1933) at Long Branch, New Jersey, using brass canisters, show that the amount of sand moved in suspension reaches 17 gm/l or 5 percent of the total load near the breaking point of the waves. Eight meters seaward of the breakpoint, suspended sediment falls to 4 gm/l. On the shoreward side the decrease is even more abrupt.

Marlette (1954), Curtis (1963) and Rudolfo (1964) used 500 ml bottles to take samples in the foreshore at the water's surface, mid-depth and 30 cm above the bottom. The results show that the weight of four replications taken over three hours increases from 0.05 to 40 gms in the swash zone. The highest concentration (79.5 gm/l) is reported in the transition zone (Little, 1968). The relative values reported by him are 1:2:4 for the plunge, surf and swash zones, respectively. Aibulatov (1966) averaged many samples that were gathered by a series of bottles lowered from a specially constructed cable car. He clearly demonstrated the effects of ridges and bars on the increase in sediment concentration in the water column at Anapa, on the Black Sea.

Inman (1949a), Terry (1951), Homma, Horikawa and Sonu, (1960) and Nagata (1961; 1964) placed multisock sediment traps in the surf zone. Nagata reported that greater concentrations move seaward in the plunge zone on steep beaches, whereas the opposite is true if the beach gradient is gentle. Homma, Horikawa and Sonu placed multidirectional, 0.075 mm mesh traps in the nearshore zone in two small bights. In the breaker zone, sediment entrapment is inversely proportional to the height above the bottom. In the surf zone 70 gm were collected in both bights near the bottom. This decreased, near the top of the water column, to 60 gm/day in one bay and increased to 450 gm/day in the other. Fukushima and Mizoguchi (1958) sampled the surf zone with 5 meter bamboo poles that were integrated into 17 positions by 5 cm holes cut between the joints. Seaward of the breaker zone the suspended load is relatively small but particles reach 150 cm above the bottom. In the breaker zone the samplers caught 233 gm in four days. No sediment was collected above 1 meter. In the surf zone the sediment distribution is bimodal showing a larger peak approximately 1 m above the bottom. Basinski and Lewandowski (1975) obtained similar results.

At Mission Bay, California, Watts (1953) took many samples of 5 minutes duration at different depths in the surf zone. He utilized a suction pump mounted in a vertical position. The greatest volume of sand (7.9 gm/l) was shoreward of the plunge point (only one sample was taken in the transition or swash zones) and 20 cm above the bottom in waves 90 cm high. Fairchild (1973) used a similar device at Ventnor, New Jersey, and found sediment concentrations from 0.025 to 2.6 gm/l. His observations were made 7.5 cm above the bottom in the surf zone with breakers 30-90 cm high. The maximum concentration was at a breaker to depth ratio of 0.8; concentration decreased abruptly seaward and gradually in the landward direction. The concentration increased again toward the transition zone. The concentrations also decreased upward in the water column but there is much scatter in the data points. The size of the suspended sand is essentially constant with respect to elevation, but the distribution curve of sediment becomes flatter with higher or more intense wave activity.

SUSPENSION OF SEDIMENT: STATE OF THE ART

Rouse (1937, 1964) derived the distribution for the vertical suspended load based on the diffusion dispersion process such that:

$$\frac{C}{C_a} = \left(\frac{h-y}{y}\,\frac{a}{h-a}\right)^z, \; z = \frac{v_s}{K u_*} \qquad (1)$$

where a = arbitrary distance from the bottom, C = concentration by volume, h = depth of water, y = vertical distance from reference level, v_s = settling velocity of grains, K = von Karman's constant, u = flow velocity, and u_* = viscous shear stress. This equation can be used to calculate the concentration of a given grain size at any distance, y, from the bed if a reference concentration (C_a) at a distance, a, above the bed is known. The equation can be integrated over the region where suspension occurs, from a to h so that:

$$g_s = \int_a^h Cu\,dy \qquad (2)$$

where C and u are functions of both y and g_s, where g_s is that suspension rate in weight per unit time and unit width.

This is the suspended load rate in weight per unit time and width and is an appropriate approximation. But the equation is limited, for it can predict the relative concentration in the vertical but not its absolute value (Einstein and Abel-Aal, 1972). Bijker (1971), however, has applied the equation to suspended sedimentary computations in both wave and current combination and found good agreement between the vertical and actual concentrations.

Vanoni and others (1963) and Kennedy and Locher (1972) have shown that in the presence of sediment the K constant used to calculate z decreases rapidly. This had prompted Einstein (1950) and Einstein and Chien (1954) to express the water velocity with the logarithmic velocity distribution so that:

$$g_s = \int_a^h C_a \left(\frac{h-y}{y} \frac{a}{h-a} \right)^z 5.75\, U'_* \log \frac{(30.2y)}{\Delta}\, dy \qquad (3)$$

where U'_* is the shear velocity due to sand grains only. The integration indicated is impossible, but iteration for various a/h values and z value gives two graphical solutions such that

$$g_s = 11.6\, C_a U'_*\, a \left(2.303 \log \frac{(30.2h)}{\Delta} I_1 + I_2 \right) \qquad (4)$$

Three other analytical approaches exist to predict the suspended sediment distribution under waves; however, a satisfactory solution does not yet exist (Das, 1973).

Bagnold (1963, 1966) approached the sediment transport problem somewhat differently. He argues that it is the existence of lifting forces equal to the immersed weight of the sand that is paramount in sediment movement. The immersed weight $m'g = (\rho_s - \rho)/\rho_s\, mg$. Therefore, the total load transported per unit width is:

$$i = i_b + i_s = \frac{\rho_s - \rho}{\rho_s} mg\bar{U} = m'_b g\bar{U}_b + m'_s g\bar{U}_s \qquad (5)$$

where i = dynamic transport rate of sediment, i_b = fraction of bedload in a given grain size, i_s = fraction of suspended load in a given grain size, ρ = density, ρ_s = effective density of sediment grains, and $\bar{U}$ = average water velocity.

This dynamic transport rate has to be converted to a work rate since the stress and velocity are not in the same direction. Therefore, the bedload fraction is multiplied by the tangent of the slope, tan a, and the suspended load by $v_s/\bar{U}_s$. This yields $i_b + i_s = m'_b g\bar{U}_b \tan a + m'_s g\bar{U}_s (v_s/\bar{U}_s)$. Assuming any energy losses are due solely to fluid drag on the bottom, then the available energy in the water column to transport sediment is: $w = \rho g h S \bar{U}$, where w is available power attributable within unit boundary area and S is the energy gradient due to slope. If the power expended in transporting the bedload is e_b, the remaining power $(1 - e_b)$ is available to maintain the suspended load. If e_s is the suspension efficiency, the total transport rate can be written as:

$$i = i_b + i_s = w \frac{e_b}{\tan a} + \frac{e_s \bar{U}_s}{v_s}(1 - e_b) \qquad (6)$$

FIG. 1.—The almometer, consisting of two parts, a high intensity light source and a series of photo-electric cells, both encased in watertight acrylic cylinders, and almometers placed in the field.

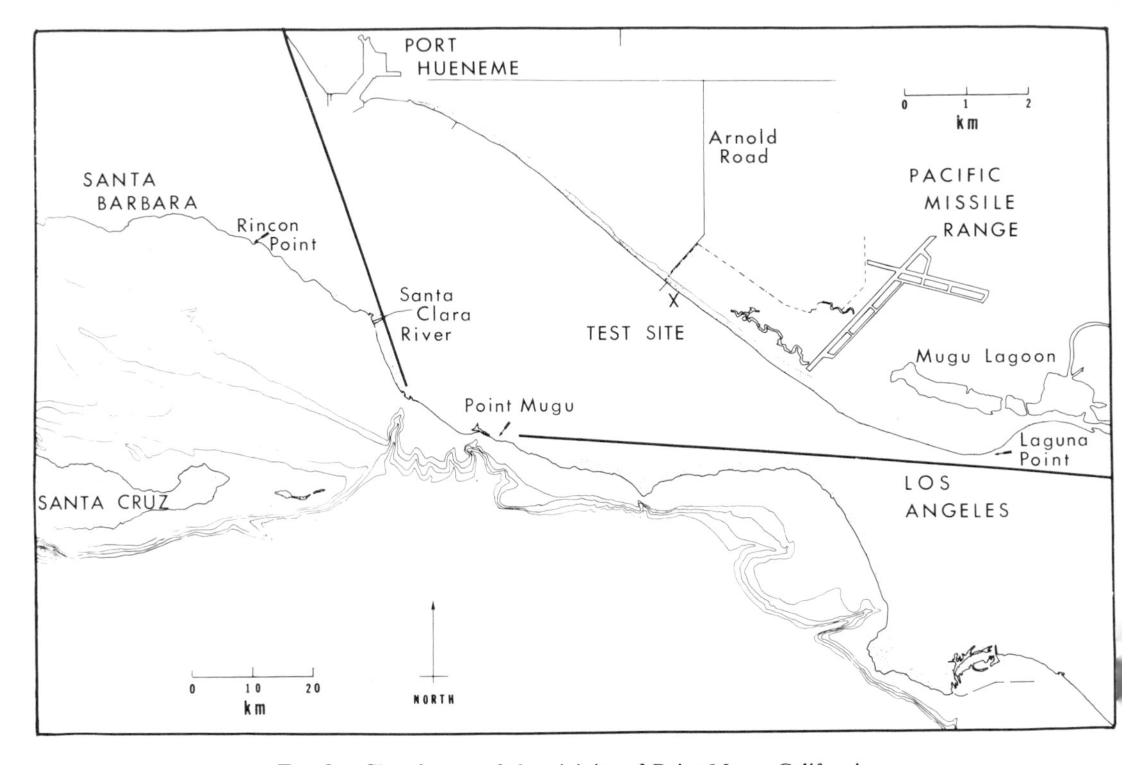

FIG. 2.—Sketch map of the vicinity of Point Magu, California.

From flume studies $e_s(1 - e_b) = 0.01$, and $U_s = U$ for true suspensions. Therefore,

$$i = \frac{we_b}{\tan a} + 0.01 \frac{\bar{U}}{v_s} \qquad (7)$$

The total sediment load can be obtained if these four parameters are known.

PRESENT FIELD STUDY

This study was undertaken at Point Mugu, California, to determine, by Eulerian means, the movement of sand normal to the shoreline and across the shoreface and beachface under the influence of breakers, surf and swash regimes on an open marine beach. A new instrument, the almometer (Fig. 1; Brenninkmeyer, 1973, 1975) was used to continuously monitor (every 1/5 of a second) the changes in topography to an accuracy of 1 cm and the density distribution of sediment in the bottom 1 meter of the water column. The instrument consists of 64 photo-electric cells placed 1-2 cm apart in a vertical column, and a high intensity fluorescent lamp. The light source was stationed at least 30 cm away from the photo-electric cells so that the opacity of the water passing between the light and sensors could be ascertained. The quantity of sand was accurately related to the opacity of the water and ranged from 10 to 900 grams per liter (Brenninkmeyer, 1976).

Three of these almometers were emplaced on the beach at Point Mugu (Fig. 2): (1) 15 cm below MLLW on the shoreface, 115 m from the baseline; (2) 30 cm above MLLW on the lower beachface, 82 m from the baseline; (3) 90 cm above MLLW on the upper beachface, 58 m from the baseline.

Sea state.—The tides were measured at Rincon Island, 19 km away. The semidiurnal minimum was 33 cm and a maximum of 225 cm with a mean of 140 cm. From the three fixed locations of the almometer, the various dynamic zones of the surf regime could be sampled as the tide rose and fell. The position of the various zones, as used in this paper, are shown in Figure 3.

Measurements of wave period, angle of wave approach, breaker height and type were recorded every two hours (Fig. 4). These were supplemented by the U.S. Navy observations of the same variables. The Navy also measured the direction and speed of longshore currents. The U.S. Army Corps of Engineers recorded deep water wave height and period. The breaker height

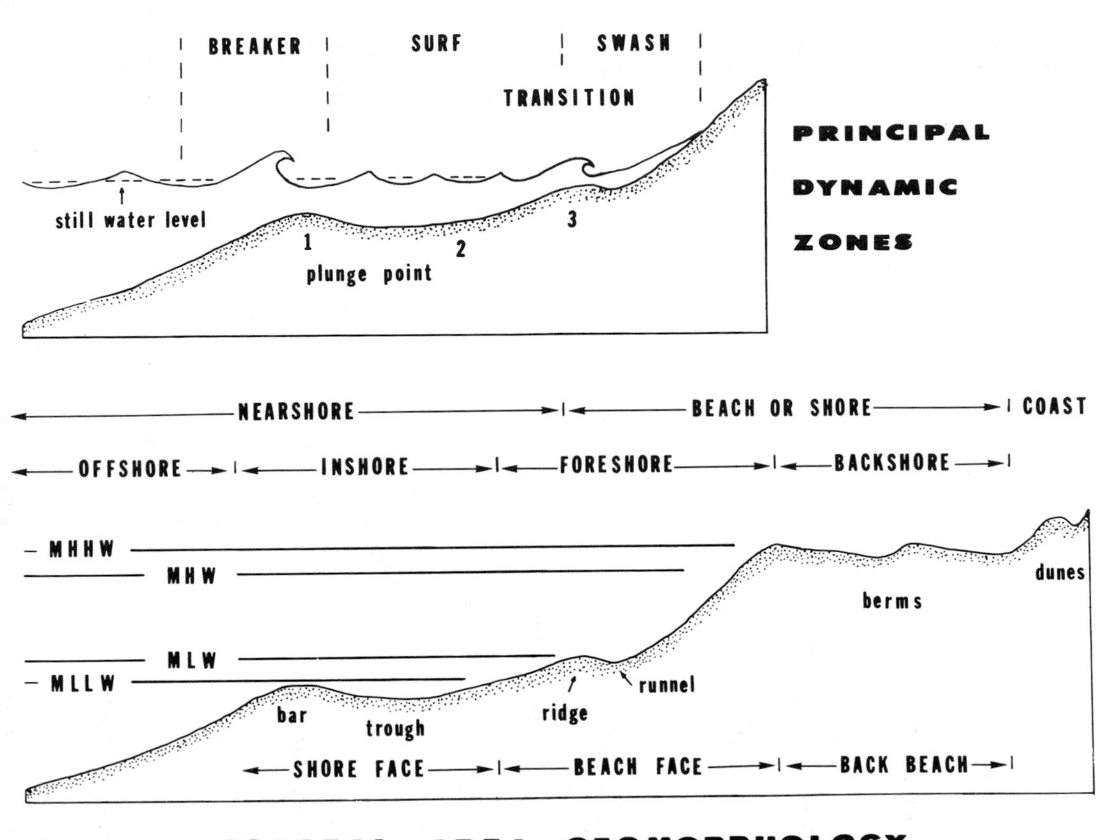

FIG. 3.—Coastal geomorphological terminology and principal dynamic zones.

averaged 75 cm and ranged from 45–105 cm. Each reading was made to the nearest 15 cm, and represents an average of measurements taken from 20 consecutive waves. The angle of incidence of the breakers showed that the waves came from both east and west in about equal frequency. This seemed to have little effect on the direction and velocity of longshore currents. The longshore drift is predominantly to the east with a velocity of up to 35 cm/sec. Only once out of 10 days during the time of the test was there a westward current; it was 10 cm/sec.

SAND FOUNTAINS IN THE SURF ZONE

Since Aibulatov's (1958) and Homma and Horikawa's (1963) studies, coastal scientists realized that suspension of sediment can occur in very short spurts under the influence of progressive waves. More recently they have been observed in standing waves (Hattori, 1969, 1975). These sudden upheavals of bottom sediment have been called sand fountains by Zenkovitch (1967; p. 121). Such fountains of sediment are quite variable and are short in duration (Fig. 5).

The three depicted smaller concentrations usually illustrated similar patterns. The 375 gm/l concentration, at times, followed a more subdued path and showed reversals not experienced by smaller concentrations. Deposition of sediment occurred more than half the time after one of these sand uplifts.

Extent and location of sand fountains.—To determine the extent, size, shape, and location of these fountains, the 135 gm/l concentrations were analyzed in detail through nine tidal cycles (Table 1). Only those times when the 135 gm/l concentration rose 15 cm above the bottom were noted. This elevation was chosen since it is the lowest elevation above the bottom at which moderately heavy sediment concentrations were intermittently thrown into suspension.

Approximately 80 percent of the 1213 explosions of the substrate occurred at the almometer on the upper beachface (Table 1). This was certainly contrary to expectation, especially in light of the fact that the almometer was above

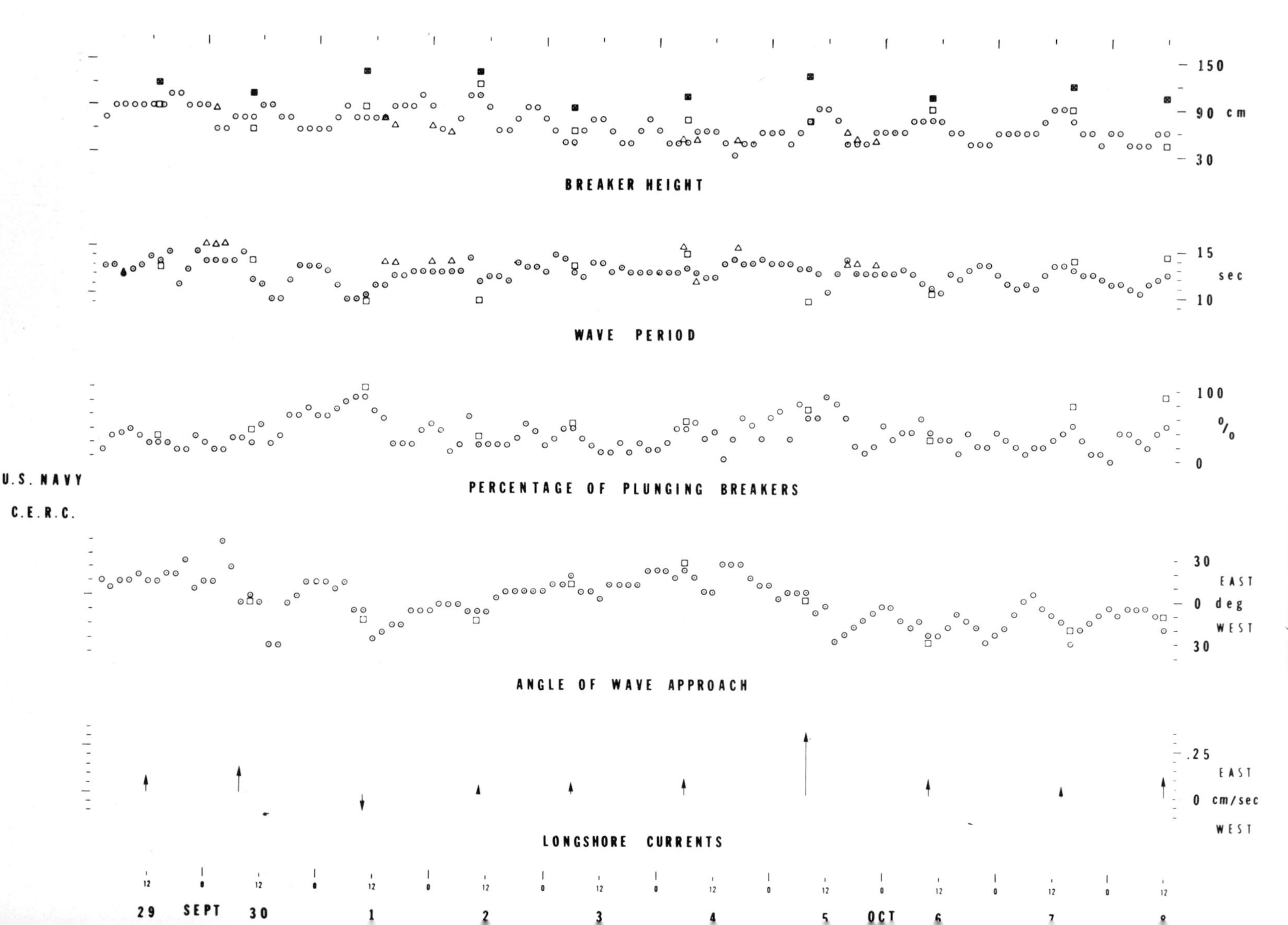
BREAKER HEIGHT
150
90 cm
30
WAVE PERIOD
15
sec
10
PERCENTAGE OF PLUNGING BREAKERS
100
%
0
ANGLE OF WAVE APPROACH
30
EAST
0 deg
WEST
30
LONGSHORE CURRENTS
.25
EAST
0 cm/sec
WEST
□ U.S. NAVY
△ C.E.R.C.
29 SEPT 30 1 2 3 4 5 OCT 6 7 8

the still water level 40 percent of the time. Part of the explanation must lie in the ability of the water flow to suspend sediment in the lower swash and especially the transition zone. Approximately 87 percent of the sand fountains occurred when the still water level was less than 30 cm above or below the elevations of the almometer stations. Inman (1949b) noted that sand of 0.18 mm diameter is the easiest to transport. Just landward of the third station the sand had a diameter of approximately 0.18 mm, whereas seaward it was finer and therefore more difficult to suspend. This does not explain the greater number of sand uplifts that occurred on the flood tide compared to the ebb tide (250% greater). It was only apparent at the third station which was above the ground water level during the flood tide, whereas, on the ebb tide it was below the ground water level. During the ebb tide, sand below the water table is somewhat consolidated due to the surface tension imparted by the ground water. The elevation of the water table lags one to three hours behind the tide (Emery and Foster, 1948; Fausak, 1970; Dominick and others, 1971) in its descent down the beach to its low water level at the lower beach face station. The surface tension, combined with a probable decrease in the roughness coefficient due to the interstitial water, makes it more difficult for the sand to be suspended.

The almometer on the lower foreshore registered the least number of sand uplifts (10 percent of the total recorded) while it was above the still water level only 14 percent of the time (14.5 hrs). Only ten percent of the explosions occurred at the almometer on the shoreface while it was completely submerged all the time. This illustrates that the predominant means of sand transport in the breaker and lower surf zones is not by means of suspension but rather by bedload. Johnson (1952) reported that when relatively flat waves are present in this zone, the major portion of sand is moved along the beach by bedload, whereas when the steepness ratio exceeds 0.025, movement is by a combination of bedload and suspension. During the present study, the waves were very flat ($H_0/L_0 = 0.002$).

The breaker zone in general, and the plunge point in particular, is difficult to locate accurately. During these tests under the prevailing wave conditions the plunge point moved past the shoreface almometer when the still water level was between 90 and 120 cm and past the lower beachface station when the tide rose above 140 cm. When the tide reached all of these levels, only 4 percent of the suspension clouds occurred.

The approach of the breaker zone, as the tide rose or fell, was signalled by a decrease in the duration of the uplifts. The average time of the uplifts dropped from about 14 seconds to 4 seconds. This dramatic decrease in the length of the fountains occurred on the lower two stations but not on the upper beachface which was not in the breaker zone during the time in question. These decreases in time were a result of the increases in water velocity near the breaker zone. Therefore, though few in number, the sand fountains occurring near the breaker zone moved considerable quantities of sand.

Overall, the sand uplifts ranged in duration from only half a second to 53 seconds, averaging 10.2 seconds. During the ebb tide the fountains lasted somewhat longer than during the flood tide. This is because the sand on the bottom of the swash zone is coarser on the ebb tide than on the flood tide, and it is negatively skewed (Brenninkmeyer, in preparation); therefore, the material in suspension should be somewhat finer. The longest uplifts occurred on the relatively flat shoreface and lower beachface during low tide. This is apparently caused by the prevalence of finer sand sizes and the slower speed of the backwash.

The tide level is relatively insignificant in effecting the heights attained by the suspension clouds. The 375 gm/l turbid zone reached 100 cm above the bottom only once during a rising tide in 110 cm of water under the breaker zone. The 40 gm/l concentration rose above the limits of measurement (112 cm) four times. The average height of the fountains was 23.6 cm on rising tide and 23.7 cm on the falling tide (Table 1). There does not seem to be any correlation between the height and the length of the sand fountains as shown in Figure 6.

The frequency of occurrence of the sand fountains was computed by averaging the time between successive suspension clouds after the first uplift per tide (Table 1). During the rising tide, when the still water level was 90–150 cm, the frequency of occurrence at the upper beachface almometer was 1.7 and 2.6 minutes. During the same tidal

←

FIG. 4.—Sea state at Point Mugu, September 29 to October 8: Breaker height, breaker period, percentage of plunging breakers, angle of wave approach and longshore currents.
⊙ Observations taken every two hours.
□ Observations taken by the U.S. Navy from the Experimental groin.
△ Recordings from the U.S. Army Corps of Engineers, Coastal Research Center deep-water wave gauge.

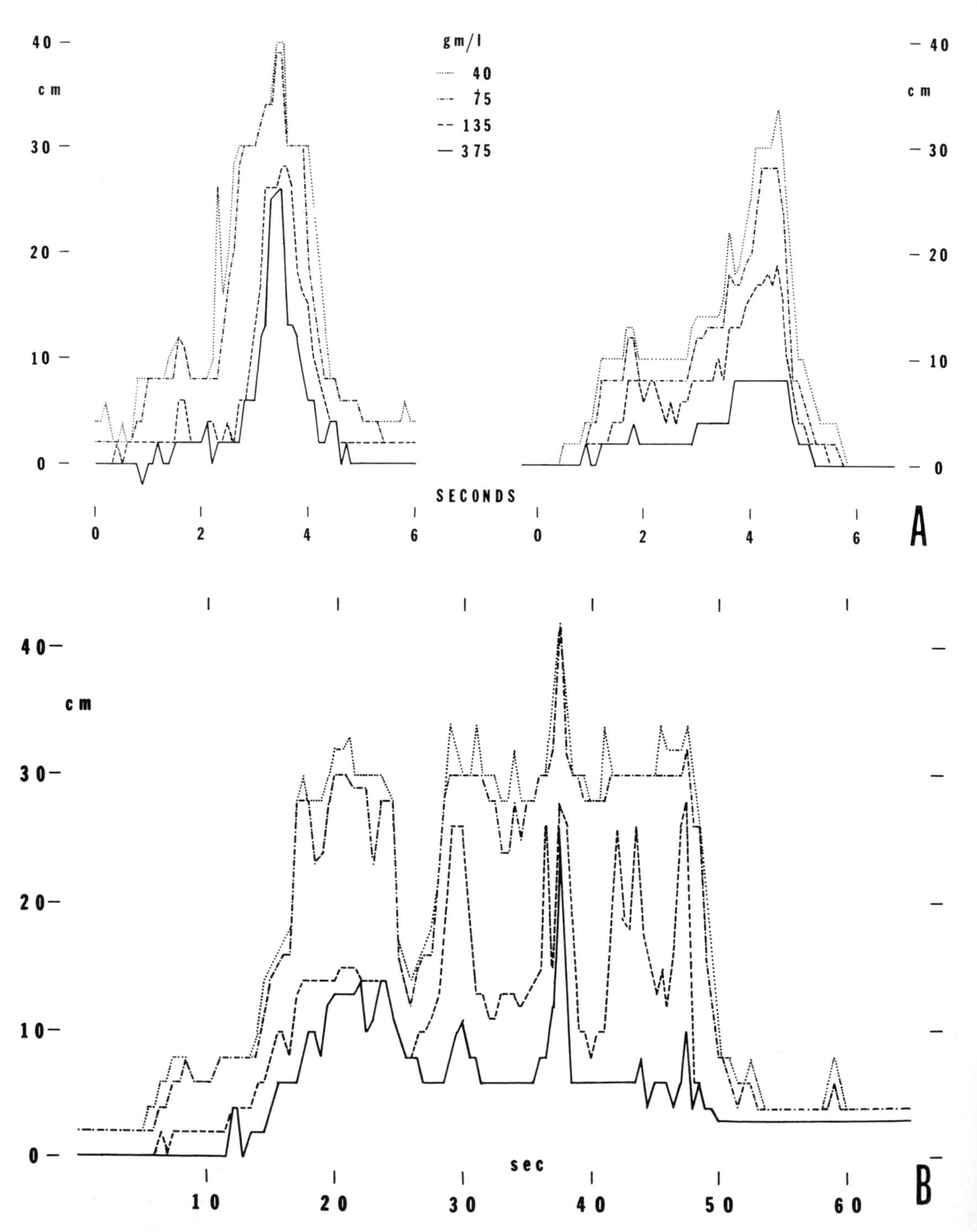

FIG. 5.—Sand fountains. *A*, Left: sand fountain on upper beachface, September 30, 22:53, ebb tide, 64 cm above MLLW, classification δAaα; right: sand fountain on lower beachface, October 1, 23:50, ebb tide, 37 cm above MLLW, classification γAbα. *B*, Sand fountain on upper beachface, September 30, 22:40, ebb tide,

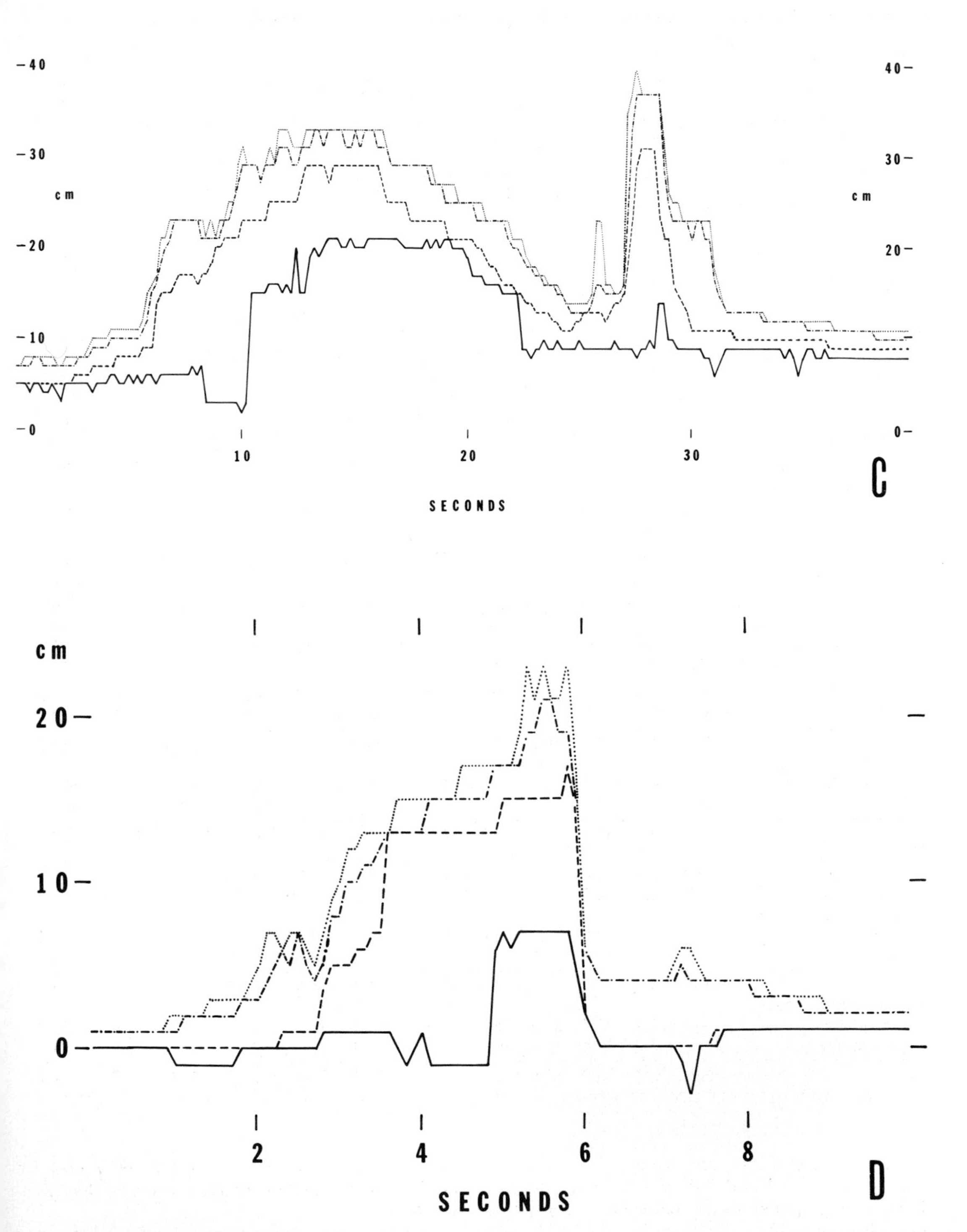

82 cm above MLLW, classification δBaα. *C*, Sand fountain on shore face, October 2, 14:36, flood tide, 21 cm above MLLW, classification βBaδ. *D*, Sand fountain on lower beachface, September 28, 22:56, ebb tide, 43 cm above MLLW, classification βAbα.

TABLE 1.—DISTRIBUTION OF SAND FOUNTAINS ACROSS THE SURFACE ZONE WITH CHANGES IN TIDAL ELEVATION DURING NINE TIDAL CYCLES

Tide level (cm)	Overall Total −30–150	Shoreface (#1) −30	0	30	60	90	120	150	Total or Average
Hours									
Rising tide	55.4	2.6	9.0	5.3	5.6	6.7	5.2	4.5	38.9
Falling tide	48.3	2.6	8.3	6.2	9.2	4.3	5.4	4.0	40.0
	103.7	5.2	17.3	11.5	14.8	11.0	10.6	8.5	78.9
Sand fountains									
Rising tide	799	5	15	4	7	7	25	16	79
Falling tide	414	1	26	9	1	6	8	2	53
	1213	6	41	13	8	13	33	18	132
Frequency (hrs)									
Rising tide	0.12	0.52	0.60	1.32	0.80	0.96	0.35	0.41	0.61
Falling tide	0.16		0.32	0.69		1.07	0.64	5.42	0.46
Length average (sec)									
Rising tide	9.2	13.8	13.7	5.8	3.9	5.7	5.3	4.6	7.4
Falling tide	11.9	15.0	14.8	14.1	4.0	3.5	4.6	3.5	11.2
	10.2	14.0	14.4	12.9	3.9	4.7	5.1	4.5	9.0
Height average (cm)									
Rising tide	23.6	18.4	18.9	23.0	17.5	29.4	21.8	20.9	20.5
Falling tide	23.7	19.8	22.2	21.7	17.0	19.7	22.9	20.0	21.7
	23.6	18.6	20.9	22.1	17.4	20.4	22.1	20.8	21.0

Tide level (cm)	Lower Beachface (#2) 0	30	60	90	120	150	Total or Average	Upper Beachface (#3) 60	90	120	150	Total or Average
Hours												
Rising tide	9.3	7.9	7.5	11.4	9.7	7.0	52.8	7.5	11.4	9.7	7.0	35.6
Falling tide	8.3	7.0	7.7	7.8	9.1	6.3	46.2	7.7	7.8	9.1	6.3	30.9
	17.6	14.9	15.2	19.2	18.8	13.3	99.0	15.2	19.2	18.8	13.3	66.5
Sand fountains												
Rising tide	5	26	7	2	1	4	45	32	407	219	17	675
Falling tide	1	50	24	1	1	0	77	7	142	109	26	284
	6	76	31	3	2	4	122	39	549	328	43	959
Frequency (hrs)												
Rising tide	1.86	0.30	1.07	5.70		1.75		0.23	0.03	0.04	0.41	
Falling tide		0.14	0.32					1.10	0.05	0.08	0.24	
Length average (sec)												
Rising tide	15.6	13.5	8.6	6.0	4.0	8.5	12.0	8.3	8.5	10.8	9.9	9.2
Falling tide	16.5	16.5	13.1	4.0	4.0		15.1	8.6	11.8	11.3	8.2	12.0
	15.8	15.5	12.1	5.3	4.0	8.5	14.0	8.4	9.3	11.0	8.9	9.8
Height average (cm)												
Rising tide	18.6	21.1	22.5	32.2	26.0	38.0	23.1	20.7	22.8	26.8	24.0	24.0
Falling tide	17.0	23.2	24.4	16.0	16.0		23.3	20.4	23.6	24.5	26.8	24.2
	18.3	22.5	24.0	26.9	22.0	38.0	23.2	20.6	23.0	26.0	25.7	24.1

stages of the ebb tide, the average time between fountains was 3.3 and 5.0 minutes respectively. Plotting the actual times between occurrences (Fig. 7) shows that during this tide level only two small peaks stand out amidst the scatter. These are at 15 and 21 seconds. The first is the predominant wave period. The second may be due to constructive interference of wave trains of shorter period or more likely, they may be due to the somewhat longer period of the backwash.

Shape of the suspension clouds.—The sand fountains were classified according to their shape. If the 135 gm/l suspensate cloud rose and dropped immediately as it moved past the almometer, it was designated as *A*; if, however, the cloud stayed

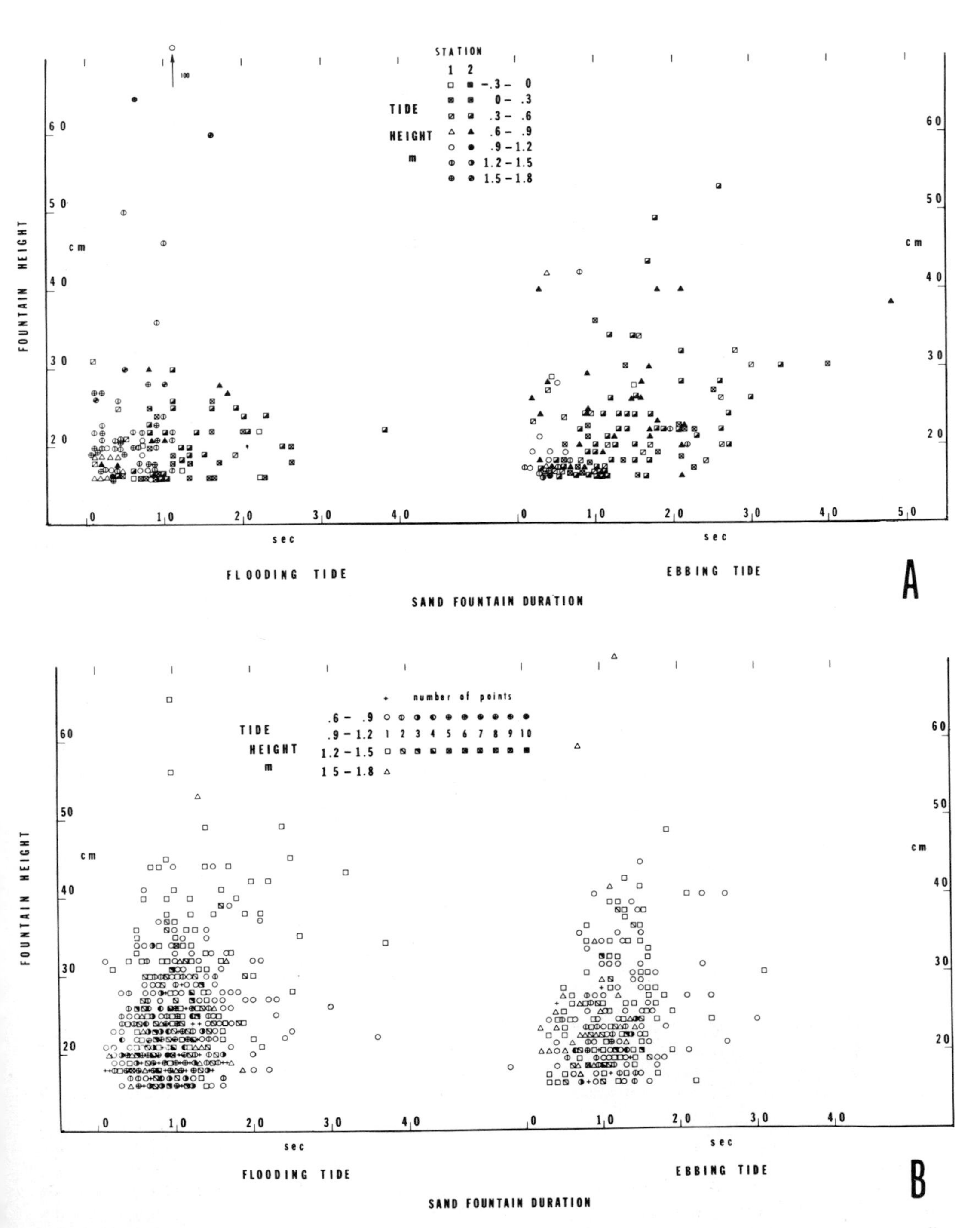

FIG. 6.—*A*, Height versus duration of sand fountains on the shoreface and lower beachface with changes in tidal elevation. *B*, Height versus duration of sand fountains on the upper beachface with changes in tidal elevation.

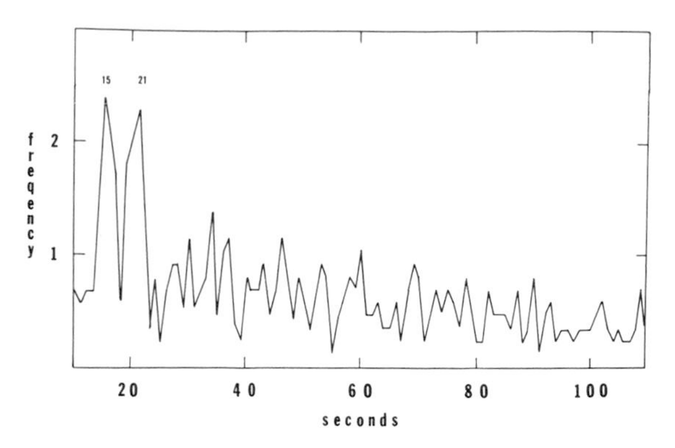

FIG. 7.—Frequency of occurrence of sand fountains on the upper beach face.

at about the same height for over 5 seconds, it was called *B*. A further subclassification delineated the types of peaks (Table 2).

TABLE 2.—CLASSIFICATION OF THE SHAPE OF SAND FOUNTAINS

A—Peaked, shorter than five seconds
B—Flat, longer than five seconds
Peak characteristics
a—peak is symmetrical, rise time equals fall time
b—rise time is longer
c—fall time is longer
f—two or more peaks, the first is higher
g—two or more peaks, the last one is higher
h—two or more peaks, both are of equal height
k—plateau in the middle
m—plateau in the middle with a peak at the end
n—plateau in the middle with a peak at the beginning
Rise and fall characteristics
α—smoothly
β—series of small steps
γ—plateau before or after main one
δ—peak before or after main one

The great majority clouds rose abruptly from the bottom on the flooding tide and died out gradually. On the ebbing tide they rose with less rapidity and waned even more slowly (Table 3). The higher the tide, the shorter the rise time. Higher up the beach there were more small plateaus and peaks before the major uplift commenced. As the tide increased in elevation, the clouds died out slowly with secondary peaks more common.

About half of the sand fountains were symmetrical. They rose and fell without any prior or posterior minor uplifts. About half the number of peaks on the flooding tide rose more slowly

TABLE 3.—DISTRIBUTION OF DIFFERENT TYPES OF SAND FOUNTAINS ACROSS THE SURF ZONE

	Flood tide			Ebb tide		
Classification	Shoreface (1)	Lower Beach Face (2)	Upper Beach Face (3)	Shoreface (1)	Lower Beach Face (2)	Upper Beach Face (3)
A	74.7	40.0	60.9	66.0	41.5	60.2
B	25.3	60.0	39.1	34.0	58.4	39.8
α	68.3	46.7	53.8	45.3	45.4	43.7
β	22.8	31.1	26.7	37.7	37.7	32.4
γ	5.0	8.9	9.9	0.9	9.1	12.7
δ	3.8	13.3	9.6	0.7	7.8	11.3
α	69.6	51.1	32.0	52.8	44.2	24.6
β	25.3	37.8	46.2	41.5	35.1	60.6
γ	3.8	8.9	11.8	3.8	14.3	4.9
δ	1.3	2.2	9.9	1.9	6.5	9.9
a	74.7	62.2	52.1	50.9	44.2	54.2
b	2.5	8.9	4.4	7.5	13.0	8.8
c	7.6	4.4	14.1	3.8	3.9	8.1
f	6.3	4.4	3.4	1.9	2.6	5.3
g	3.8		3.3	9.4	6.5	4.9
h		8.9	5.8	13.2	15.6	5.6
k		2.2	5.0		3.9	3.9
l	1.3		1.5	5.7	1.3	1.1
m	1.3	8.9	3.0	5.7	7.8	2.8
n	2.5		7.4	1.9	1.3	5.3

than on the ebbing tide. The reverse holds for the slower fall of the explosions. There were approximately twice as many sand fountains with 2 or more peaks on the ebb tide than on the flood tide. The other types of fluctuations were approximately equal for both tidal phases.

Cause of sand suspension in the surf zone.—The suspension clouds occur for at least five reasons, each of which may be valid for different parts of the coastal area. In the area of shoaling waves Inman and Nasu (1956) showed that vertical velocities up to 60 cm/sec directed downward, exist not only under wave crests but also under wave troughs. Upward vertical velocities are highest under the wave crests. However, not all wave crests possess an accompanying upward velocity. Morrison and Crooke (1953) noted that the maximum horizontal and vertical velocities occur when the wave crest approaches. When the wave is about to break, the maximum velocity is ahead of the wave front near the plunge point (Iverson, 1952). The maximum horizontal velocity is almost equal to the crest celerity just shoreward of the highest part of the crest (Ippen and Kulin, 1954) after breaking or 0.09 T after the crest (Adeyemo, 1971) depending on the beach slope. The temporal distribution of these velocities coupled with the accompanying turbulence of the breaking waves are more than sufficient to raise sediment into suspension.

Inside the surf zone, measurements by Longinov (1958, 1961, 1964, 1968) showed that discrepancies between theoretical (Airy's) and actual water velocities vary by a factor of 5 for the surface and a factor of greater than 20 for the bottom. His recordings, by means of a dynamometer, showed that positive vertical velocities, a few cm above the bottom, reached 50 percent of the horizontal velocities. Higher in the water column they reached 80 percent of the horizontal velocities (Fig. 8). Maximum vertical velocities were not randomly distributed but occurred either when the horizontal velocity was zero, changing from shoreward to seaward or just after the shoreward maximum. These vertical velocities may be due to breaker vortices (Miller, this volume) and not the orbital velocity structures of waves. Perhaps this explains the onset of suspension that has been observed when the horizontal velocity decreases (Inman and Bagnold, 1963; Palmer, 1970).

When the wave breaks, the wave front is saturated with air bubbles and foam which can penetrate the bottom at plunge point and at the inner breaker line. Fuhrboter (1971) and Miller (1972) have observed that as much as 20-30 percent air by volume may be generated within a breaking wave. The ascent of these bubbles has a suctional effect sufficient to draw up sand (Aibulatov,

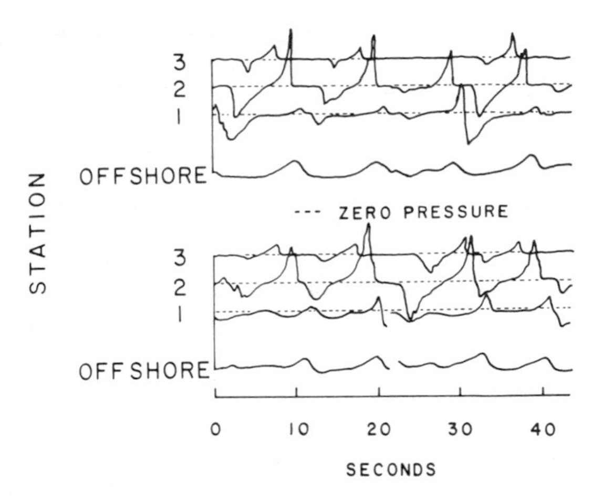

FIG. 8.—Changes in horizontal water velocity across the inshore and foreshore, 1 is seaward of the breaker zone, 2 is in the transition zone, 3 is in the swash zone. (After Longinov, 1964).

1958, 1961). Medvedev and Aibulatov (1958) and Aibulatov, Bolyrev, and Griesener (1961) noted that the descent and rise of air-laden water in this zone caused continuously changing vertical eddies. The two in combination stirred up clouds of sediment several meters wide and lasting several seconds. In the Black Sea these clouds reached a height of 20–30 cm.

The dearth of sand fountains near the primary breakers illustrates that the air bubbles do not reach the bottom but are absorbed in the surficial water layers. As Miller and others (1974) noted, the impact pressure on the bottom (caused by breaking waves) increases gradually, but closer to the water surface there is a sharp increase. The reverse is true in a bore where there is a drastic increase close to the bottom. Even secondary breakers, in the lower surf zone, did not appear to be effective lifting agents. However, close to shore, in the transition zone, the entrapment of air, seems to be very effective in lifting sand above the bottom.

Miller and Zeigler (1958, 1964), Ingle (1966) and Ippen (1966) suggested another reason for the suspension clouds. The water circulation in the surf zone depends upon the relative timing of the arrival of the incoming bore and the backrush of the previous surge. When the two are coincident there is little exchange, because the breaker is formed primarily from the seaward moving water. The deep water wave steepness has little influence on wave instability in the surf zone and the breaking limit of a solitary wave ($H = 0.78h$) is, therefore not applicable (Horikawa and Kuo, 1967). The same is true for laboratory undular bores (Miller, 1973). Thus the inner breakers may be caused by the return flow of the backwash.

Pressure profiles and electromagnetic current measurements by Schiffman (1965), Miller and others (1974) and also by Huntley and Bowen, (1975) in the surf zone show surge profiles that increase gradually, reach a maximum and then decline so fast as to be almost instantaneous (Fig. 8). If a backwash has a velocity greater than $(gh)^{1/2}$, the incoming bore can not progress against it, and produces a hydraulic jump or roll wave (Grant, 1943). If the average backwash velocity is 180 cm/sec, then the water depth at which the Froude number becomes critical is just under 30 cm.

The breaking of a roll wave appears to be similar to a standing wave (Simons and others, 1965; Allen, 1968) except that it is formed by opposite flowing water masses. In the breaking of a standing wave there is a brief suspension of much sediment, then the water and sediment movement very nearly stops (shown by zero horizontal pressures) and most of the suspended sediments fall back to the bed. The deposits of these suspended sediments may form the inner rough facies (Clifton, and others, 1971) composed of steep-sided symmetrical ripples on steep beaches and perhaps even the ridges and runnels so common on rather flat beaches. These surface protuberances on their own may cause more suspension by creating a barrier to the backwash flow and again setting up a hydraulic jump. The rate at which sediment is entrained is governed, to a great extent, by the ripple spacing (Kennedy and Locher, 1972).

Some of the suspension clouds may be computor introduced, due to the mode of digitization. The highest sensor registering a chosen concentration was utilized. If a sediment-laden backwash moved over a clean incoming bore, the digitized results would indicate an upheaval of bottom sediment, which was not the case. It belied the assumption used in digitization that the sediment concentration increased downward. Overtopping by the backwash has been observed on beaches (Miller and Zeigler, 1964). In reviewing four original analog tapes, 10-15 percent of the so-called sand fountains resulted from an increase of the 135 gm/l concentration at a level somewhere above the bottom and a decrease in sediment concentration below it (Brenninkmeyer, 1976). Overtopping, of course, is not the only means for this inverse concentration gradient. A roll wave may give a similar effect.

Finally, some of the suspension clouds may have been due to the stationary position of the almometers. The suspension in a primary bore, carried past the station, was counted as a sand fountain. When the velocity reversed, the sediment cloud entrained during the incoming wave may not have settled back to the bed and was swept back in the opposite direction again being counted. Relatively few of the reported sediment clouds were due to this double counting. The average size of the sand on the foreshore (0.16 mm) falls almost 22 cm in a second in quiet sea water (Gibbs, Matthews and Link, 1971). It would be expected, therefore, that most of the sand in a suspension cloud would fall back to the bottom during the time it took for the water to reverse direction. A reversal of water motion from on shore to offshore is the probable cause of the sand fountain shown in Figure 5c.

SUSPENDED SAND MOVEMENT ACROSS THE SURF ZONE

In order to show the respective modes of sediment transportation by bores and backwash, several examples will be cited. The similarity between A and B in Figure 9 is clear. If the initial disturbances are taken for comparison, the backwash traveled 175 cm/sec. Either this velcity was not sufficient or there was a lack of turbulence, for on the upper beachface (#3) the 135 gm/l suspension cloud reached 38 cm above the

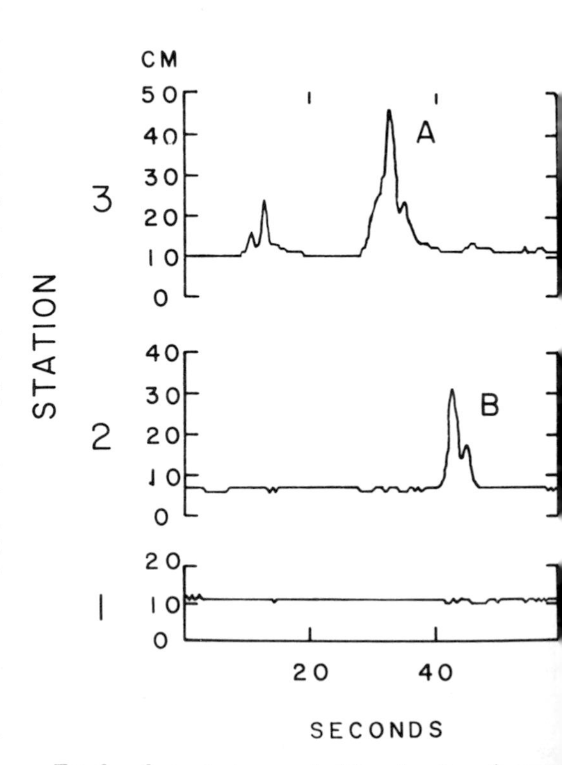

FIG. 9.—One minute record of the elevation changes across the foreshore (#1), lower beach face (#2) and upper beach face (#3) of the 135 gm/l concentration during the seaward traverse of a backwash. October 1, 22:00, Ebb tide, 100 cm above MLLW, breaker height 105 cm, 13 second period, 40% plunging, approaching land at a 5° angle from the west.

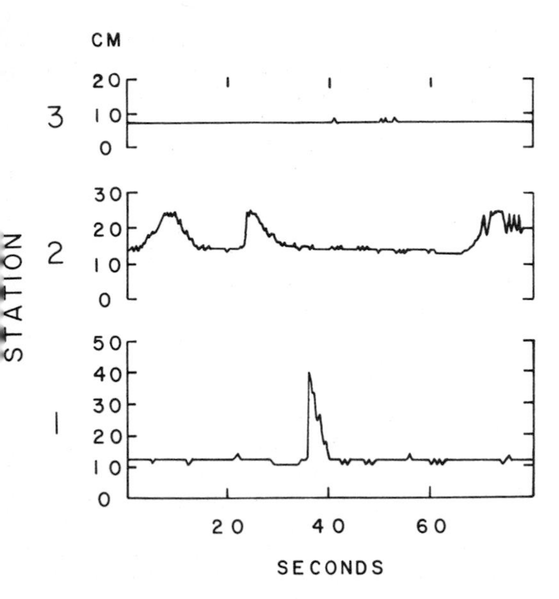

FIG. 10.—Eighty second record of the elevation changes across the foreshore (#1), lower beach face (#2) and upper beach face (#3) of the 135 gm/l concentration showing an increase in suspension as a backwash moves seaward. September 30, 23:39, ebb tide, 42 cm above MLLW. Breaker height 60 cm, period 12.5 seconds, 60% plunging, approaching land at a 10° angle from the east.

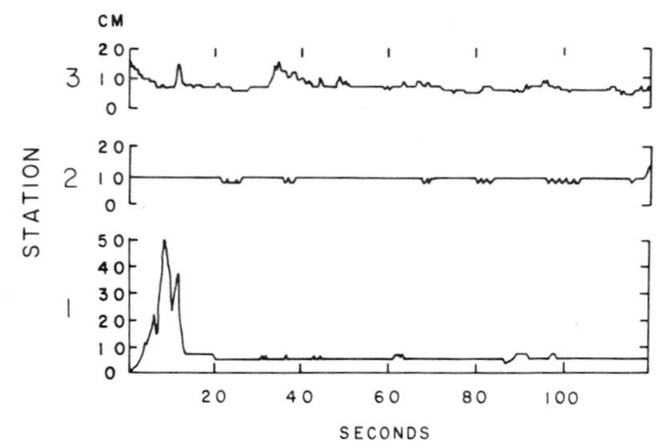

FIG. 11.—Two minute record of elevation changes across the foreshore (#1), the lower beach face (#2) and the upper beach face (#3) of the 135 gm/l concentration after an incoming wave. October 3, flood tide, 104 cm above MLLW. Wave height 60 cm, 14 second period, 40% plunging, approaching land at a 10° angle from the east.

bottom. By the time it reached the lower beachface (#2) it was only 23 cm above the bottom.

In Figure 10, the elevation of suspended sediment increased from 13 to 29 cm above the substrate, but the water velocity was at least 250 cm/sec. The sand fountain on the foreshore (#1) was apparently caused by the backwash, because both clouds start abruptly and have a slower denouement. It is conceivable that a surge forming the cloud at station one was the same one that produced the cloud (33 seconds later) near station two, but the shapes do not match and the water velocity would be 110 cm/sec. This is a little slow considering that the still water level was 60 cm above the bottom at the foreshore station.

Only occasionally can a single surge be followed across a portion of the surf zone. The same bore was rarely detectable with certainty across all three stations. The inability of tracing suspension plumes across all three stations may be due to (1) the breaking of the waves between the stations and thereby losing their identity, (2) sediment not being readily moved throughout the breadth of the surf zone, or (3) the sediment can be in such a state of upheaval that the juxtaposition is obscured. Figure 11 is representative of this lack of correlation. In the breaker zone (#1), 135 gm/l of sand was lifted at least 50 cm above the bottom. By the time this bore reached station 2 (33 meters away) in the middle of the surf zone, no trace of the suspended sediment motion was ascertained. In the lower swash zone (#3), the reformed wave lifted some sediment 8 cm above the bottom 32 seconds later.

Figure 12 shows the most common occurrence of suspended sediment in the surf zone. Although #1 is in the breaker zone and some movement of sediment does take place, all of it is very close to the bottom. There is little, if any, motion in the middle of the surf zone. It is only at the transition zone (#3) that suspension is frequent.

The identity of the incoming surge was fre-

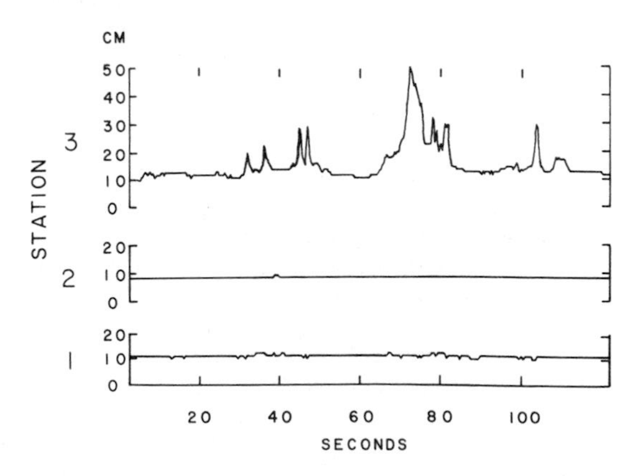

FIG. 12.—Two minute record of elevation changes across the breaker zone (#1), the middle surf zone (#2) and the transition zone (#3) of the 135 gm/l concentration. October 1, 07:02, flood tide, 100 cm above MLLW. Breaker height 90 cm, 9 second period, 80% plunging, approaching land at a 10° angle from the east.

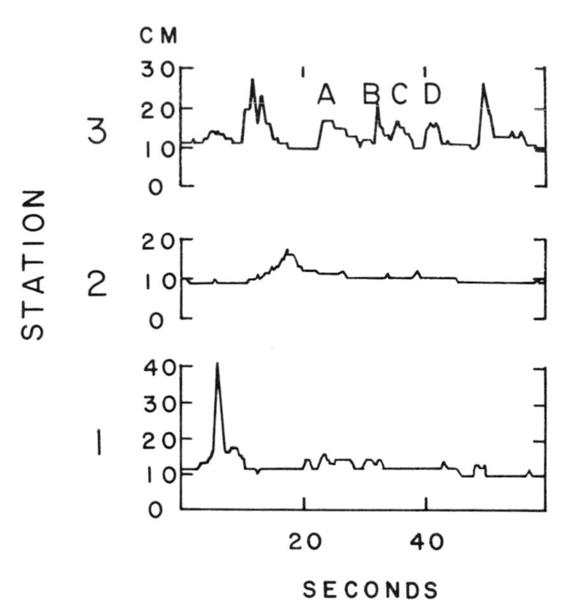

FIG. 13.—One minute record of the elevation changes across the foreshore (#1), lower beach face (#2) and upper beach face (#3) of the 135 gm/l concentration under an incoming bore. September 30, 05.24, flood tide, breaker height 75 cm, period 13 seconds, 25% plunging, approaching land at 20° angle from the east.

quently lost in transport across the inshore and foreshore. In Figure 13 an incoming bore spurted sand into a cloud 30 cm high and lasting 8 seconds. Eight seconds later when the surge reached the second station the height of the 135 gm/l load had decreased to 9 cm but increased to 9 seconds in duration. After the surge had passed this point, a 2 cm layer was deposited. By the time the bore reached the upper beach its density was masked. If its influence is first recognized at A, then the water velocity would be 290 cm/sec. This seems fast because the velocity on the lower foreshore was 305 cm/sec. If on the other hand, its influence is detected at B then the average velocity of the surge between stations two and three is half that between one and two. If the wave hits station three at C, its velocity is 110 cm/sec; if at D, 85 cm/sec. Each of these is possible, but 110 cm/sec is favored because the cloud commencing at C has a slow demise which, as seen, tends to indicate a backwash phenomenon. The same is not really true for the one beginning at B. Note that the agitation of the sand surface was continuous and pronounced at the upper beachface, whereas, upheavals were sporadic and subdued on the lower beachface and shoreface.

The inequality of sediment movement across a beach is brought out even more clearly when different concentrations are considered. In the two-minute record depicted in Figure 14, only one major disturbance in the seaward station was recorded, while at least nine are shown in the shoreward station. This commonly occurred when the upper beachface was in very shallow water. The still water level was 10 cm below the sand level at station 3. The average swash reached a height of 25 cm above it. During the same time the 135 gm/l sand concentration attained a height of 19 cm above the bottom. These records illustrate that the entire water column in the transition and lower swash zones can be saturated with sediment.

In the inshore area (#1) under the breaker zone the 375 gm/l concentration, although changing constantly, normally varied within a prism 6 cm thick. Note that in the sand fountain, the upheaval of the largest concentration reached a height almost equal to that of the smaller concentration.

At the same time the lower beachface was more passive. The uplift of sand was barely differentiable 16.3 seconds after it had occurred at the seaward almometer (a velocity of 195 cm/sec). Aside from that instant, the 375 gm/l concentration fluctuated only 2 cm. Four to six cm above it the lower concentration (40 gm/l) showed variations of approximately the same magnitude.

At the upper beachface (#3) nine uplifts of sand occurred, however, the heavy concentration (375 gm/l) was generally not affected and was lifted 19 cm above the bottom only once. Erosion occurred prior to and following the uplifts. Just before the arrival at A there was erosion of the substrate to a depth of 6 cm without any concomitant increase in the height of the lighter concentration. It was apparent that this erosion was due to a secondary breaker, for after its passage, a cloud of suspended sediment billowed upward. The time of its uplift and demise was symmetrical. Seven and one-half seconds later another suspension cloud (B) moved past the station. Now the cloud tapered off more gradually. Apparently this sequence of A and B represents an incoming bore and its backwash. The same is true of the couplet beginning at 37 seconds.

At high tide (Fig. 15) the record looked different. Outside the breaker zone (#1) disturbances were 13 seconds apart and represented each passing wave. There were several times when up to 3 other uplifts occurred between the wave period. These may be a response to shorter period waves. The uplifts also may have been a response to bottom roughness, because this zone was in the outer rough facies (Clifton and others, 1971) composed of lunate megaripples. Homma and others (1965) have shown, that usually 3 suspended peaks per wave exist close to the bottom above a rippled surface. Note that the two concentrations

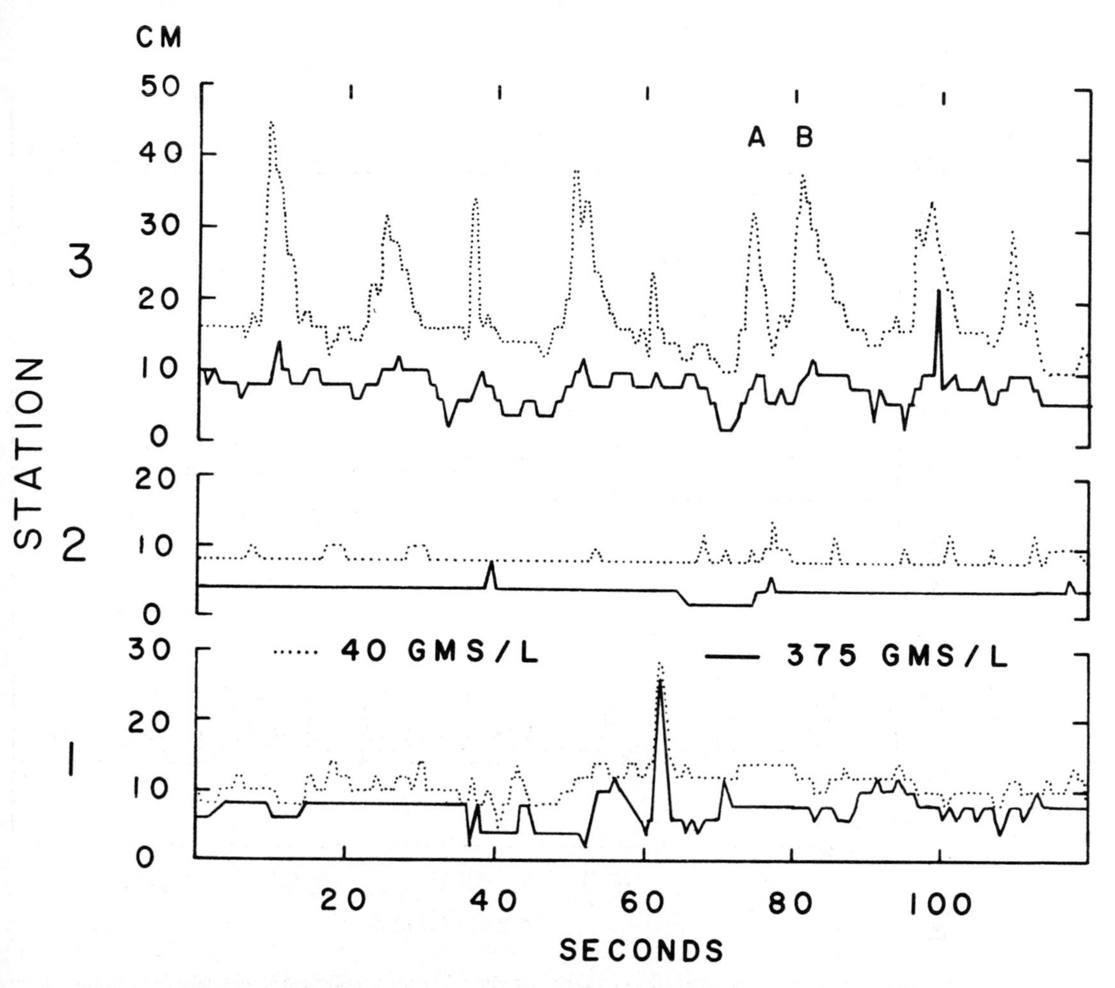

FIG. 14.—Two minute record of the elevation changes of the 375 gm/l (solid line) and 40 gm/l (dotted line) at mid tide. (#1) is in the breaker zone, (#2) is in the middle of the surf zone, (#3) is in the transition zone. September 29, 05:56, flood tide, 110 cm above MLLW. Breaker height 90 cm, period 12 seconds, 35% plunging, approaching land at a 10° angle from the east.

were seldom more than 4 cm apart and had similar patterns of movement.

On the lower foreshore (#2), this time located in the outer surf zone, 6 cm generally separated the 40 gm/l and the 375 gm/l concentrations. In this zone there was no strong suggestion of sand transport associated with the wave period.

In the inner part of the surf zone (#3), uplifts were commencing. The difference in elevation between the measured concentrations was never more than 4 cm. The 375 gm/l surface between the time of uplifts stayed almost constant whereas in the transition zone (#3, Fig. 14) that concentration showed rapid elevation changes.

At low tide (Fig. 16) the center of suspension shifted to the lower beachface. Station two experienced many explosions. At station three, located at the very top of the swash zone, the small drops in elevation of the sand surface illustrated that the incoming swash changed the packing density occasionally and that the sand surface rebounded to its previous position almost immediately.

On the upper foreshore (#1), which was under the middle surf regime, the two concentrations are almost coincident. Even in the transition zone (#2) the levels of the different concentrations were approximately the same between the times of the sand fountains.

Quantity of sand moved in suspension.—The amount of sand moved by these sand fountains is considerable. To determine the quantity of sand involved, the area occupied by each of the measured concentrations was determined by planimetry for 200 randomly selected explosions. The

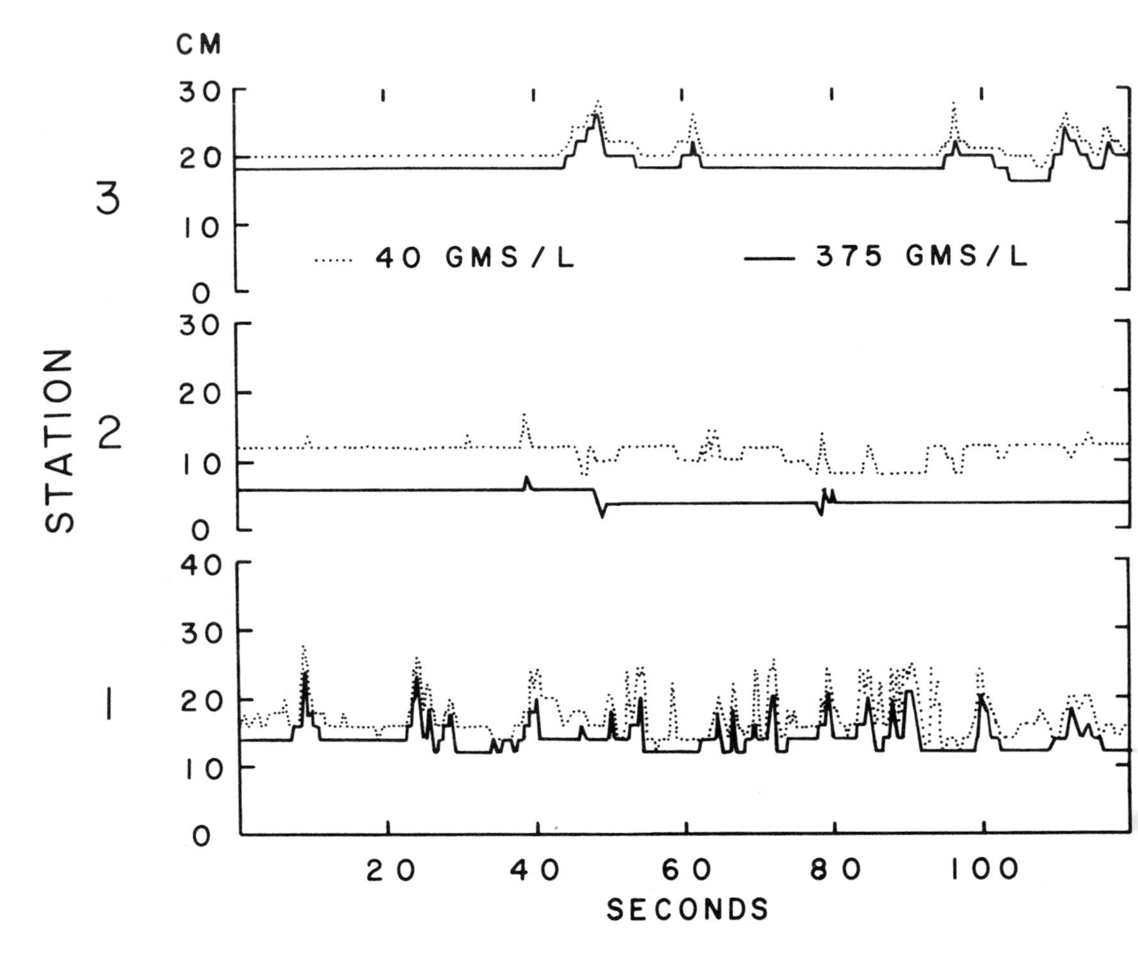

FIG. 15.—Two minute record of the elevation changes of the 375 gm/l (solid line) and 40 gm/l (dotted line) at high tide. (#1) is outside the breaker zone, (#2) is in the outer surf zone, (#3) is in the inner surf zone. September 30, 19:54, high tide, 160 cm above MLLW. Breaker height 60 cm, period 13 seconds, 60% plunging, approaching parallel to shore.

area was multiplied by the concentration and the assumed velocity of the water. The velocity of the water was taken as 180 cm/sec, an average of that reported for the inner surf zone and swash zone under similar wave conditions and beach slope by Demarest (1947), Miller and Zeigler (1958), Dolan and Ferm (1966) and Dolan, Ferm and McArthur (1969). The product of the area, concentration and velocity produced a mean of 189,783 gm per sand fountain or approximately 0.08 m^3. A 120 cm/sec average velocity, as reported by Schiffman (1965) and Brenninkmeyer (1968) yielded 127,154 gm or 0.05 m^3 per lofted cloud. This latter velocity is approximately the same as the water velocity based on solitary wave theory (Munk, 1949). Measurements by Inman and others (1971) showed that at the bottom, the velocity of a water particle under a bore was nearly equal to that predicted by solitary theory, whereas, at the surface it may be up to 8 times faster.

If the number of explosions occurring per tide were cumulatively added, then between 4.9 and 8.1 m^3 are moved in suspension 15 cm above the bottom on the lower beach face per strip 30 cm wide. This is 3-4 times the amount reported by the U.S. Army Corps of Engineers, Beach Erosion Board (1933) for Pensacola, Florida, but about the same as for Long Branch, New Jersey.

After the data for each almometer were plotted between the constraints of the limit of the upper swash and the unknown amount on the seaward side of the breaker zone, then between 150 and 300 m^3 of sand were placed in suspension per 30 cm wide zone per tide. The amount on the outer foreshore dropped to almost nothing

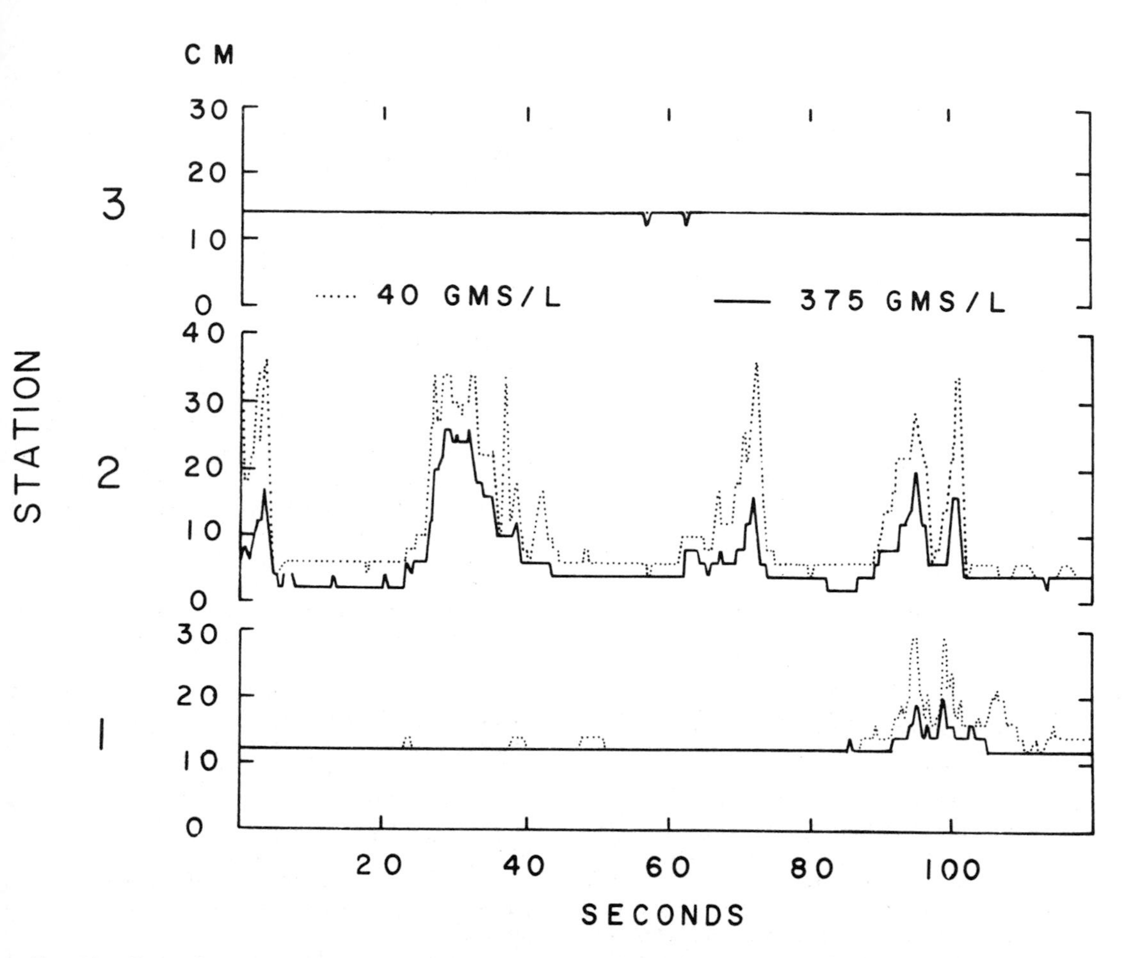

FIG. 16.—Two minute record of the elevation changes of the 375 gm/l (solid line) and 40 gm/l (dotted line) at low tide. (#1) is in the middle surf zone, (#2) is in the transition zone, (#3) is in the upper swash zone. October 2, 00:02, low tide, 30 cm above MLLW. Breaker height 90 cm, period 13 seconds, 50% plunging, approaching land at a 10° angle from the west.

seaward of the lower tide plunge point (U.S. Army Corps of Engineers, Beach Erosion Board, 1933). The curve between these points is, of course, somewhat arbitrary but should have the general shape shown in Figure 17.

By far the largest amount of sand is transported in suspension shoreward of the high water plunge point. The shape of the curve is determined by the position of the transition zone. Since this zone at Point Mugu is commonly between 60 and 120 cm above MLLW, this gives the cumulative curve its leptokurtic shape.

In the vicinity of Point Mugu approximately 45×10^3 m^3 were moved yearly by longshore transport between the berm and MLLW (U.S. Army Corps of Engineers, Los Angeles District, Hydrographic Surveys, D. O. Sheets: 3348-3350C, 2434-2437D, D2466-2471, E2411-2414, D2466-2471-70). The quantity of sand thrown into suspension per year, 15 cm above the bottom is $105\text{–}210 \times 10^3$ m^3 or approximately 2–5 times that moved by longshore currents. Thus more than two-thirds of the sand in suspension at Point Mugu is not involved in longshore transport but moves normal to the shoreline. This is somewhat lower than Inman's and Bagnold's (1963) tenfold figure, but the amount obtained should increase tremendously if the bottommost 15 cm are included.

On the other hand, if there were strong longshore currents and their progress was not checked by the nearby impervious experimental groin, then their velocity profile (Komar, 1971; Jonsson and others, 1975) would be strikingly similar to that of Figure 17. Observations by Inman and Quinn (1952) and theoretical relations (Bowen, 1969; Longuet-Higgins, 1970, 1972) showed that the longshore velocities had approximately the same asymmetrical shape as the suspended sediment

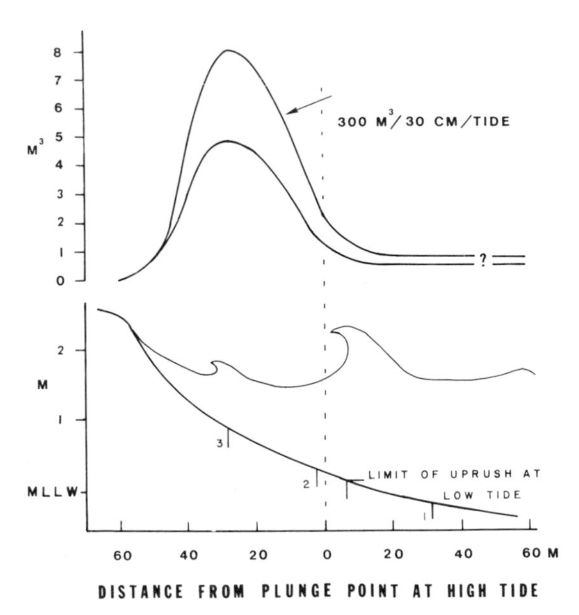

FIG. 17.—Quantity of sand transported in suspension per strip one foot wide per tide. The upper curve assumes an average inner surf and swash velocity of 180 cm/sec. The lower curve is for 120 cm/sec.

distribution. One is drawn to conclude that a large fraction of the longshore transport is by suspension. This conclusion is also supported by some field and laboratory investigations (Galvin, 1973). Sands caught in bedload traps are only a tenth to a hundredth of the value expected from longshore transport rate-energy flux correlations.

CONCLUSIONS

Modes and frequency of sand transport differ within each of the dynamic zones of the coastal area. Outside the breaker zone, sand moves predominantly by bedload. Movement of sand more than 15 cm above the bottom is rare. The average change in sand surface elevation per wave ranges from 0–6 cm.

Inside the breaker zone sand moves more rapidly. The sand surface changes swiftly with differences in elevation of more than 6 cm per wave not uncommon. Sand is rarely thrown into suspension. The suspension clouds that do occur have a rapid symmetrical rise and fall and are of very short duration.

Sedimentation is small in the outer surf zone after the transformation of the incoming wave to a bore. The sand surface changes no more than 4 cm with the passage of a bore. Suspension of sand more than 15 cm above the bottom is almost nonexistent. The distance between the bottom and the lighter sediment concentration seems to be dependent on the ground water level in the beach. If the outer surf zone is below the water table, the various concentrations are almost coincident. If the outer surf zone is above the water table, the 40 gm/l concentration is approximately 6 cm above the bottom.

In the inner surf zone suspended sand transport increases in frequency, elevation and duration because in this area the incoming bore breaks and the zone is influenced by the backwash. The bottom elevations are affected by these suspension clouds.

In the transition zone, at the still water level, sand movement by suspension becomes predominant. Upheavals in the sediment are frequent, show a variety of different shapes and are of various durations. The frequency of occurrence of these sand fountains shows a considerable spread. Two periods stand out at Point Mugu. The first is a little longer than the prevailing swell period. The second, somewhat longer, may be due to interference or reflection or, more likely, to the fact that only the longest waves can reach this zone and also to the longer period of the backwash.

In the swash zone, sand movement reverts back to bedload. Transport of any quantity of sand 15 cm above the bottom is exceedingly rare.

Sediments are suspended by:

1. The horizontal and vertical water velocities and fluctuations in velocity are sufficient to suspend even the coarsest sand.
2. Entrapped air in a breaking wave has a suctional effect after the bubbles penetrate the bottom.
3. A hydraulic jump or roll wave is created if the backwash velocity is sufficiently high or if ripples are present.

Twice as many suspensions of sediment occur on the rising tide as on the ebb tide due to the lag of the ground water table behind the tidal elevation. The tidal level is not significant in influencing the height the sand fountains attain. At Point Mugu they reach an average height of 24 cm and last 10 seconds. The amount of sand per tide thrown into suspension 15 cm above the bottom for a strip 30 cm wide is between 150–300 cubic meters. This amounts to 105–210 × 10^3 m^3 per year which is 2–5 times that moved by longshore currents.

ACKNOWLEDGEMENTS

The author is indebted to the Department of Geological Sciences at the University of Southern California whose faculty, staff and students aided in the construction of the almometer and in the field tests. The U.S. Navy through its Pacific Missile Range at Point Mugu provided the recording facilities and computor reduction of the data. The work of Messrs. Frank Agapoff, Del Pierce and Clifford LeMieux is appreciated.

The critical remarks of Drs. Donn Gorsline, Robert Miller, and George Oertel markedly improved the manuscript. This project was supported by grants from the Office of Naval Research, NONR (6)-00013-72 and N00014-67-A-0269-0009C and the National Science Foundation GA22842 and GB6319.

REFERENCES

Adeyemo, M. D., 1971, Velocity fields in the wave breaker zone: Am. Soc. Civil Engineers, Proc. 12th Conf. on Coastal Eng., v. 1, p. 435-460.

Aibulatov, N. A., 1958, Novye issledovaniya vdol'beregovogo peremescheniya peschanykh nanosov v more [New investigations of longshore migration of sandy sediment in the sea]: Okeanograf. Komiss. Akad. Nauk Byeloruss. SSR, v. 1, p. 72-80.

——, 1961, Nalyudyeniya za vdolbegobyn peremescheniem peschanykh nanosov u otmelogo akkulyativnogo merega [Observations of the longshore sand drift along a shallow coast of accumulation]: Akad. Nauk SSSR, Inst. Okeanol., Trud., v. 53, p. 3-18.

——, 1966, Issledovanie vdol'beregovogo permeslichenia peschanykh nanosov v more [Investigations of longshore migration of sand in the sea]: Nauka Pub., Moscow, 159 p.

——, Bolyrev, V. L., and Griesener, H., 1961, Das Studium der Sediment Bewegung in Flussen und Meeren mit Hilfe von Luminerszierenden Farbstoffe und radioactiven Isotopen: Petermanns Geog. Mitt., Jg. 105, p. 117-186, 254-263.

Allen, R., 1968, Current ripples: North-Holland Pub. Co., Amsterdam, 433 p.

Bagnold, R. A., 1963, Beach and nearshore processes: Part I, Mechanics of marine sedimentation: *IN* M. N. Hill (ed.), The Sea: Interscience Pub., New York, v. 3, p. 507-528.

——, 1966, An approach to the sediment transport problem from general physics: U.S. Geol. Survey Prof. Paper 422-J, 37 p.

Basinski, T., and Lewandowski, A., 1975, Field investigation of suspended sediment: Am. Soc. Civil Engineers, Proc. 14th Conf. on Coastal Eng., p. 1096-1108.

Bhattacharya, P. K., Glover, J. R., and Kennedy, J. F., 1969, An electrooptical probe for measuring of suspended sediment concentration: Internat. Assoc. for Hydraulic Res., Proc. 13th Congr., v. 2, p. 241-250.

Bijker, E. W., 1971, Longshore transport computations: Am. Soc. Civil Engineers Proc., Jour. Waterways, Harbors and Coastal Div., v. 97, WW4, p. 687-702.

Bowen, A. J., 1969, Rip currents: Jour. Geophys. Res., v. 74, p. 5467-5478.

Brenninkmeyer, B. M., 1968, Sedimentation patterns in the lower swash zone: Univ. Southern California, Unpub. rep. in sedimentation, 18 p.

——, 1973, Synoptic surf zone sedimentation patterns: Ph.D. Dissertation, Univ. Southern California, 274 p.

——, 1975, Mode and period of sand transport in the surf zone: Am. Soc. Civil Engineers, Proc. 14th Conf. on Coastal Eng., 812-827.

——, 1976, In site measurements of rapidly fluctuating high sediment concentrations: Marine Geol., v. 20, p. 117-128.

Clifton, H. E., Hunter, R. E., and Phillips, R. L., 1971, Depositional structures and processes in the non-barred high-energy nearshore: Jour. Sed. Petrology, v. 41, p. 651-670.

Cook, D. O., 1970, Sand transport by shoaling waves: Ph.D. Dissertation, Univ. Southern California, 148 p.

Curtis, C. M., 1963, Suspended sediments in a breaking wave: Univ. Southern California, Unpub. rep. in sedimentation, 19 p.

Das, M. M., 1973, Suspended sediment and longshore sediment transport data review: Am. Soc. Civil Engineers, Proc. 13th Conf. on Coastal Eng., p. 1027-1048.

Demarest, D. F., 1947, Rhomboid ripple marks and their relationship to beach slope: Jour. Sed. Petrology, v. 17, p. 18-22.

Dolan, R., and Ferm, J. C., 1966, Swash processes and beach characteristics: Prof. Geographer, v. 18, p. 210-213.

——, —— and McArthur, D., 1969, Measurements of beach process variables, Outer Banks, North Carolina: Louisiana State Univ., Coastal Stud. Inst. Tech. Rep. 64, 79 p.

Dominick, T. F., Wilkins, B., and Roberts, H., 1971, Mathematical model for beach groundwater fluctuations: Water Resources Res., v. 7, p. 1626-1635.

Einstein, H. A., 1950, The bed-load function for sediment transportation in open channel flow: U.S. Dep. Agriculture, Tech. Bull. 1026, 70 p.

—— and Abdel-Aal, F. M., 1972, Einstein's bedload function at high sediment rates: Am. Soc. Civil Engineers Proc., Jour. Hydraulics Div., HY1, v. 98, p. 137-151.

—— and Chien, N., 1954, Second approximation to the solution of the suspended-load theory: Univ. California, Inst. Eng. Res. Rep. 3, 45 p.

Emery, K. O., and Foster, J. F., 1948, Water tables in marine beaches: Jour. Marine Res., v. 3, p. 644-654.

Fairchild, J. C., 1959, Suspended sediment sampling in laboratory wave action: U.S. Army Corps of Engineers, Beach Erosion Board Tech. Memo. 115, 25 p.

——, 1973, Longshore transport of suspended sediment: Am. Soc. Civil Engineers, Proc. 13th Conf. on Coastal Eng., p. 1069-1088.

Fausak, L. E., 1970, The beach water table as a response variable of the beach-ocean-atmosphere system: M. S. Thesis, Virginia Inst. Marine Sci., 52 p.
Federal Interagency Sedimentation Project, 1970, Project accomplishments during the calendar year 1969: Mimeographed report.
Fuhrboter, A., 1971, Air entrainment and energy dissipation in breakers: Am. Soc. Civil Engineers, Proc. 12th Conf. on Coastal Eng., p. 391-398.
Fukushima, H., and Mizoguchi, Y., 1958, Field investigation of suspended littoral drift: Coastal Eng. in Japan, v. 1, p. 131-134.
Galvin, C. J., Jr., 1973, A gross longshore transport rate formula: Am. Soc. Civil Engineers, Proc. 13th Conf. on Coastal Eng., p. 953-970.
Gibbs, R. J., Matthews, M. D., and Link, D. A., 1971, The relationship between size and settling velocity: Jour. Sed. Petrology, v. 41, p. 7-18.
Grant, U. S., 1943, Waves as a sand transporting agent: Am. Jour. Sci., v. 241, p. 117-123.
Hattori, M., 1969, The mechanics of suspended sediment due to standing waves: Coastal Eng. in Japan, v. 12, p. 69-81.
——, 1975, Delay distance in suspended sediment transport: Am. Soc. Civil Engineers, Proc. 14th Conf. on Coastal Eng., p. 1086-1095.
Homma, M., and Horikawa, K., 1963, A laboratory study of suspended action due to wave action: Internat. Assoc. for Hydraulic Res., Proc. 10th Congr., v. 1, p. 213-220.
——, —— and Kajima, R., 1965, A study of suspended sediment due to wave action: Coastal Eng. in Japan, v. 3, p. 85-103.
——, —— and Sonu, C., 1960, A study of beach erosion at the sheltered beaches of Katasi and Kamakura, Japan: Coastal Eng. in Japan, v. 3, p. 101-122.
Horikawa, K., and Kuo, C., 1967, A study of wave transformation in side surf zone: Am. Soc. Civil Engineers, Proc. 10th Conf. on Coastal Eng., p. 217-233.
—— and Watanabe, A., 1970, Turbulence and sediment concentration due to waves: Am. Soc. Civil Engineers, Proc. 12th Conf. on Coastal Eng., p. 751-766.
Huntley, D. A., and Bowen, A. T., 1975, Field measurements of nearshore velocities: Am. Soc. Civil Engineers, Proc. 14th Conf. on Coastal Eng., p. 538-557.
Ingle, J. C., 1966, The movement of beach sand: Elsevier, Amsterdam, 221 p.
Inman, D. L., 1949a, Sediment trap studies of suspended material near the surf zone: Scripps Inst. Oceanography, Quart. Prog. Rep. to U.S. Army Corps of Engineers Beach Erosion Board, No. 2, p. 5-6.
——, 1949b, Sorting of sediment in the light of fluid mechanics: Jour. Sed. Petrology, v. 19, p. 51-70.
—— and Bagnold, R. A., 1963, Beach and nearshore processes, Part II: Littoral processes: *In* M. N. Hill (ed.), The Sea: Interscience Pub., New York, v. 3, p. 529-553.
—— and Nasu, M., 1956, Orbital velocities associated with wave action near the breaker zone: U.S. Army Corps of Engineers, Beach Erosion Board Tech. Memo. 79, 43 p.
—— and Quinn, W. H., 1951, Currents in the surf zone: Univ. California Council on Wave Res., Proc. 2nd Conf. Coastal Eng., p. 24-36.
——, Tait, R. J., and Nordstrom, C. E., 1971, Mixing in the surf zone: Jour. Geophys. Res., v. 76, p. 3493-3514.
Ippen, A. T., 1966, Estuary and coastline hydrodynamics: McGraw-Hill Book Co., New York, 744 p.
—— and Kulin, G., 1954, The shoaling and breaking of the solitary wave: Univ. California Council on Wave Res., Proc. 5th Conf. Coastal Eng., p. 27-47.
Iverson, H. W., 1952, Laboratory studies of breakers: Natl. Bur. Standards Circ. 521, p. 9-32.
Johnson, J. W., 1952, Final report on sand transportation by wave action: Univ. California, Dep. Engineering, Wave Res. Lab. Tech. Rep., Ser. 14, No. 9, 6 p.
Jonsson, I. G., Skovgaard, O., and Jacobreu, T. S., 1975, Computation of longshore currents: Am. Soc. Civil Engineers, Proc. 14th Conf. on Coastal Eng., p. 699-714.
Kennedy, J. F., and Locher, F. A., 1972, Sediment suspension by water waves: *In* R. E. Meyer (ed.), Waves on beaches: Academic Press, New York, p. 249-295.
Komar, P. D., 1971, The mechanics of sand transport on beaches: Jour. Geophys. Res., v. 76, p. 713-721.
Little, R. D., 1968, Suspended sediment in the surf zone: Univ. Southern California, Unpub. rep. in sedimentation, 28 p.
Longinov, V. V., 1958, Nekotorye dannye nablyudenii nad gorizontal' nymi volnovymi davleniyami v pridonnom sloye beregovoi zony v prirodnykh usloviyakh [Some data of field observations on horizontal wave pressures in the bottom layer of the shore zone]: Akad. Nauk SSSR, Inst. Okeanol., Trud., v. 28, p. 100-156.
——, 1961, Kopryedyelyeniyu maksimalnkh volnovkh skorostyey v beregovoi zone morya [The determination of maximum wave velocities in the shore zone of the sea]: Akad. Nauk SSSR, Inst. Okeanol., Trud., v. 48, p. 287-327.
——, 1964, Some aspects of wave action on a gently sloping sandy beach: Internat. Geol. Rev., v. 6, p. 212-227.
——, 1968, Opredelynye kontsentratii peschanukh vzvesei v beregovoi zony pri volnenii fotoelektricheskim metodom [Determination by a photoelectric method of sand suspension concentration in the littoral zone during waves]: Gosudarstvennyi proektno-knostruktorskii i Nauchno-issledovatel' skii Inst. Morskogo Transporta [Leningrad], v. 20/26, p. 82-92.
Longuet-Higgins, M. S., 1970, On the longshore currents generated by obliquely incident sea waves: Jour. Geophys. Res., v. 75, p. 6778-6801.

——, 1972, Recent progress in the study of longshore currents: *In* R. E. Meyer (ed.), Waves on beaches: Academic Press, New York, p. 203-248.

MARLETTE, J. W., 1954, The breakwater at Redondo Beach, California, and its effects on erosion and sedimentation: M. S. Thesis, Univ. Southern California, 82 p.

MEDVEDEV, V. S., AND AIBULATOV, N. A., 1958, Izuchenye dinamiki otmelogo peschanogo berega pomoshch'yu lyuminoforov i podvesnoi dorogi [Dynamic studies of a shallow sandy shore by means of lumenophores and aerial tramway]: Akad. Nauk SSSR, Inst. Okeanol., Trud., v. 28, p. 37-55.

MILLER, R. L., 1972, Study of air entrainment in breaker waves: Am. Geophys. Union Trans., v. 53' p. 426.

——, 1973, The role of surface tension in breaking waves: Am. Soc. Civil Engineers, Proc. 13th Conf. on Coastal Eng., p. 433-449.

——, LEVERETTE, S., SULLIVAN, J. O., TOCHKO, J., AND THERIAULT, K., 1974, The effect of breaker shape on impact pressures in the surf: Univ. Chicago, Fluid Dynamics and Sediment Trans. Lab. Tech. Rep. 14, 45 p.

—— AND ZIEGLER, J. M., 1958, A model relating dynamics and sediment pattern in equilibrium in the region of shoaling waves, breaker zone and foreshore: Jour. Geology, v. 66, p. 417-441.

—— AND ——, 1964, A study of sediment distribution in the zone of shoaling waves over complicated bottom topography: *In* R. L. Miller (ed.), Papers in marine geology: Macmillan Co., New York, p. 133-153.

MORRISON, J. R., AND CROOKE, R. C., 1953, The mechanics of deep water, shallow water, and breaking waves: U.S. Army Corps of Engineers, Beach Erosion Board Tech. Memo. 40, 14 p.

MUNK, W. H., 1949, The solitary wave and its application to surf problems: New York Acad. Sci, Annals, v. 51, p. 376-401.

NAGATA, Y., 1961, Balance of the transport of sediment due to wave action in shoaling water, surf zone on foreshore: Recs. Oceanog. Works Japan, v. 6, p. 53-62.

——, 1964, Deformation of temporal pattern of orbital wave velocity and sediment transport in shoaling water, in breaker zone and on foreshore: Jour. Oceanog. Soc. Japan, v. 20, p. 57-70.

PALMER, H. D., 1970, Wave-induced scour around natural and artificial objects: Ph.D. Dissertation, Univ. Southern California, 172 p.

ROUSE, H., 1937, Modern conceptions of the mechanics of turbulence: Am. Soc. Civil Engineers Trans., v. 102, p. 463-543.

——, 1964, Sediment transport mechanics: Suspension of sediment: A discussion: Am. Soc. Civil Engineers Proc., Jour. Hydraulics Div., HYl, v. 90, p. 361-363.

RUDOLFO, K., 1964, Suspended sediments in California waters: M. S. Thesis, Univ. Southern California, 91 p.

SCHIDELER, G. L., AND MCGRATH, D. G., 1974, Submarine strain-gage instrumentation for monitoring diurnal beach sediment migration: Jour. Sed. Petrology, v. 44, p. 1305-1309.

SCHIFFMAN, A., 1965, Energy measurements of the swash-surf system: Limnology and Oceanography, v. 10, p. 255-260.

SHINOHARA, K., TSUBAKI, T., YOSBITAKA, M., AND AGEMORI, C., 1958, Sand transport along a model sandy beach by wave action: Coastal Eng. in Japan, v. 1, p. 111-130.

SIMONS, D. B., RICHARDSON, E. V., AND NORTON, C. F., 1965, Sedimentary structures generated by flow in alluvial channels: *In* G. V. Middleton (ed.), Primary sedimentary structures and their hydrodynamic interpretation: Soc. Econ. Paleontol. Mineral., Spec. Pub. 12, p. 34-52.

TERRY, R. D., 1951, Suspended sediment study of surf at Huntington Beach, California: Univ. Southern California, Unpub. rep. in sedimentation, 21 p.

U.S. ARMY CORPS OF ENGINEERS, BEACH EROSION BOARD, 1933, Interim Report: April 15, 1933.

U.S. INTERAGENCY COMMITTEE ON WATER RESOURCES, 1963, Measurements and analysis of sediment loads in streams: U.S. Interagency Comm. on Water Res., Rep. 14, 151 p.

UNIVERSITY OF IOWA, 1970, Laboratory investigations of suspended sediment samplers: Univ. Iowa, Hydraulics Lab. Rep. 5.

VANONI, V. A., AND OTHERS, 1963, Sediment transport mechanics: Suspension of sediment: Report of Task Committee on Sedimentation Manual: Am. Soc. Civil Engineers Proc., Jour. Hydraulics Div., HY5, v. 89, p. 45-76.

WATTS, G. M., 1953, Development and field tests of a sampler for suspended sediment in wave action: U.S. Army Corps of Engineers, Beach Erosion Board Tech. Memo. 34, 41 p.

WENZEL, D., 1975, Measuring sand discharge near the sea-bottom: Am. Soc. Civil Engineers, Proc. 14th Conf. on Coastal Eng., p. 741-755.

ZENKOVICH, V. P., 1967, Processes of coastal development: Interscience Pub., New York, 738 p.

ROLE OF VORTICES IN SURF ZONE PREDICTION: SEDIMENTATION AND WAVE FORCES

ROBERT L. MILLER
University of Chicago, Chicago, Illinois 60637

ABSTRACT

There are difficulties in the creation of numerical prediction models in the surf zone. Specific examination of fundamental breaker mechanics is necessary, if successful prediction is to be achieved. Investigation of the internal velocity field indicates the importance of "breaker vortices" whose size and strength are a function of breaker shape. Numerous experiments have provided data for a detailed analysis of the vortices. As a result, a first approximation model is discussed, although further work is necessary. The effect of breaker shape and breaker vortices is indicated in three applications; (1) bedforms in the surf zone, (2) impact pressures due to breaking waves, (3) the interaction of post-breaking bore and foreshore. In each case the reliability of prediction is considered.

INTRODUCTION

One of the most complicated dynamic regions of the coast is the surf and run-up zone. This is conveniently defined as the region where incoming waves break and eventually run up the foreshore of the beach, if one is present. The width of the surf zone may vary from tens of meters to as much as a thousand meters, depending on sea state and the morphology of the bottom. A number of geomorphological features are formed, at least in part, in this region. Natural features such as sea cliffs and beaches, and also man made structures undergo extreme impact or erosional forces during storms. The mechanisms of these processes is not well understood.

It is not surprising that under these circumstances there are no generally accepted numerical prediction models for sediment transport mechanics and associated bedforms in the onshore-offshore direction in surf zone and foreshore.[1] The general problem is, however, under active investigation. Two approaches have been employed in recent publications. Fox and Davis (1973) have developed a format employing the deductive method. Field observations are taken from meteorological conditions and sea state during a storm. The bedforms and bottom configuration are measured before and after the the storm. Particular effects are thus deduced from general observations. A second approach employs the inductive method. Attention is focused on the details of the internal fluid dynamics of the breaker, and on the sediment transport resulting from the force distribution. Understanding gained in this way should eventually form a sound basis for numerical prediction models. The present paper deals with recent results in the context of the second approach.

Within the last decade, significant advances have been made in classification and prediction of various properties of breaking waves. The following is not intended to present a complete discussion, but to emphasize those properties which form the context of the present paper. Reviews of the status of knowledge about various aspects of breaking waves are found in Wiegel (1964), Galvin (1968), and Miller (1972a, 1972b), and of sediment transport under nonbreaking waves by Kennedy and Locher (1972), and Einstein (1972).

A number of studies has been made of the dimensions of breaking waves. For example, prediction of breaker height in terms of deep water wave height has been published by Iverson (1952), Wiegel and Beebe (1956), Komar and Gaughan (1972), and Weggel (1972). The distance of breaker travel has been examined by Galvin (1969), and the effect of surface tension on the breaking process by Miller (1972a).

The internal velocity field in breakers has been studied in the laboratory by Iverson (1952), Ippen and Kulin (1954), and Wiegel and Skjei (1958), and in the field by Inman and Nasu (1956), Miller and Ziegler (1964, 1967), Wood (1973), and Thornton and Krapohl (1974). Classification of breakers by external shape has also been reported by several investigators including Patrick and Wiegal (1955) and Galvin (1968). There appears to be general agreement that the major breaker types range in a continuous "spectrum" of shapes from spilling to plunging and with a third category, the surging breaker, forming in some manner an extension of the plunging end of the array. Steep waves on low angle beach slopes tend to form spilling breakers, whereas less steep waves on

[1] It is interesting to note that in contrast, the longshore component has been thoroughly studied and several widely used models have been published. A recent review is given by Longuet-Higgens (1972).

higher angle beach slopes tend to form plunging breakers according to laboratory observations by Patrick and Wiegel (1955) and by Galvin (1968). Deep water steepness H_O/L_O is the most widely used parameter to predict the breaker shape. Galvin (1968) has also considered the applicability of "breaker steepness". The key to the problem at this time lies in devising a successful quantitative parameter or set of parameters to replace the qualitative terms "spilling" or "plunging" breaker. At present all studies in the literature are at best semiquantitative, in that graphs are presented of various parameters such as H_O/L_O, beach slope α, as opposed to strictly descriptive terms, e.g., "plunging breaker," "spilling breaker" or "surging breaker." Once these descriptive terms are successfully replaced by adequate quantitative parameters $\beta_1, \beta_2 \ldots$, it will be possible to consider the transfer from incoming swell spectrum to breaker type and thus to bedform. In the following discussion attention is focused on the easily distinguished portion of the array of breaker shapes from spilling at one end to plunging at the other. The less common surging breaker form does not actually "break" in the same sense, and the internal mechanics are omitted in the present paper.

BREAKING MECHANISM

At present, the available shallow water wave equations have been extended to include surf and run-up on the beach face. A comprehensive review is given by Meyer and Taylor (1972). The theory has not been extended to include the internal mechanics of fluid motion during the breaking process. This is needed to predict a number of natural phenomena associated with the surf zone. Thus a necessary input to prediction models, at this stage of knowledge, is a detailed semiquantitative description of the breaking process, especially of the internal fluid motions. The description of the breaking process in this section is based on experimental studies in the Fluid Dynamics and Sediment Transport Laboratory at The University of Chicago.

The first instability in the plunging breaker occurs on the free surface at the crest. It takes the form of a local steepening of the forward faces which passes through the vertical plane, and subsequently (when viewed in the plane parallel to wave advance) takes the form of a jet which progressively extends forward and down until it closes upon the forward slope of the wave (Figs. 1, 2). At this point a vortex is formed,

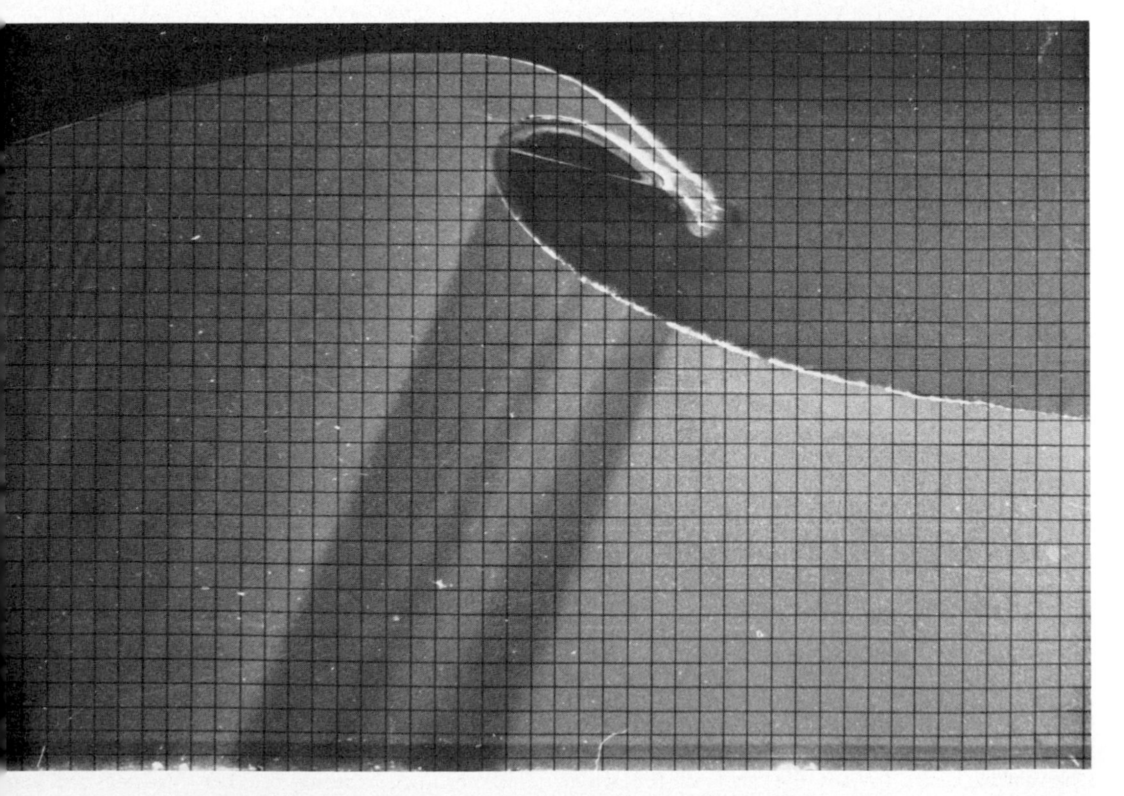

FIG. 1.—Jet stage of breaking process for plunging breaker in collimated light. Grid scale = 1 cm.

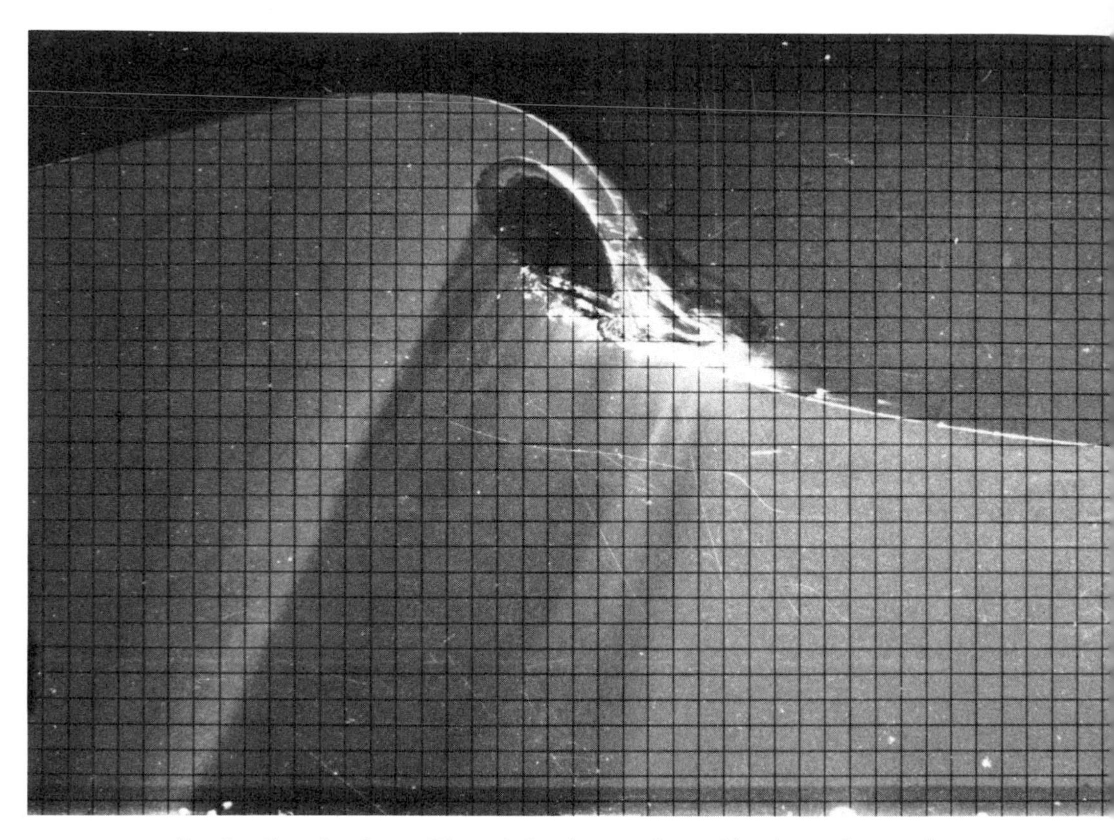

Fig. 2.—Same breaker as Figure 1. Jet closes on front of breaker to form cavity.

with the circulation around the cavity in the sense of the wave advance. When the tip of the "jet" strikes the forward free surface, the resulting splash forms a second vortex in front of the first one (Fig. 3). This process has been observed to repeat, resulting in as many as five successive vortices. After the second, however, each successive vortex is weaker than the preceeding one (Fig. 4). In this manner, a series of large, single-impulse eddies are introduced from the surface, and contribute to the rapid decay of the wave during the breaking phase. As each vortex moves forward and down into the interior of the wave, a mixing process takes place inside the cavity. Eventually the cavity is completely replaced by a zone of concentration of bubbles. During the early stage of the vortex, a large number of bubbles is formed in this manner. These are entrained in the circulation and serve to indicate the details of the local motion. In some cases the rapid compression forces the air upward through the free surface in the form of a spout (Fig. 5). The forward velocity of the vortex is necessarily slower than the wave velocity. As the vortex drifts toward the back of the wave it expands and slows down. Figures 6 and 7 illustrate the trajectories of the vortices and the expansion phase before decay. The complete decay of the vortex is indicated by the bubbles which no longer follow a circulation path but instead rise directly to the free surface at some distance behind the breaker crest at less than or equal to, the "rise velocity" of bubbles. When the bubbles reach the free surface they generate "foam."

This process eventually transforms the incoming wave into a bore mode which then progresses up the foreshore. If however, the Froude number of the broken wave is not sustained above critical due for example to the water depth, then a new wave will form and go through the breaking process once more, but closer to the shore. A discussion of waves reforming after breaking is given in Horikawa and Kuo (1966), and Wood (1973). Variations in magnitude of breaker vortices are correlated with gradation from plunging to spilling breaker shape.

Repeated observations indicate that the breaker vortex system which is generated on a large scale in the plunging breaker, also occurs in the same manner in the spilling breaker. Vortices are generated on a much smaller scale in the spilling

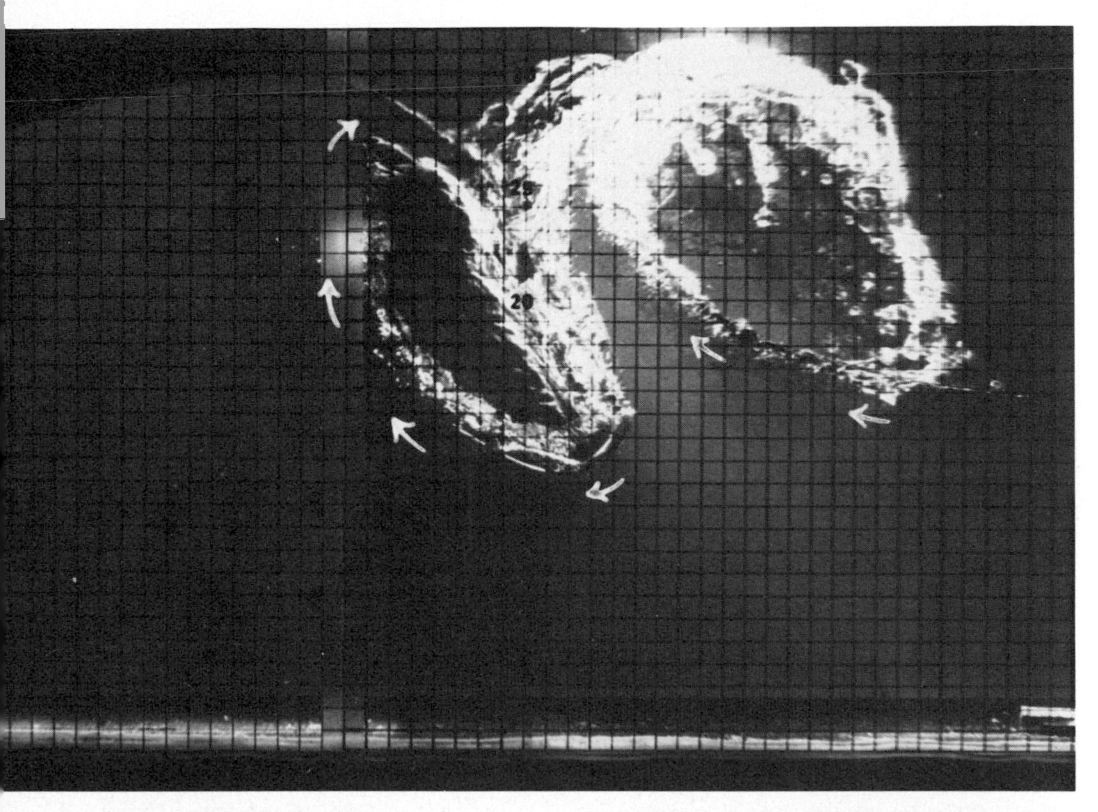

FIG. 3.—Plunging breaker. Formation of second vortex in front of first one. Circulation pattern as shown. Grid scale = 1 cm.

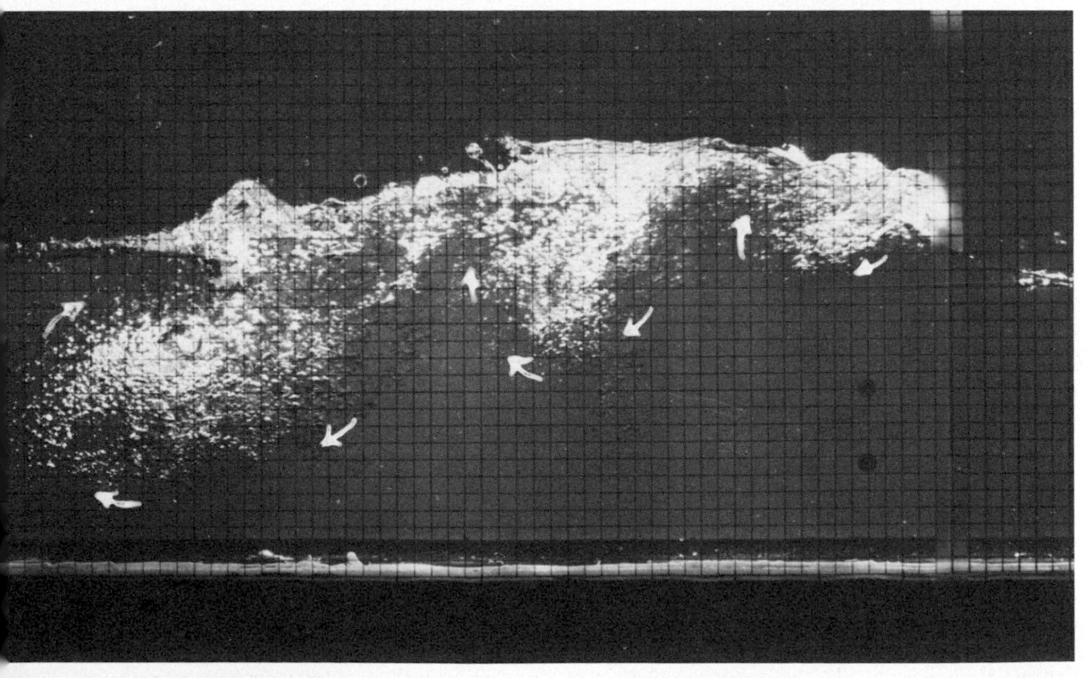

FIG. 4.—Plunging breaker sequence of three successive vortices indicated by bubbles. First vortex approaching bottom.

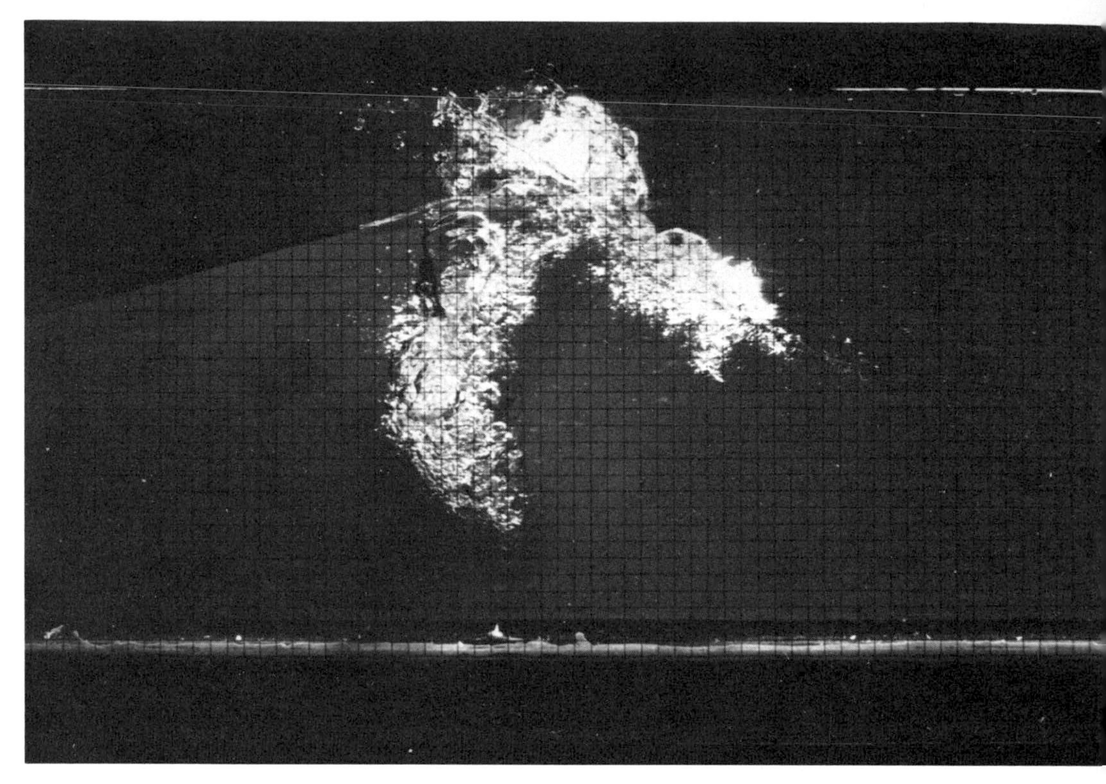

FIG. 5.—Plunging breaker with rapid compression of first vortex cavity forcing air through free surface in form of a spout.

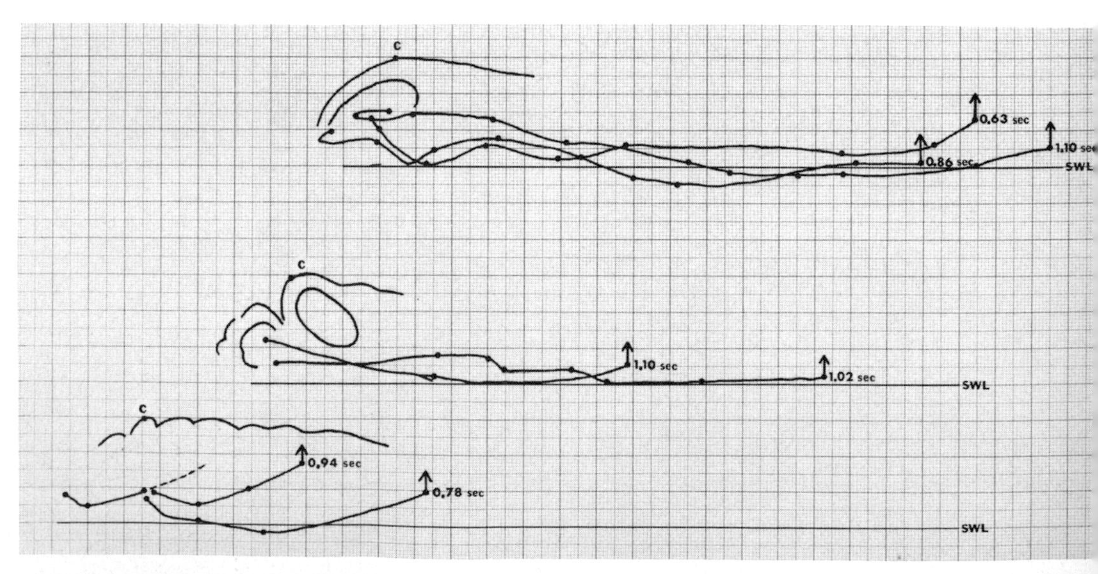

FIG. 6.—Trajectory of vortex centers relative to the crest, for plunging breakers. *Top*, trajectory of first vortex for three different breakers. *Middle*, trajectory of second vortex for two different breakers. *Bottom*, trajectory of third vortex for three different breakers. Ordinate is vertical distance relative to crest at *C*. Abscissa is position of crest in time units relative to zero at crest. Data taken with camera moving with crest at 64 frames/sec. Arrow indicates complete decay of vortex with bubbles rising vertically. Total elapsed time given for each sample.

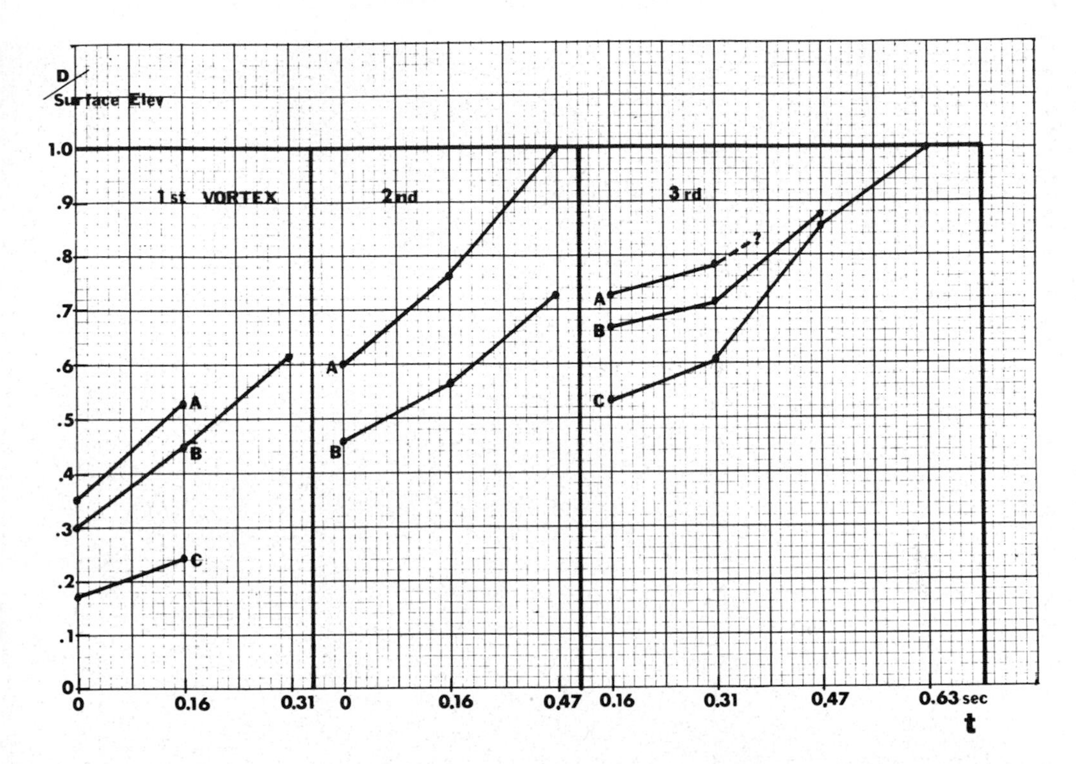

FIG. 7.—Rate of expansion of first three successive vortices in the plunging breaker. Ordinate is ratio of vortex diameter (as indicated by entrained bubbles) to surface elevation above vortex center. At D/surface elevation = 1.0, the vortex extends from surface to bottom. Abscissa is elapsed time. Several examples, A, B, C, are given for each vortex.

breaker and are confined to the near region of the free surface (Fig. 8). The rate at which the breaker height decreases reflects the magnitude of penetration of the vortex system. This rate of decrease in crest height, η, during breaking may be interpreted as representing a record of "energy transfer." Because total energy must be conserved, variation in the pattern of $d\eta/d_x$ yields insight into the way the energy may be transferred to the bottom, absorbed in internal friction (molecular viscosity), and small scale turbulence, and by the breaker vortices. Figures 9a and 9b illustrate the change in η which occurs just after the initial break point for plunging breakers. This region coincides with the generation of a series of very large vortices. When the vortices decay the fluid motion approximates a weak bore. This metamorphosis is illustrated in another way in Figure 10a. In contrast, the decrease in η takes place gradually for the spilling breaker (Fig. 9b). This initiation of small vortices at initial breaking continues on a small scale as the fluid movement gradually enters a bore mode (Fig. 10b).

It is reasonable to infer that significant transfer of momentum to the bottom sediment is accomplished through a large portion of the total shoreward progression of the slowly decaying spilling breaker.

The conclusion is drawn that breaker shape is a continuous gradation in which each shape undergoes the same processes, but differs in magnitude of the vortices. The two extremes in this gradation are the plunging breaker and the spilling breaker. At the plunging breaker extreme, the vortices[2] are large relative to the water depth (Fig. 4). The second and third vortices may extend surface to bottom before complete decay. At the

[2] The presence of an air pocket trapped by the upper portion of a plunging breaker has been noted by a number of investigators, for example Evans (1940), Kuelegan (1948), and Galvin (1968). However, the observation and analysis of a series of vortices grading in magnitude with breaker shape and their role in surf dynamics, for example bedform generation and impact pressures, originate with the present author.

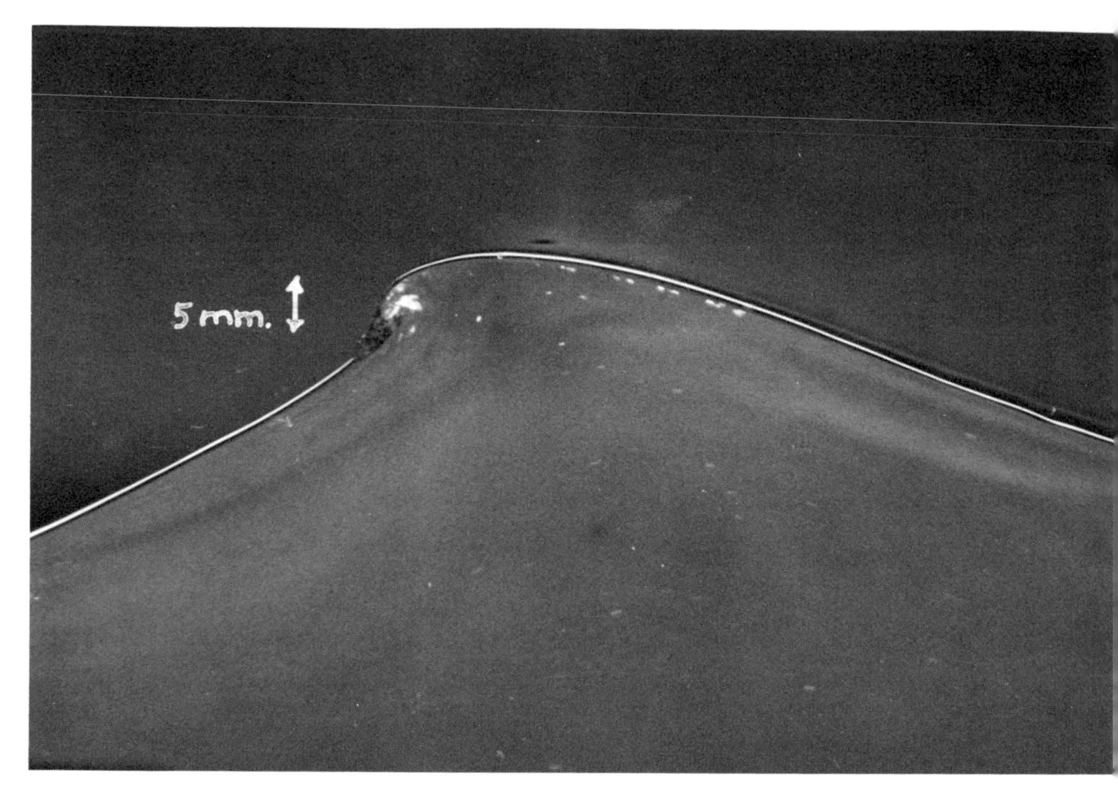

FIG. 8.—Initial breaking in spilling breaker, enlarged by use of telephoto lens.

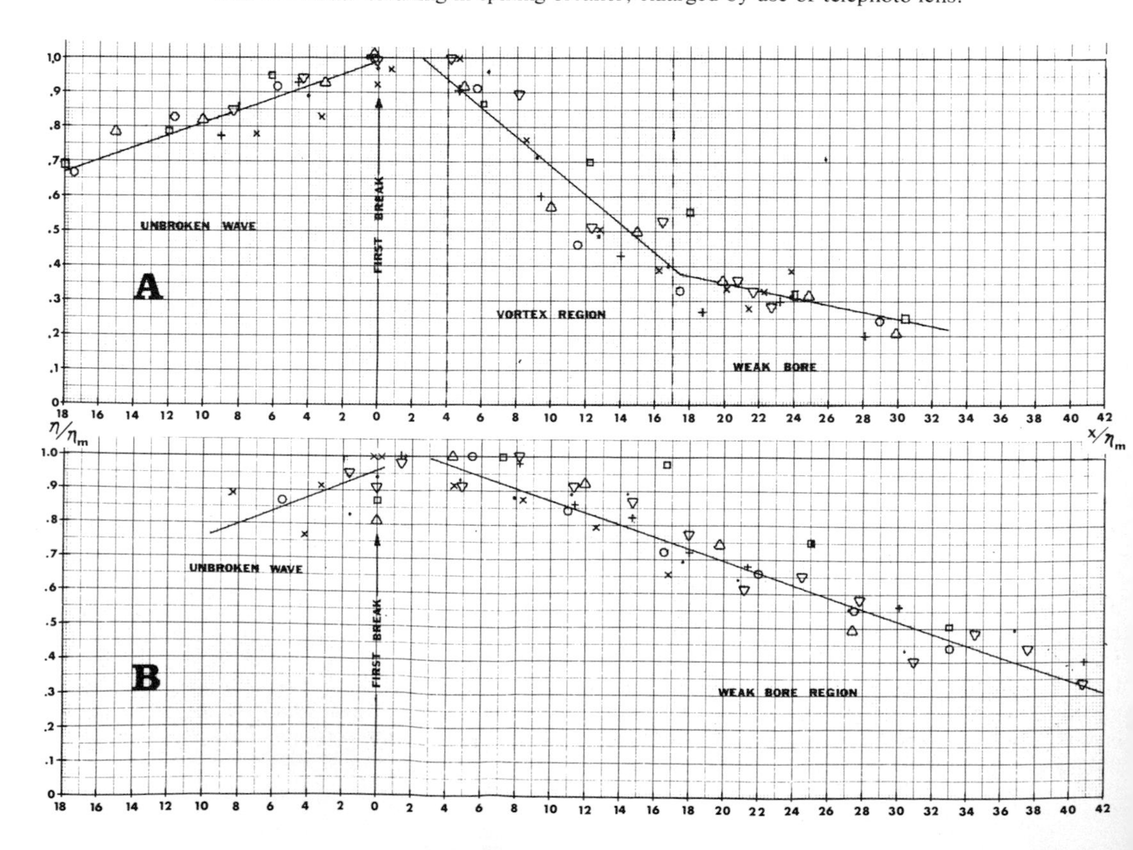

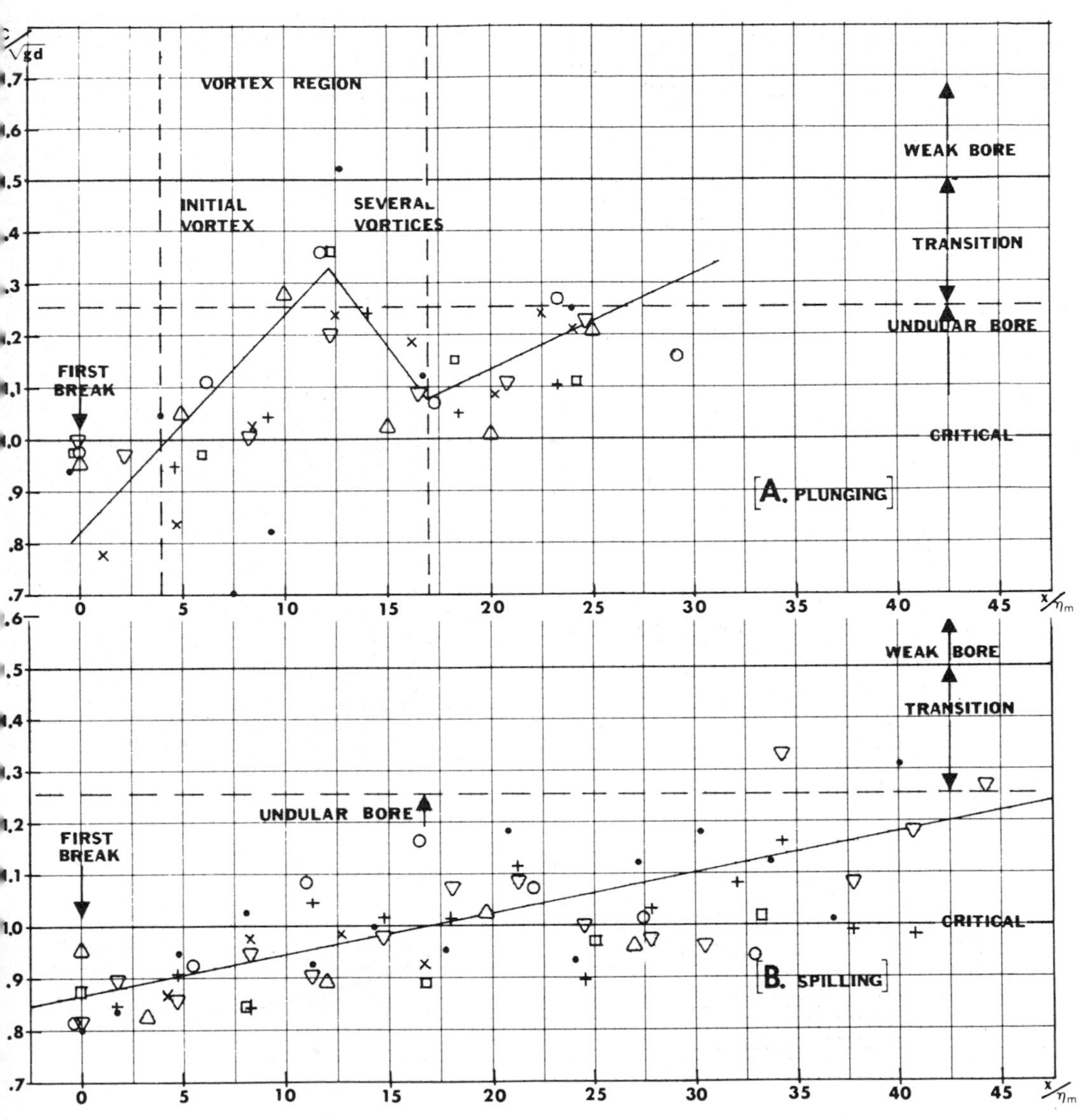

FIG. 10.—Froude number change during breaking. *A*, Plunging breaker. Data obtained from same source as in Figure 9. The ordinate is a dimensionless Froude number $c/\sqrt{gd}$ where c is crest velocity and d is distance from S.W.L. to sloping bottom. Abscissa is same as in Figure 9. Designation of horizontal regions by Froude number, based on Miller (1968). These are intended for comparison only—undular bores were not observed in the present study. Vertical regions obtained in the same manner as for Figure 9. *B*, Spilling breaker. Data obtained from same source as in Figure 9. Designated regions as in *A*, above.

FIG. 9.—Change in η, crest height above still water level, during breaking. *A*, Plunging breaker. Data from various breakers ranging from T = 2.5 to 4.3 sec, S.W.L. = 22 to 26 cm. Ordinate is dimensionless ratio of η to η_{max}. Abscissa is dimensionless ratio of distance from initial break point, X, to η_{max}, positive in direction of wave advance. Designation of regions based from 16 mm motion picture film with camera moving at speed of wave crest. *B*, Spilling breaker. Data from various breakers ranging from T = 1.0 to 2.9 sec, S.W.L. from 22 to 26 cm. Ordinate and abscissa as in A.

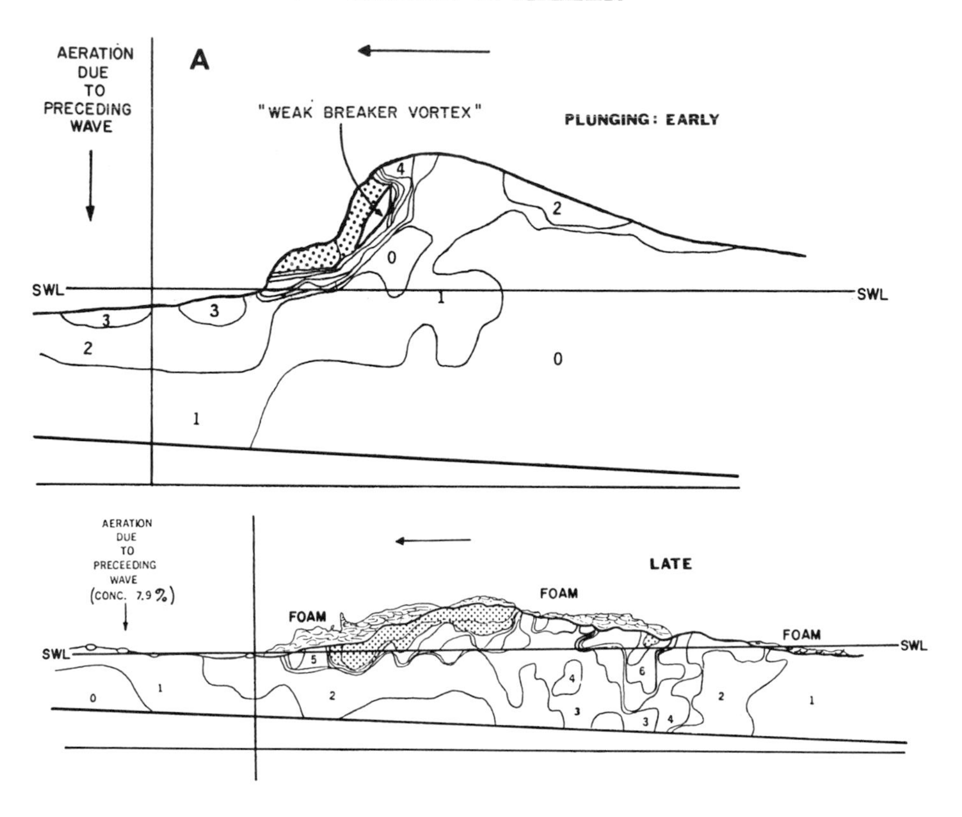
AERATION DUE TO PRECEDING WAVE
A
"WEAK BREAKER VORTEX"
PLUNGING: EARLY
SWL
AERATION DUE TO PRECEEDING WAVE (CONC. 7.9 %)
LATE
FOAM
FOAM
FOAM
SWL

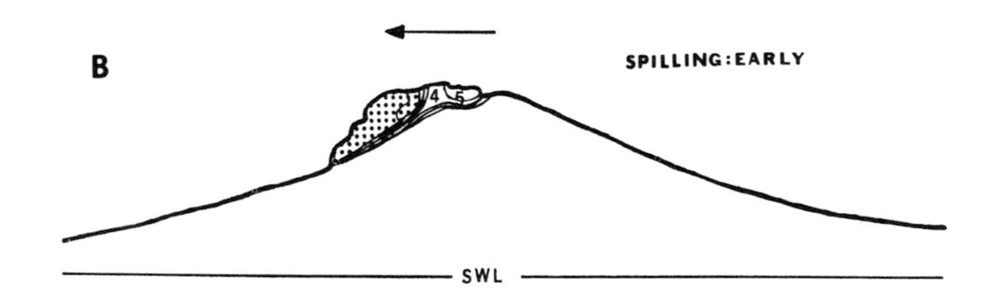
B
SPILLING:EARLY
SWL

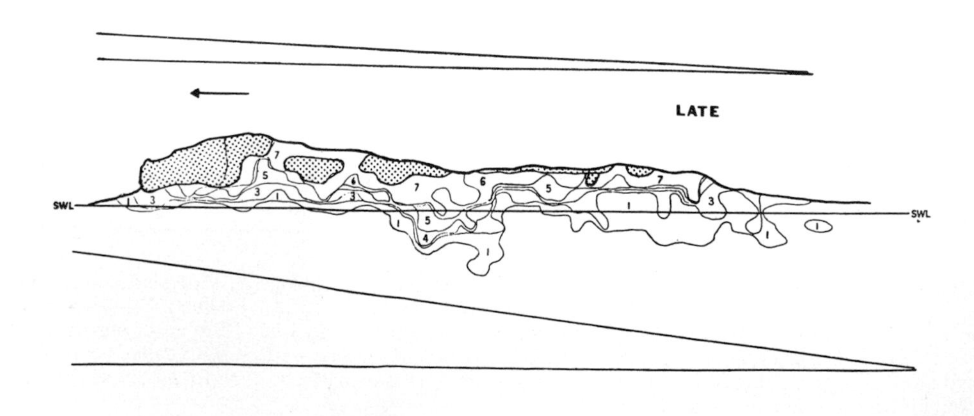
LATE
SWL
SWL

pilling breaker extreme, the vortices are confined o the region near the surface of the fluid and sually do not extend below still water level.

The mechanism by which air is entrained in he fluid during wave breaking also seems clearly llustrated by the preceding discussion. Contour naps of bubble density were constructed (Miller, 972b) and indicate that the distribution of bubbles 'aries according to breaker shape (Figs. 11, 12). An important consequence is the variation of mass y volume in the breaker face which has a ignificant effect on impact forces in surf. This opic is discussed in a succeeding section of the resent paper.

In order to establish a firm basis for a numerical rediction model, the gross aspects of fluid motion n the breaker must be ascertained. In principal, he response of the bottom sediment to the fluid an then be established on a quantitative basis, nalogous to sediment transport predictions in iver flow.

The experimental observation of internal fluid novements in the breaking wave described pre- 'iously suggests that the major components of luid motion consist of the following: (1) a horeward horizontal drift of the fluid particles, lways present and important, (2) a superimposed et of single impulse transient vortices or eddies, vhose magnitude and strength are a function of reaker shape, (3) small scale turbulence distrib- ted throughout the fluid, and (4) a bottom bound- ry layer, where the velocity gradient rapidly lecreases; the fluid in this region contains a ignificant percentage of sediment. As shown in he bedform experiments described later, the eturn flow may under certain circumstances be significant secondary component (Russell and Osirio, 1958). It is not clear to what extent this henomenon occurs in the field.

The tangential velocity of the inertial vortex n the vicinity of the bottom is of particular interest ince it is in this region that the shear stresses re applied to the individual grains. The mecha- nism of formation of the several successive vor- ices is such that large numbers of air bubbles re entrained. Several attempts in the laboratory sing approximately neutrally buoyant particles o mark the local velocity field were unsuccessful, ecause the bubble concentration effectively ob-

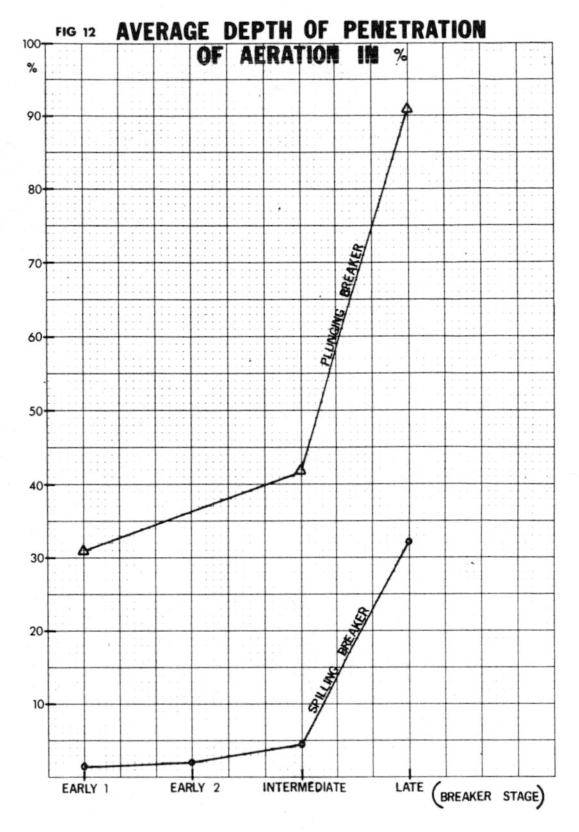

FIG. 12.—Average depth of penetration of air bubbles for plunging vs. spilling breakers. Data from photo analysis, as in Figure 11.

scured the tracers and interfered with their move- ments. Accordingly it was decided to use the bubbles themselves to obtain an order of magni- tude estimate of the near bottom tangential veloc- ity of the vortex, and the resultant near bottom velocity under the breaker.

A large number of photographs were taken through the glass wall of the wave tank. The camera was fixed in position so as to include the vortex as it came in contact with the bottom. Collimated strobe lighting directed downward from above illuminated a narrow 3 cm portion along the axis of channel in an otherwise darkened room. In this way, wall effects were eliminated,

FIG. 11.—Contour maps of bubble concentration in breakers. Enlarged photos were overlain by a 1 cm grid. Each cm square was subdivided into 10 × 10 mm units. A count of the number of 1mm units containing 1 or more bubbles yields a percentage estimate for the square, on a scale from 1 to 10. (5 means 50% of the mm units contain bubbles.) The dotted region indicates 100% bubble concentration. *A*, Plunging breaker. Early stage: average concentration = 8.7%, $\eta b/Db$ = 0.77. Late stage (weak bore): average concentration = 31.4%, $\eta b/Db$ = 0.58. *B*, Spilling breaker. Early stage: average concentration = 1.5%, $\eta b/Db$ = 0.68. Late stage: average concentration = 18.8%, $\eta b/Db$ = 1.11.

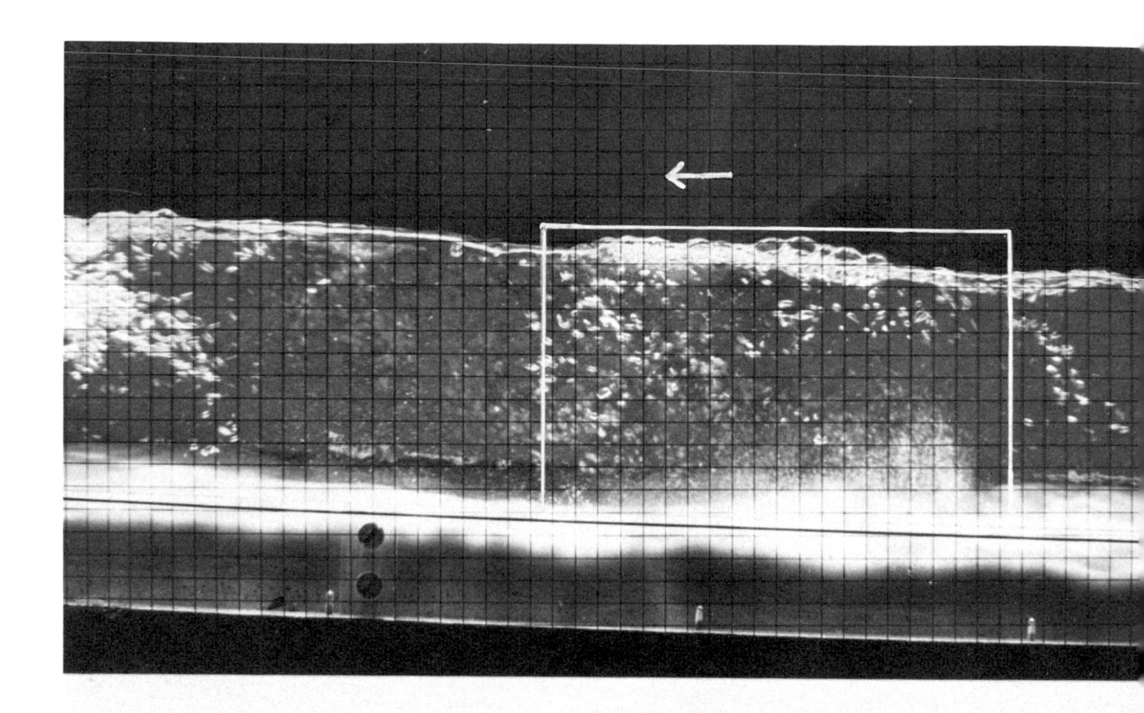

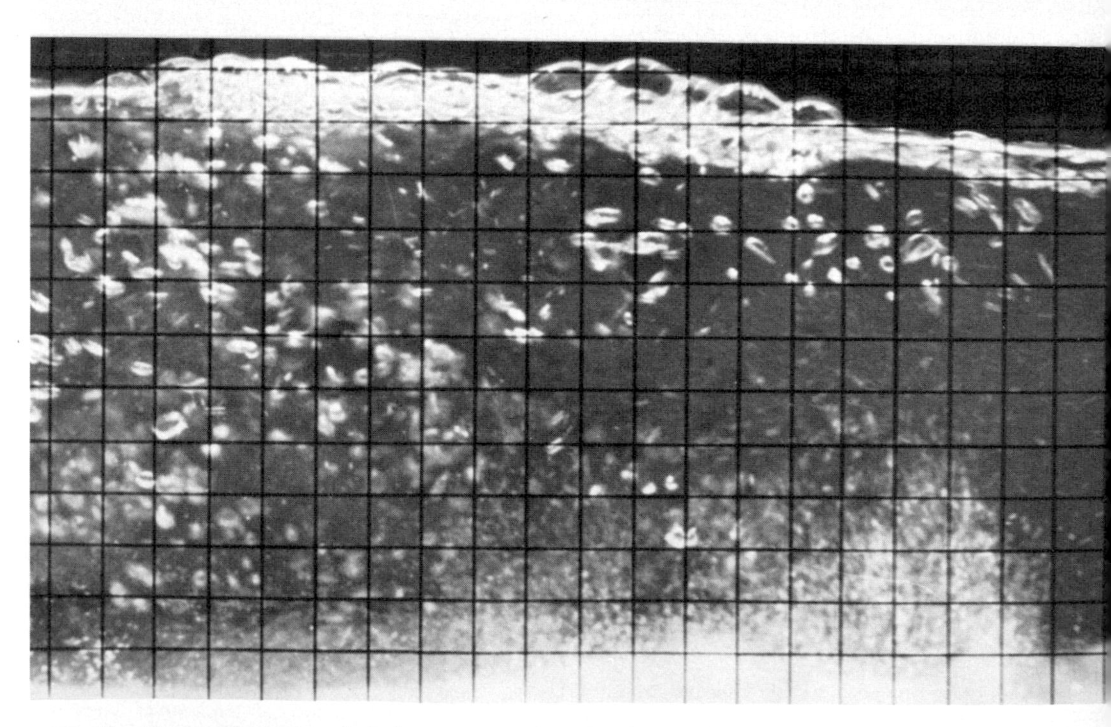

FIG. 13.—Top: open shutter photo in collimated light showing bubble streaks as vortex acts on sand bottom. Note burst of sand particles on seaward side of hollow. Bottom: close-up of section indicated in upper photograph. Grid scale is 1 cm.

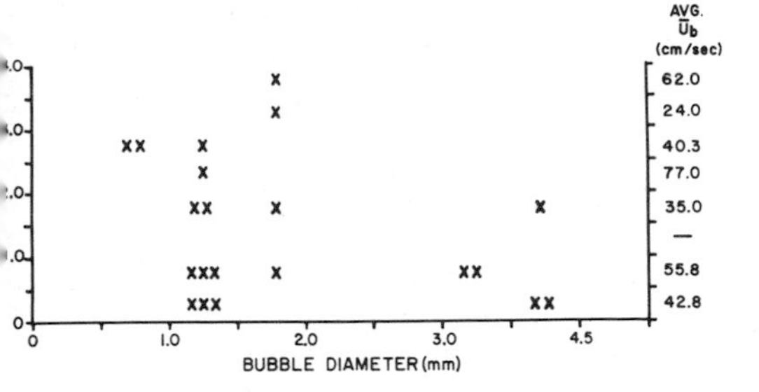

FIG. 14—Near bottom net horizontal velocity with ımera fixed. Positive numbers indicate shoreward and ɛgative numbers indicate seaward. The right hand ɔlumn indicates average $\bar{U}_b$ for each distance above ɔttom.

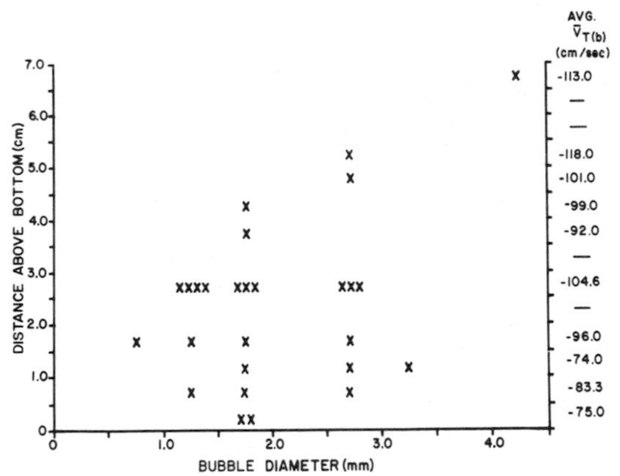

FIG. 15.—Near bottom tangential velocity of breaker vortex with camera moving with the wave. Positive numbers indicate shoreward and negative numbers indicate seaward velocities. The right hand column indicates average $\bar{V}_{T(b)}$ for each distance above bottom.

nd in effect a plane of bubbles appeared in sharp ɔcus on the film (Figs. 13 a, b).

Velocity estimates were obtained by employing bubble streak technique. Camera shutter speeds ınging from 1/250 to 1/60 sec gave the best esults. For a given wave resulting in essentially ne same plunging breaker-breaker vortex seuence, the variation in estimating bubble speed vas small and did not vary systematically with hutter speed in the 1/250 to 1/60 sec range (Fig. 4).

The bubbles entrained in the vortex reflect its notion but differ from a fluid particle in that here is an additional positive buoyancy compoent. This difficulty was met by choosing only orizontal streaks, and by noting the bubble size as evidenced by the width of the streak). Analysis f data indicates that systemic variation in bubble ize vs. horizontal speed does not appear. This s not surprising in view of the relatively small ange in terminal rise velocity and drag coeffiients for the bubble sizes indicated in Figures 4 and 15 (see Table 1).

The fixed camera analysis described above epresents the combined shoreward (+) component due to the wave speed, and the seaward −) component due to the tangential velocity of he vortex in the vicinity of the bottom. The net motion is always shoreward indicating the wave velocity is greater than the opposing tangential vortex velocity at the bottom. This was determined independently by analysing bubble motion at the bottom tangential to the vortex, using motion picture strips and a Lafayette single frame film analyzer.

An additional series of photographs was taken with the camera mounted on a monorail parallel to the glass side of the tank and moving with the wave speed. The camera was equipped with an electrically driven "fast-back" mechanism which advances the film to the next exposure at a rate of 6 per second (Fig. 15).

An equation relating the various horizontal velocity components may be written as

$$C - \bar{V}_{T(b)} = \bar{U}_B \tag{1}$$

where C is the wave velocity measured at the crest (Fig. 16). It is assumed that the horizontal velocity component due to the progressive wave is essentially constant, surface to near bottom,

TABLE 1.—TERMINAL VELOCITY (V_T), REYNOLDS NUMBER (R_e), AND DRAG COEFFICIENT (C_D) FOR VARIOUS BUBBLE DIAMETERS

Bubble diameter (mm)	Terminal velocity (V_T)	Reynolds number (R_e)	Drag coefficient (C_D)
1.0	7-14 cm/sec	40-400	1.05-0.80
1.5	14-16	425-500	0.80-0.90
2.0	16-20	575-600	0.95-1.00
2.5	21-22	620-700	1.00-1.01
3.0	22-23	720-790	1.01
3.5	23	800-890	1.02
4.0	23	900-1000	1.03
4.5	23	1100-1400	1.10-1.15

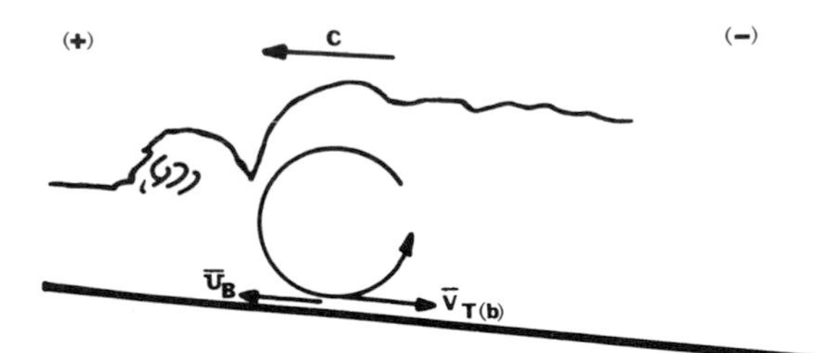

FIG. 16.—Schematic of major velocity components inside breaker. See text for notation.

and is equivalent to C; $\bar{V}_{T(b)}$ is the tangential velocity of the vortex in the vicinity of the bottom as estimated by horizontal-entrained bubble streaks in a reference frame moving with the wave crest; $\bar{U}_B$ (reference frame fixed) is the net horizontal velocity component in the vicinity of the bottom particles. The sign is (+) for shoreward direction and (−) for seaward direction. The bar above U_B and $V_{T(b)}$ indicates that these are average velocities over local fluctuations due to turbulence.

Table 2 shows estimates of $\bar{V}_{T(b)}$ and $\bar{U}_B$ from bubble streak measurements for a range of bubble sizes. The fourth column represents the sum of $\bar{U}_B$ *and* $\bar{V}_{T(b)}$ as an estimate of C. For comparison, C is noted just after the first vortex is formed and an average for three waves is 137 cm/sec. The sums of $\bar{U}_B$ and $\bar{V}_{T(b)}$ in the fourth column compare well with an average value of 137 cm/sec for C, for this set of experiments.

Preliminary measurements indicate that for a breaker height of 35 cm and a crest velocity of 140 cm/sec, the net horizontal velocity of the plunging breaker within several centimeters of the bottom, is on the order of 55 cm/sec. This is more than adequate to move sediment under unidirectional flow conditions. Velocity of the breaker was taken from motion picture-Lafayette analyzer data. Other properties were obtained in a similar way such as the depth of penetration of a vortex, the number of vortices per breaker and the associated decrease in breaker height a vortices are generated, which was described earl er.

A FIRST APPROXIMATION MODEL

At this stage it is premature to construct a mode which would adequately represent the contrit utions of the four components of motion whic were listed earlier. However, as a first approx mation, a combination of (1) the horizontal dri and associated pressure gradient and (2) the sup erimposed vortex system will represent the majc components and also reflect the gradation fro plunging to spilling breaker shape. Figure 1 illustrates the first approximation model.

The horizontal drift profile is represented a a gradient in $\bar{U}$, which is non-zero at the sed ment-water interface. It is generally recognize that in shallow water, just before breaking, th solitary wave gives a more satisfactory approx mation than Stokes waves. The solitary wave a limiting case of the more general nonline cnoidal wave theory due originally to Kortewe and DeVries (1895). Laitone (1960), has develope a second approximation for cnoidal and solitar waves. The horizontal water particle velocit equation has the form,

$$\frac{\bar{U}}{\sqrt{gd}} = \frac{H}{d}\left[1 + \frac{1}{4}\frac{H}{d} - \frac{3}{2}\frac{H}{d}\frac{y^2}{d^2}\right]\operatorname{sech}^2(A\omega x - Ct) + \frac{H^2}{d}\left[-1 + \frac{9}{4}\frac{y^2}{d^2}\right]\operatorname{sech}^4(A\omega x - Ct) \qquad ($$

where

$$A\omega x = \frac{x}{d}\frac{3}{4}\frac{H}{d}\left[1 - \frac{5}{8}\frac{H}{d}\right], \qquad ($$

H = maximum wave height, d = undisturbe water depth, and C = wave celerity. Masch ar Wiegel (1961) have published tables for solutio

TABLE 2.—COMPARISON OF AVERAGE TANGENTIAL VELOCITY ($\bar{V}_T$) AND NET VELOCITY ($\bar{U}$) FOR VARIOUS BUBBLE DIAMETERS

Bubble diameter (mm)	$\bar{V}_{T(\text{bottom})}$	$\bar{U}_{\text{bottom}}$	$\Sigma(\bar{V}_{T(b)} + \bar{U}_b)$
0.6–1.0	91 cm/sec	52 cm/sec	143.0 cm/sec
1.1–1.5	84	51.1	135.1
1.6–2.0	93.2	53.7	146.9
2.1–2.5	125	—	—
2.6–3.0	92.4	51	143.4
3.1–3.5	58	83.3	141.3
3.6–4.0	—	37	—
4.1–4.5	113	—	—

According to equation (1) the sum of $\bar{V}_T + \bar{U}$ shown in the last column should be equivalent to wave celerity at the crest.

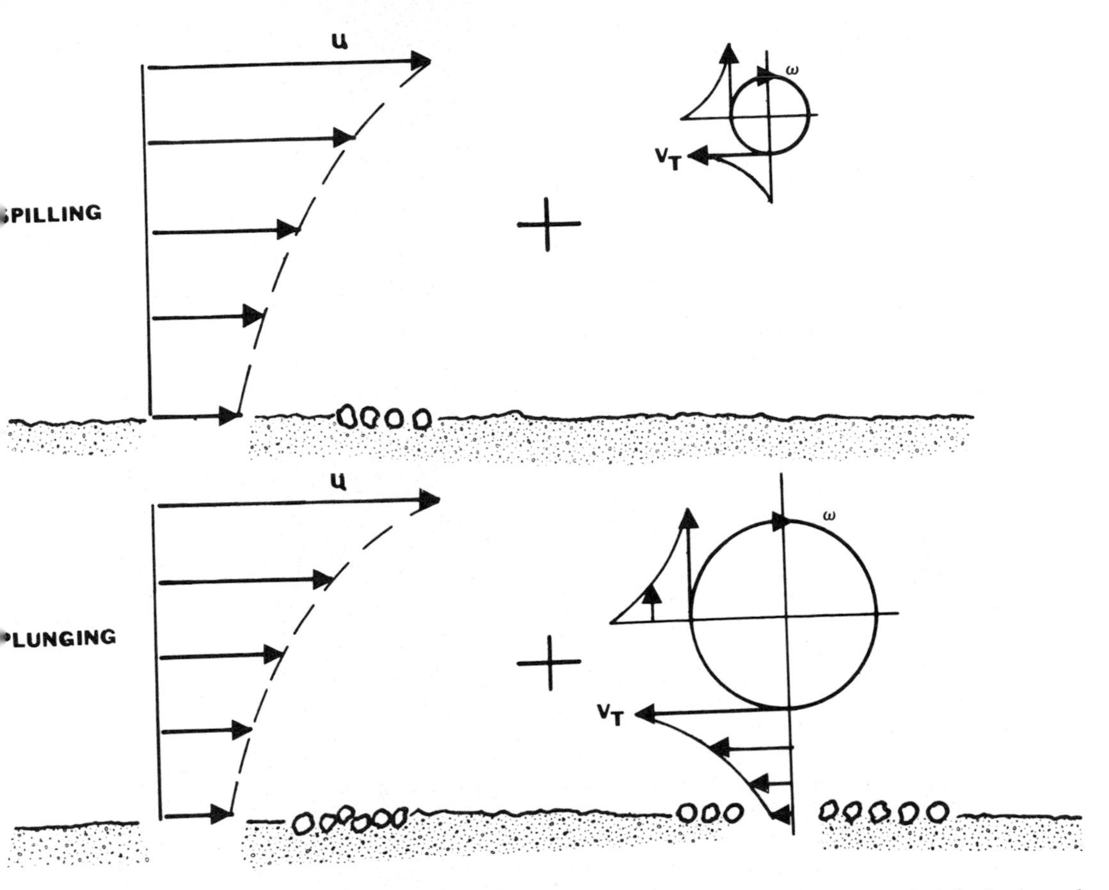

FIG. 17.—Schematic of first approximation model for internal velocity in breaking wave. At left, horizontal drift; at right, vortex representation. ω is angular velocity of vortex, V_T tangential velocity, and U horizontal shoreward drift component. Summation for spilling case, at top and plunging case, at bottom.

to a similar equation for $\bar{U}$ (due to Kuelegan and Patterson, 1940) suitable for application to a numerical model, and have tabulated most of the necessary functions used to obtain numerical answers. In addition, an expression for the pressure is available in a form suitable for computation. Thus, the horizontal drift component and associated pressure gradient appear adequately represented by available theory.

The vortex representation however, is not at this time available in equation form. The problem is under investigation. A promising approach is through an extension of circular Couette flow. In that case, consideration is made of laminar, steady, axisymmetric flow between two concentric cylinders. The Navier-Stokes equations in cylindrical coordinates, reduce to

$$-\frac{V^2}{r} = -\frac{1}{\rho}\frac{dp}{dr} \text{ in the } r \text{ direction} \qquad (4)$$

and

$$0 = \nu\left(\frac{\partial^2 v}{\partial r^2} + \frac{\partial 1}{\partial r}\frac{v}{r} - \frac{v}{r^2}\right) \text{ in the } \theta \text{ direction.} \qquad (5)$$

Solutions have been obtained for the velocity gradient, dynamic pressure gradient and bottom shear stress. If the solid inner cylinder is rotating and is taken to be the core of the vortex, then a solution with proper boundary conditions should yield a useful representation of the vortex component of motion. The proper boundary conditions include a free surface above the cylinder, a solid, impervious boundary below the cylinder, and infinite fluid extent in the horizontal. An analytic solution appears to be very difficult. Accordingly an analogue has been constructed in the laboratory. Controlled angular velocity of a lucite cylinder will generate the required pressure and velocity gradients and the bottom shear stress. It is hoped that numerical values will be obtained for the velocity gradient by streak photography of suspended aluminum particles, and the pressure at the bottom can be determined by fast response pressure gauges. Preliminary trials are encouraging.

ROLE OF BREAKER SHAPE IN THE FORMATION AND ERASURE OF SURF ZONE BED FORMS

As indicated earlier, observations of the internal mechanics of breakers were made using optical techniques with collimated light. These clearly indicated that the first and second breaker vortices in plunging breakers significantly affected the bottom flow. It was thus inferred that the concentrated energy loss in the plunging breaker could be attributed to a large, transient eddy viscosity and a transfer of momentum to the bottom sediment. It seemed apparent that in nature this plunging breaker process should affect the bottom sediment in a manner different from that of the spilling breaker where the vortices were confined to the uppermost region of the breaking wave.

Accordingly, a series of experiments was made to determine the effect of the end members of the plunging-spilling array on the bottom sediment. In order to look particularly at the origin of bed forms, the initial condition included a smooth sand bed at a slope commonly found in the nearshore (2°) and carefully controlled single, repeatable, breaker types. The optical laboratory techniques were used to measure the internal velocity field including that due to the vortices.

A partition running the full length of the wave tank facility at the University of Chicago, formed a channel with the dimensions; depth = 1 m, width = 36 cm, length = 25 m. A fixed 2° slope was constructed at the down-wave end. The outer wall of the channel is glass throughout the full length. A pan, 10 cm deep, extending the full width of the channel was inserted in the slope and filled with well rounded beach sand (median size = 1/2 mm; range, 1/4 mm to 3/4 mm). The sand area extended from the vicinity of the toe of ramp to the intersection with still water level.

Suitable breakers were generated in the following manner. The pan section of the ramp was covered with a metal plate to create a uniform impermeable slope. Various settings of the periodic wave generator were tried until a strong plunging breaker was formed, approximately in the middle of the pan section. Settings for a very gentle spilling breaker were established in the same way. The metal plate protecting the sand was then removed and the channel was ready for the periodic wave experiments.

SERIES A, (NO RETURN FLOW)

First experiment.—Initial conditions consisted of a smooth sand surface with a periodic wave of $\lambda = 6.4$ m and $T = 3.58$ sec. The breaker type was plunging with a breaker height of 26. cm. Still water level at the toe of slope was 27. cm with a slope angle of 2°. The large vortices (Fig. 4) created by the plunging breaker quickl excavated the trough and formed a weak ridg on the seaward side. In profile, the resultin bedform has the sinusoidal appearance of a offshore bar. The experiment was stopped afte 5 minutes of continuous operation; approximatel 84 identical plunging breakers.

Second experiment.—Initial conditions consisted of a partial sinusoidal form created by th plunging breaker sequence with a periodic wav of $\lambda = 5.30$ m and $T = 2.81$ sec. Breakers wer spilling, with a breaker height of 29.0 cm. Stil water level at toe of slope was 27.5 cm with slope angle of 2°. The vortices created by th spilling breaker were confined to the region nea the surface. The region near the bottom experienced strong horizontal velocity in the directio of the shore as under an unbroken wave o translation in shallow water, but with little o no reversal in flow direction between successiv breakers. The ridge portion of the bar migrate shoreward into the trough portion with a significant erasure of the sinusoid bar profile previousl created by the plunging breaker sequence of th first experiment (Fig. 18). The experiment laste for 10 minutes of continuous operation; approximately 210 identical spilling breakers.

Third experiment.—Initial condition consiste of a smooth sand surface. The breaker type wa spilling with all wave characteristics the same a in second experiment. The sinusoidal bar profile was not formed. Instead the sand migrated steadily up the slope in the form of small asymmetric ripples as found in unidirectional flow. This indicated that the spilling breaker does not form bars when initial condition is smooth slope. Running time was 1 hr; 1280 spilling breakers.

Conclusions.—The experiments suggest that bars are formed when the large scale vortices created by plunging breakers are present, and are erased by the translation effect of spilling breakers when the vortices are confined to the vicinity of the surface. A reasonable inference is that migration is caused by a series of breakers somewhere between the end members of the plunging-spilling shape spectrum. There is convincing evidence that the strength of breaker vortices which vary directly with breaker shape is a significant factor in the generation of bed forms in the surf zone (Fig. 18).

SERIES B, (WITH RETURN FLOW)

The first set of experiments eliminated the return flow component by using a short ramp with the uprush trapped on the other side. Backwash was thus negligible. By extending the ramp

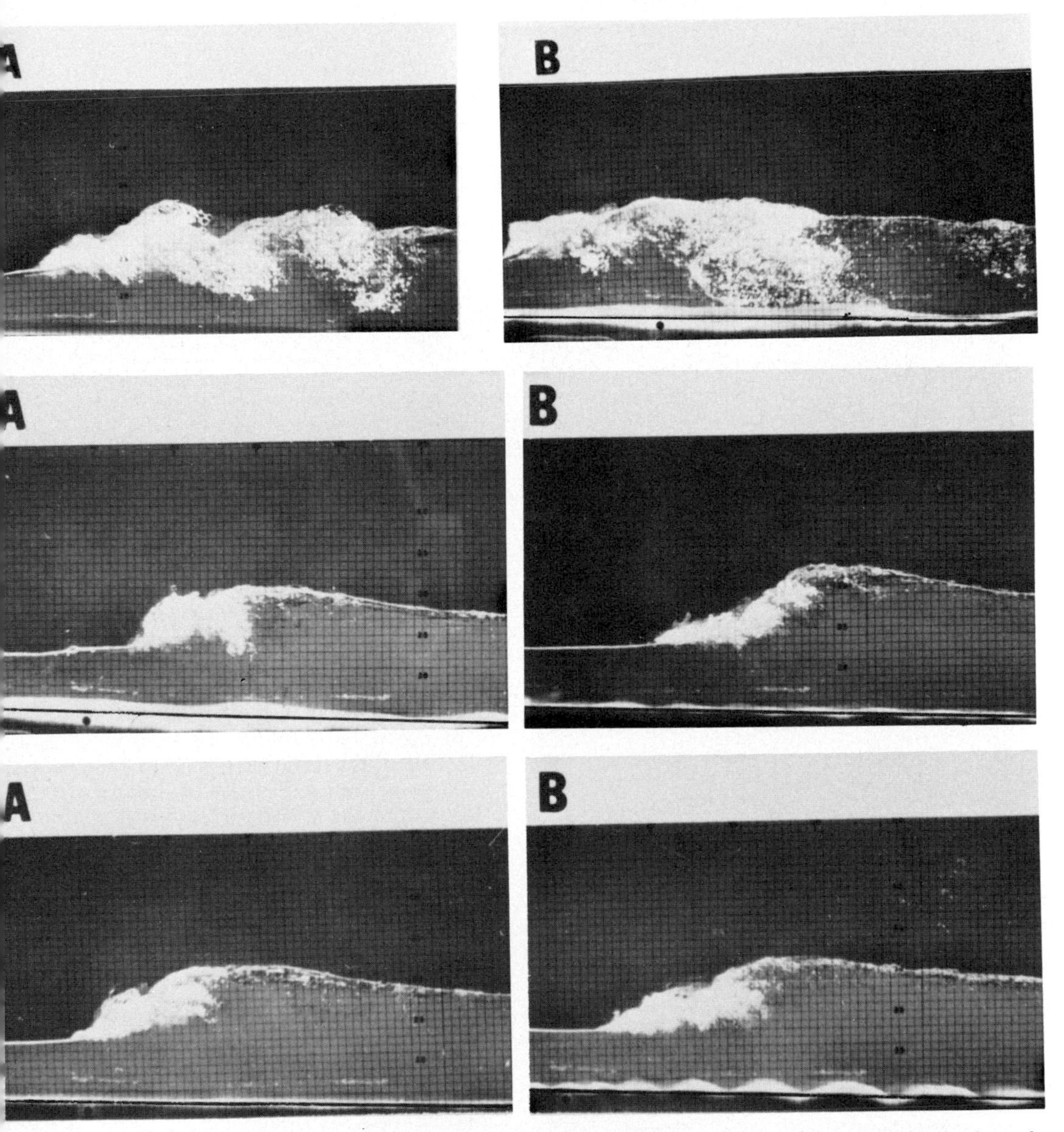

FIG. 18.—Photographs illustrating the experiments on bar formation. Straight line shows original smooth sand level. *Top A*, Plunging breaker over smooth sand surface (initial conditions). *Top B*, Formation of shoreward "hollow" of bar profile by plunging breaker. *Middle A*, Spilling breaker over initial condition, the previously formed bar. *Middle B*, "Hollow" filled in by spilling breaker. *Bottom A*, Spilling breaker over smooth sand (initial conditions). *Bottom B*, After extensive series of spilling breakers, bar profile does not appear.

the normal uprush and backwash were present. In this closed system a seaward velocity component, the "return flow," was present (Russell and Osirio, 1958).

The set of experiments in A were repeated. This time, as expected with the addition of the seaward directed return flow component, the ridge region was significantly more pronounced, with a profile quite close to the classical sinusoid found in nature (Fig. 19). The spilling breaker series once again tended to erode the ridge and fill the trough but not as markedly as in the previous set where the seaward return flow component was eliminated.

Conclusions.—Addition of the return flow adds a seaward velocity component to the system. The result is that sediment thrown into suspension in the vicinity of the ridge tends to settle in that

FIG. 19.—Classic bar profile created by plunging breaker series. Straight line shows original smooth sand level.

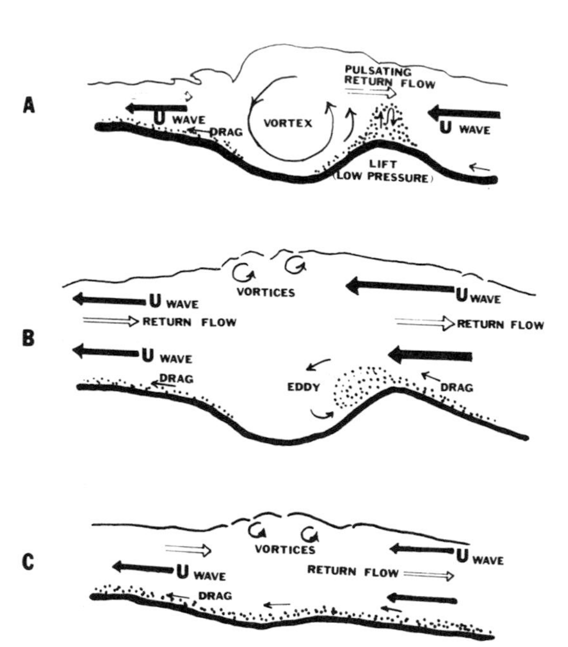

FIG. 20.—Schematic of fluid-sediment interaction. *A*, The proposed distribution of velocities as the plunging breaker creates the bar profile. The velocities are, in order of magnitude, (1) the horizontal component of the incoming wave and/or breaker, *U*, (2) the angular velocity ω of the vortex, (3) the pulsating return flow, and (4) the small scale turbulence (not specifically shown in the figure). Supposed regions of predominant drag or lift are indicated. *B*, The fluid motions as the spilling breaker erases the bar. The vortices are small and confined to the upper portion of the wave. The dominant flow is *U*. An eddy forms shoreward of the ridge. *C*, The final appearance of the profile as continued spilling breaker velocity field transports sand shoreward.

region rather than to be transported over the trough toward the beach. In general, the experiments suggest that the interaction of plunging breaker vortices with the bottom will create a bar profile and that the spilling breaker tends to erase the bar. It is recognized that a simplified set of controlled experiments in a wave tank, on a scale of the order of 1/5 that of the natural case does not in itself justify immediate extrapolation to the field.[3] However, this is definitely a promising area for detailed experimental study of sediment transport mechanics in the surf zone, and should lead to a better understanding of interaction of breaking waves and the bottom. Figure 20 is a schematic interpretation of the fluid-sediment interaction.

EFFECT OF BREAKER SHAPE ON IMPACT PRESSURES IN SURF

"Impact pressure" is defined as the impact of a translatory mass of water against a rigid surface. The pressure peak is of large magnitude but very short duration. The phenomenon is also known as "shock pressure." A one dimensional analysis (Weggel, 1968) leads to the relationship

$$p_{max}\, t_{max} = -2\rho \frac{\forall}{A} U_o \qquad (6)$$

where $p = p_{max}$ at $t = t_{max}$; at moment of impact p = pressure, t = time, ρ = density of fluid, A = cross sectional area of impact, $\forall$ = volume of fluid with front face equal to A, U_o is horizontal

[3] The presence of sand ripples must be carefully investigated to determine whether they are an artifact of the experimental conditions or an important secondary effect in the system.

velocity component at t_o, an arbitrary time before impact.

The right side of the equation is expected to be constant, and the minus sign shows that velocity and resultant force are in opposite directions. It is assumed that the horizontal velocity, U, is reduced to zero simultaneously, on the front face of the wave, and that the horizontal velocity is the same everywhere. Actually the variation in breaker geometry is quite large, but it appeared to me that if breaker type is taken into account, the variation within types may be small.

Another difficulty in consideration of surf is the presence of entrained air. The equation given above (6) predicts that $p_{max} \rightarrow \infty$ as $t_{max} \rightarrow o$, as Weggel (1968) has pointed out. A more realistic approach to the theoretical upper limit for p_{max} is some function of the local acoustic velocity of the fluid.

$$p_{lim} = \rho\, U_o C_a \qquad (7)$$

where C_a = local acoustic velocity, taken as a constant. However, Gibson, (1970) has shown that the velocity of sound can be significantly reduced by addition of small amounts of air in the fluid. For example, for an air-water mix of less than 1% by volume, there is a 90% decrease in the velocity of sound. In real breaking waves, Fuhrboter (1970) and Miller (1972b) have observed as much as 20 to 30% of air by volume. Because the degree and distribution of air entrainment depends on breaker type (Miller, 1972b and section 1 of this paper), the magnitude of impact pressure, as it is affected by aeration, should also vary by breaker type.

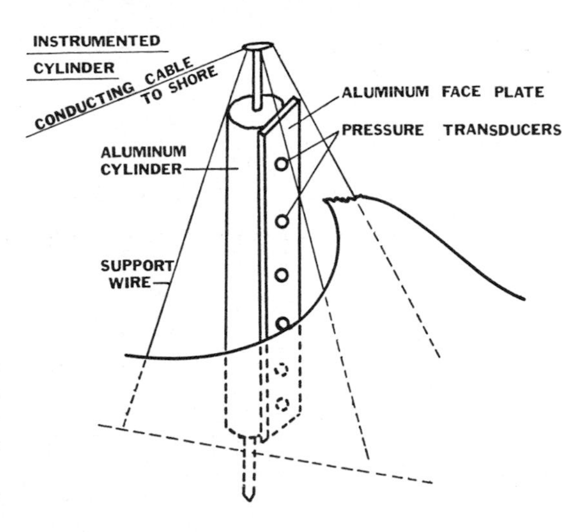

FIG. 21.—Diagram of facility for field measurement of impact pressure in surf.

FIELD STUDY

In view of the anticipated importance of breaker shape in the problems discussed above, a series of field experiments was carried out on full scale breakers (Miller and others, 1974). The strategy was to place a vertical cylinder seaward of the foreshore, on the tidal flat. The strain gage instrumented cylinder (Fig. 21) was emplaced at low tide, when the tidal flat was fully exposed. As the tide came in, the cylinder recorded the first arrivals as post-breaking bores. The vertical array of sensors yielded the data as a gradient from near bottom to free surface. As the tide continued to rise, the direct impact of spilling or plunging breakers was recorded. Finally at high tide, the near breaking waves were measured.

Results.—Figure 22 illustrates the results obtained from the study. Simultaneous traces are taken from the oscilloscope for a typical example from each of the four wave or breaker types. The figure indicates that the bore formed just after the collapse of the plunging breaker generated the largest impact pressures. Weak bores generated by spilling breakers, on the other hand, do not generate large impact pressures. This results in a wide scatter of data for this breaker type. The maximum impact pressure in the vertical gradient is at the base of the bore. The next largest impact pressures are due to plunging breakers, but the gradient is reversed, with maximum at the top. The spilling breaker is similar in vertical gradient, but smaller in magnitude. Although the near-breaking wave generates the smallest impact pressures, impact "spikes" are noted when the wave surface strikes the sensor face.

The role of entrained air and the angle of impact of the free surface, are reasonably explained when breaker shapes are taken into account. The air bubbles are concentrated near the top of the bore front with relatively little air at the base of the bore front. This explains the reversal of the impact pressure gradient from that found in the plunging and spilling breakers. In these cases the mass per unit volume is also a variable, rather than a constant as assumed in the introductory analysis. Comparison is made of the field data with published papers on both field and laboratory studies. Interest is focused on the role of breaker shape in improving prediction of impact pressures. Figure 23 illustrates the use of breaker height to predict impact pressure. It is clear that there is no clear-cut linear relationship, even with the plot of a family of curves of the form $p = K(H)$ where K is a constant. In addition the laboratory data gives consistently higher values of pressure for a given breaker height than the field data. Predicting dynamic pressure from wave velocity (Fig. 24) gives fair agreement (provided breaker types are differentiated) with a family of curves

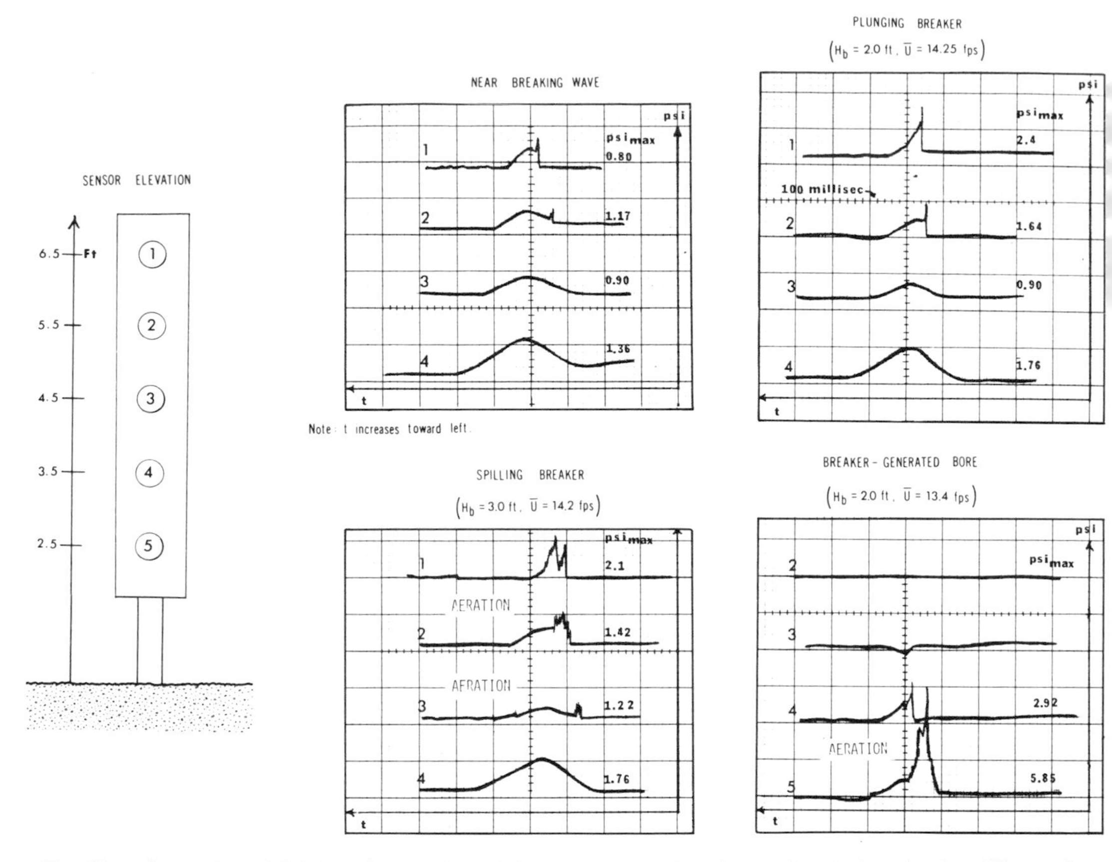

FIG. 22.—Illustration of field results. Ordinate is impact pressure in psi. Abscissa is duration in milliseconds.

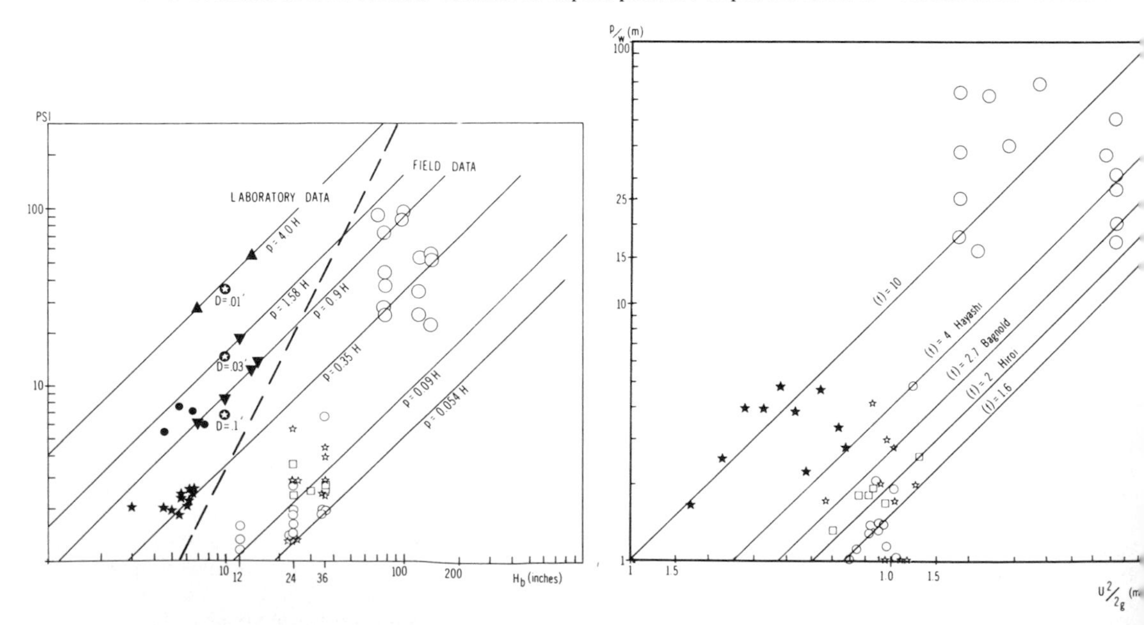

FIG. 23.—Impact pressure is psi vs. breaker height, H_b. Open symbols = field data, solid symbols = laboratory data.

FIG. 24.—Dynamic pressure, p/w vs. velocity $U^2/2g$; p/w is pressure head, $U^2/2g$ is head of fluid at impact, f is a coefficient, w is unit weight of water.

of the form $p/w = (f)\, U^2/2g$. However, bores appear to follow a different relationship and do not fit the curves shown on the diagram.

SUMMARY

Breaker shape is shown to be a significant factor in the mechanics of impact pressures, and hence in prediction of this property in the surf zone. The effect of breaker shape on the instant of impact and on the distribution of density per unit volume in the bubble zone is indicated.

INTERACTION OF POST-BREAKING BORE WITH THE FORESHORE

In this section we consider the post-breaking bore as it progresses up the foreshore. Interaction of fluid and sediment in this region is very complicated. The key to the sediment transport mechanics lies in observation of the internal kinematics in the vicinity of the sand bed, but observations of this kind are difficult to obtain in the field. Accordingly a laboratory study was made (Nelson and Miller, 1975) in which a single

TILTING FLUME APPARATUS

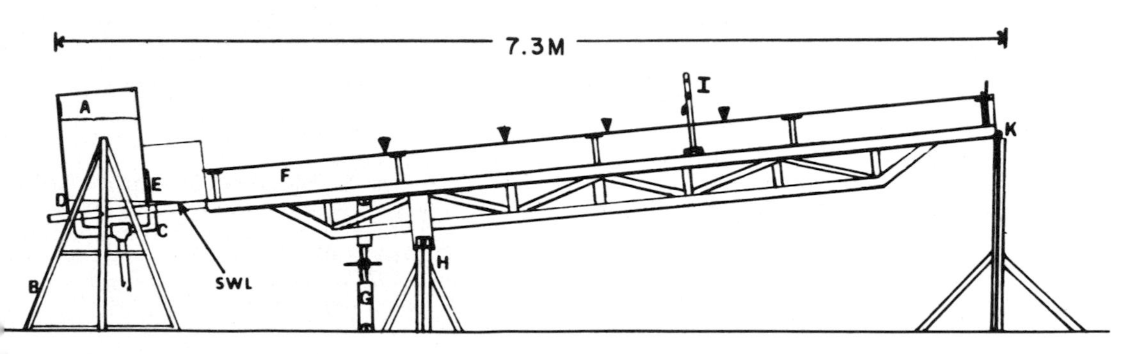

A. Head Tank
B. "A Frame" Support
C. Drain Pipes
D. False Floor
E. Sliding Gate
F. Glass Wall
G. Double Thread Jack
H. "Anti-Sway" Support
I. Traversing Mechanism
J. Glass Wall Support
K. Hinge and Support
▼ Stations (probes)
SWL: Still Water Level

HEAD TANK APPARATUS

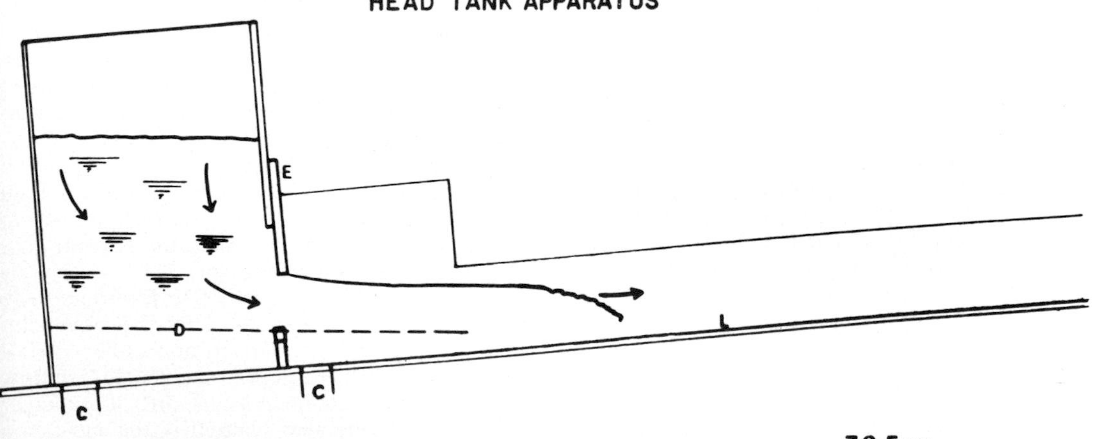

FIG. 25.—Tilting flume designed for foreshore studies. Head tank: 30 cm wide, 60 cm long and 90 cm high. Flume: 30 cm wide, 22 cm high and 7.3 m long.

uprush and backwash cycle was isolated, and the sediment transport mechanics observed through the glass wall of a specially designed tilting flume.

Experimental design.—It is widely recognized that, regardless of the wave type approaching the foreshore, after breaking the wave is usually a bore. A backward tilting flume was constructed at the University of Chicago (Fig. 25). A charge of water in a standing head tank at the lower end of the tilted flume is suddenly released by raising a gate following the method of Prins (1958). The result is a bore of controlled strength, which propagates upslope over a sand bed. Eventually the flow velocity decreases to zero. The fluid then accelerates from rest back down the slope. When it reaches the base of the slope a bypass at the bottom of the head tank is opened to prevent a secondary reflection. In this way a single uprush and backwash cycle is isolated for study.

Initial bore heights range from 12 to 26 cm. St. Peter sand was used (well rounded, Md = 0.5 mm). Since the bore heights and sand sizes approximated those found in nature, scaling was not considered to be necessary.

Measurement and observation.—Because of the violent and complex flow patterns, a Lagrangian frame of reference was not attempted. Accordingly, utilizing the Eulerian approach, a number of fixed stations were established at intervals from the base to the top of the sand slope. At each station the fluid profile was recorded by a capacitance probe. The velocity was recorded by elapsed time between probes, and by motion picture film. The sediment bedforms, suspended sediments trajectories, and the internal structures of the fluid, were obtained by both still and motion picture films. Streak analysis was used to obtain velocities of sediment and of bubbles, which represented the local fluid velocity. An illustration of measurement technique is in the determination of the average flow velocity, $\bar{U}$, a necessary parameter in many of the analyses. A parabolic velocity profile as in river flow, is assumed for the flow region behind the turbulent front. $\bar{U}$ is taken to be the local horizontal velocity at 0.6 of the full depth measured down from the free surface. At this level, analysis of bubble streaks, or fine sediment particles on a frame to frame basis, is made from film records. The velocity at the free surface, $\bar{U}_s$, is also obtained in this way. The shear velocity, U_*, is then estimated from a relationship suggested by Southard and others (1971),

$$U_S - \bar{U} = 2.5\, U_* \qquad (8)$$

and the bottom shear stress from the definition of U_*

$$\tau_o = U_*^2 \qquad (9)$$

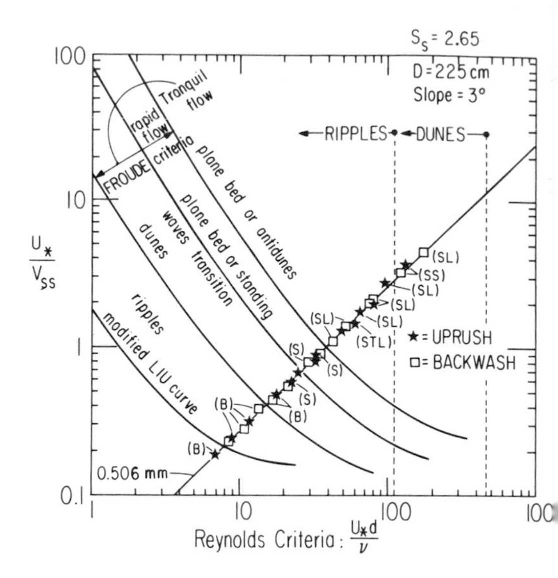

FIG. 26.—Simons' (1961) bedform diagram. Ordinate U_*/V_{ss} where $U_* = \sqrt{\tau_o/\rho}$ Abscissa, $U_* d/\nu$, where d = grain diameter, ν = viscosity. Experimental data from Nelson and Miller (1975), where B = bed load, S = saltation, STL = saltation layer, SL = suspension layer, SS = sand spurt, and SP = sand plume.

Comparison of sediment transport modes (bedforms) during the uprush—backwash cycle with published bedform criteria for unidirectional flow.—Good agreement is found for the flow region behind the turbulent front during the uprush and for the entire backwash. Figure 26 shows a plot of bedform data obtained during the experimental study on the bedform criteria diagram of Simons, Richardson and Albertson (1961). The ordinate is the ratio of U_* to V_{ss}, the fall velocity of the sediment particle, and the abscissa is a Reynolds number, $U_* d/\nu$, where d is the grain diameter and ν is the kinematic viscosity of the fluid. Surprisingly good agreement is indicated in both uprush and backwash. In another relevant diagram, Bogardi (1965) indicates expected bedform regions. The ordinate is the inverse squared Froude number, gd/U_*^2, and the abscissa is grain diameter. Figure 27 shows a plot of data from Nelson and Miller (1975), along the 0.5 mm line. Again there is good agreement.

Relevance of breaker shape to foreshore sediment transport mechanics.—A number of conclusions were drawn by Nelson and Miller (1975), regarding the nature of uprush and backwash flow and the resulting sediment transport. Of particular interest in the present context is the indication from the experimental data, that low energy initial bores tend to transport sediment upslope, and then percolate into the bed leaving a deposit of

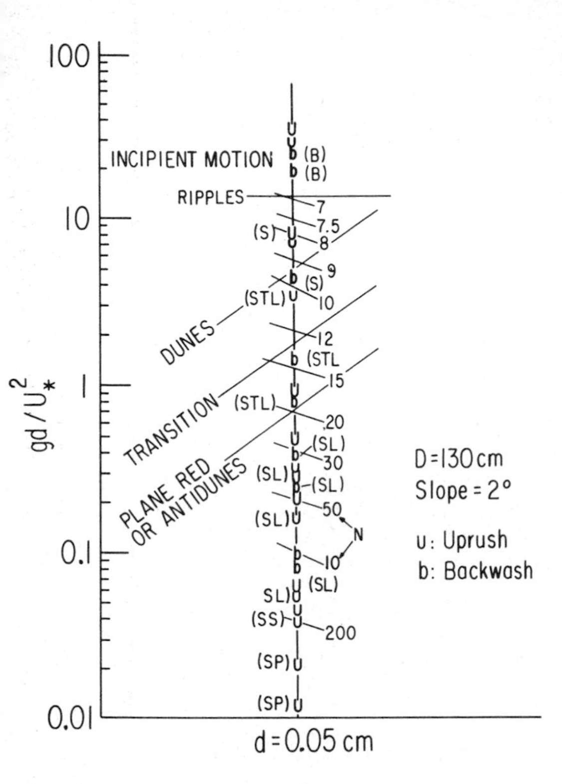

FIG. 27.—Bogardi's (1965) bedform diagram. Notation as in Figure 26. D = distance from head tank.

transported sediment. In contrast, the high energy high volume bores tend to have large backwash modes; the uprush-transported sediment is moved back down slope. Insight into the erosion-deposition balance on the foreshore is thus gained through analysis of initial bore strength. Initial bore strength in turn appears to be a function of breaker shape, as indicated in the present paper under Breaking Mechanism.

CONCLUSIONS

Analysis of the internal components of motion in breaking waves was made using the optical laboratory technique described earlier.

The results, especially with respect to the role of breaker vortices, seem sufficient to attempt a first approximation prediction model. Although further work is needed, preliminary trials appear promising. Insight into various phenomena associated with the surf zone is gained when breaker shape and associated breaker vortices are taken into account.

ACKNOWLEDGEMENTS

The optical techniques were developed with the close cooperation of Robert V. White, Associate Engineer at the Fluid Dynamics and Sediment Transport Laboratory at the University of Chicago. Robert White also did the photography displayed in this paper. My thanks also to Barry Lesht, University of Chicago, and William Wood, Purdue University, for a critical reading of the manuscript. This study was entirely supported by the Office of Naval Research, Geography Programs, through Contract N00014-75-C-0931.

REFERENCES

BOGARDI, J. L., 1965, European concepts of sediment transportation: Am. Soc. Civil Engineers Proc., Jour. Hydraulics Div., HY1, No. 4195, p. 29-54.

EINSTEIN, H. A., 1972, A basic description of sediment transport on beaches: *In* R. E. Meyer (ed.), Waves on beaches: Academic Press, New York, p. 53-94.

EVANS, O. F., 1940, The low and ball of the eastern shore of Lake Michigan: Jour. Geology, v. 48, p. 476-511.

FOX, W. T., AND DAVIS, R. A., JR., 1973, Simulation model for storm cycles and beach erosion on Lake Michigan: Geol. Soc. America Bull., v. 84, p. 1769-1790.

FUHRBOTER, A., 1970, Air entrainment and energy dissipation in breakers: Am. Soc. Civil Engineers, Proc. 12th Conf. on Coastal Eng., p. 391-398.

GALVIN, C. J., JR., 1968, Breaker type classification on three laboratory beaches: Jour. Geophys. Res., v. 73, p. 3651-3659.

——, 1969, Breaker travel and choice of design height: Am. Soc. Civil Engineers Proc., Jour. Waterways and Harbors Div., No. WW2, p. 175-191.

GIBSON, F. W., 1970, Measurements of the effect of air bubbles on the speed of sound in water: Acoustical Soc. America Jour., v. 48, p. 1195-1197.

HORIKAWA, K., AND KUO, C. T., 1966, A study of wave transformations inside surf zone: Am. Soc. Civil Engineers, Proc. 10th Conf. on Coastal Eng., v. 1, p. 217-233.

INMAN, D. L., AND NASU, N., 1956, Orbital velocity associated with wave action near the breaker zone: U.S. Army Corps of Engineers, Beach Erosion Board Tech. Memo. 79, 43 p.

IPPEN, A., AND KULIN, G., 1954, The shoaling and breaking of the solitary wave: Am. Soc. Civil Engineers, Proc. 5th Conf. on Coastal Eng., p. 27-49.

IVERSON, H. W., 1952, Laboratory study of breakers: Natl. Bur. Standards Circ. 521, p. 9-32.

KENNEDY, J. F., AND LOCHER, F. A., 1972, Sediment suspension by water waves: *In* R. E. Meyer (ed.), Waves on beaches: Academic Press, New York, p. 249-295.

Komar, P. D., and Gaughan, M. K., 1972, Airy wave theory and breaker height prediction: Am. Soc. Civil Engineers, Proc. 13th Conf. on Coastal Eng., p. 405-418.
Korteweg, D. J., and de Vries, G., 1895, On the change of form of long waves advancing in a rectangular canal, and on a new type of long stationary waves: Phil. Mag., Ser. 5, v. 39, p. 422-443.
Kuelegan, G. H., 1948, An experimental study of submarine sand bars: U.S. Army Corps of Engineers, Beach Erosion Board Tech. Rep. 3, 40 p.
——, and Patterson, G. W., 1940, Mathematical theory of irrotational translation waves: Natl. Bur. Standards Jour. Res., v. 24, p. 47-101.
Laitone, E. V., 1960, The second approximation to cnoidal and solitary waves: Jour. Fluid Mechanics, v. 9, p. 430-444.
Longuet-Higgins, M. S., 1972, Recent progress in the study of long-shore currents: *In* R. E. Meyer (ed.), Waves on beaches: Academic Press, New York, p. 203-248.
Masch, F. D., and Wiegel, R. L., 1961, Cnoidal waves: Tables of functions: Eng. Found. Council on Wave Res., Berkeley, California, 129 p.
Meyer, R. E., and Taylor, A. D., 1972, Run-up on beaches: *In* R. E. Meyer (ed.), Waves on beaches: Academic Press, New York, p. 357-412.
Miller, R. L., 1972a, The role of surface tension in breaking waves: Am. Soc. Civil Engineers, Proc. 13th Conf. on Coastal Eng., p. 433-449.
——, 1972b, Study of air entrainment in breaking waves: Am. Geophys. Union Trans., EOS, v. 53, p. 426.
——, Leverette, S., O'Sullivan, J., Tochko, J, and Theriault, K., 1974, Field measurements of impact pressures in surf: Am. Soc. Civil Engineers, Proc. 14th Conf on Coastal Eng., p. 1761-1777.
—— and Zieglier, J. M., 1964, The internal velocity field in breaking waves: Am. Soc. Civil Engineers, Proc. 9th Conf. on Coastal Eng., p. 103-122.
Nelson, C. L., and Miller, R. L., 1975, The interaction of fluid and sediment on the foreshore: Univ. Chicago, Dep. Geophys. Sci., Fluid Dynamics and Sed. Trans. Lab. Tech. Rep. 15, 176 p.
Patrick, D. A., and Wiegel, R. L., 1955, Amphibian tractors in the surf: Am. Soc. Nav. Architechts and Marine Engineers, Proc. 1st Conf. on Waves, p. 397-422.
Prins, J. E., 1958, Characteristics of waves generated by a local disturbance: Am. Geophys. Union Trans., v. 39, p. 865-874.
Russell, R. C. H., and Osirio, J. D. C., 1958, An experimental investigation of drift profiles in a closed channel: Am. Soc. Civil Engineers, Proc. 6th Conf. on Coastal Eng., p. 171-193.
Simons, D. B., Richardson, E. V., and Albertson, M. L., 1961, Flume studies using medium sand: U.S. Geol. Survey Prof. Paper 1498-A, 76 p.
Thornton, E. B., and Krapohl, R. F., 1974, Water particle velocities measured under ocean waves: Jour. Geophys. Res., v. 79, p. 847-852.
Weggel, J. R., 1972, Maximum breaker height for design: Am. Soc. Civil Engineers, Proc. 13th Conf. on Coastal Eng., p. 419-432.
Wiegel, R. L., 1964, Oceanographical engineering: Prentice-Hall, Inc., Englewood Cliffs, New Jersey, 532 p.
—— and Beebe, K. E., 1956, The design wave in shallow water: Am. Soc. Civil Engineers Proc., Jour. Waterways Div., v. 82, No. WW1, Paper 910, 21 p.
—— and Skjei, R. E., 1958, Breaking wave force prediction: Am. Soc. Civil Engineers Proc., Jour. Waterways and Harbors Div., v. 84, No. WW2, Paper 1573, 14 p.
Wood, W. L., Jr., 1973, A wave and current investigation in the nearshore zone: Purdue Univ., Dep. Geol. Sci. Final Rep., O.N.R., NONR-N00014-68-A-0109-0002, 148 p.

SWASH—GROUNDWATER—BEACH PROFILE INTERACTIONS

EVANS WADDELL
Purdue University, West Lafayette, Indiana 47906

ABSTRACT

Because features of an uninterrupted swash appear primarily dependent on only breaker height and beach slope, duration of an undisturbed swash can and often will exceed breaker period. When this situation occurs, collisions between successive swashes expend energy on the lower beach face thereby curtailing or inhibiting some uprushes. Magnitude of wave energy reflection and hence generation of leaky mode, nearshore standing waves appears in part to be conditional on the intensity of swash interaction. As a result, collisions between swashes can have a significant influence on nearshore and beach processes.

Swash and standing waves cause corresponding oscillations of the beach water table in the vicinity of the beach face by transmission mass flux through the saturated portion of the beach. As a result of low pass filter characteristics of the beach matrix, groundwater movement induced by lower frequency standing waves is the more significant of these two discrete frequency forcing functions. It is possible for these water table fluctuations to periodically alter the upper beach face environment so that sequential deposition and erosion result.

High frequency oscillations of beach elevation (on the order of 40 seconds or greater) have been measured on the upper and mid region of the beach face. These regular changes in sand level are produced by action of multiple swashes and appear to be the result of sequential deposition of material moved upslope as suspended load and subsequent erosion as this material moves downslope as bedload.

INTRODUCTION

Studies of the subaerial beach have had a pattern of examining sequentially higher frequency phenomena. Each level of temporal resolution (e.g. annual response, seasonal response, . . . , tidal cycle response) has supplied some further identification of relationships pertinent to the system. These studies, however, have not generally supplied information regarding the fundamental mechanics which govern this particular environment. Consequently, at times it may be possible to anticipate net response of the system (Harrison, 1969), but it has not been possible to state specifically why this response occurred. Fortunately, with the recent use of high speed photography (Miller, 1968; Nelson and Miller, 1974) and continuously recording electronic sensors (Machemehl and Herbich, 1970; Waddell, 1973), it is now possible to isolate some salient and fundamental characteristics and interactions of this dynamic environment.

One prior study which showed considerable insight into this environment was by Emery and Gale (1951) who made use of visual and photographic observations to infer qualitative interrelationships between incident waves, swash, and ground water. They described (1) small changes in beach water table induced by individual swashes, and (2) evolution of across beach swash profiles during the uprush-backwash cycle. Emery and Gale also noted that swash moved up the beach with a small, steep-faced front, and, that during backwash, it retreated in such a way that the swash mass thinned without significant changes in location of the leading edge until most of the water had moved down the beach face. In describing the combined shorebreak-uprush sequence Emery and Gale observed that smaller, high frequency breakers disappeared prior to moving up the beach face as swash. Because of this, they suggested that the beach acts as a filter which permits passage of only larger or longer waves. They also observed that the period of swash was always larger than that of input waves just seaward of the shorebreak. It was also noted that with gentler beach slopes there was a greater difference between periods of swashes and input waves.

Specific attempts have been made to establish the influence of groundwater on beach changes. Emery and Foster (1948) found that during ebbing tide, the water table near the beach face sloped seaward while during flooding tide the water table sloped landward. They indicate that response of the beach groundwater to tide was inversely proportional to the distance from the beach face. They reasoned that "vertical permeability (of the beach) must be lower than horizontal permeability because beaches contain thin alternating coarser and finer layers" (Emery and Foster, 1948, p. 648). Horizontal mass flux was associated with movement of the tidal wave through the beach matrix behind the swash slope.

Grant (1948) recognized that height of the groundwater table was related to whether the beach was prograding or eroding. The model which Grant presents for this relationship indicates that a dry beach facilitates deposition on the foreshore by reducing backwash flow velocity and thus prolonging existence of laminar flow. On a saturated beach, backwash is supplemented by outflow through the zone of effluent which makes backwash turbulent earlier in the cycle. Outflow also dilates sand grains in this region, which further encourages erosion.

Longuet-Higgins and Parkin (1962, p. 196) found that when an impermeable roofing-felt was inserted in the beach about 3 inches below the sand surface, "the waves quickly eroded the shingle overlying the roofing-felt but disturbed only to a lesser extent the shingle on either side." These investigators attributed these results "to the fact that over the roofing-felt, the backwash could not penetrate to a depth of more than three inches, and so, the backwash there was relatively undiminished."

Duncan (1964) investigated combined influences of semidiurnal tidal and ground water fluctuations on deposition and erosion in the swash zone. During flooding tide, sand that was deposited on the beach slope above the water table was taken from the region below the water table. This deposition was associated with loss of swash mass due to infiltration and consequent decrease in seaward acting momentum. As infiltration continued, the groundwater table elevated, thus causing the lower boundary of the zone of infiltration to migrate landward. There was also evidence of an associated outflow lower on the beach face. This outflow dilated sand grains to encourage further erosion on that portion of the slope. On the other hand, during falling tide, part of the zone of infiltration was removed from swash influence, and consequently little mass was lost to infiltration. Backwash was, in fact, enhanced by outflow of ground water causing a pad of material to move downslope until the profile approximated the initial low tide configuration.

Recently, Pollack and Hummon (1971) identified four beach zones based on both degree of saturation and on spatial and temporal changes of water content. The zones in order of occurrence down the beach face are (1) zone of dry sand, (2) zone of retention, (3) zone of resurgence, and (4) zone of saturation.

The zone of dry sand was influenced by swash action only during occasional tidal cycles. The zone of retention was influenced by swash for some period during all tides; therefore, sand in this region was never completely dry. During time of exposure, water content was controlled by gravity. This zone was porous and experienced a great deal of inflow and outflow. The zone of resurgence was characterized by intensive mass flux throughout the tidal cycle. This zone expands and contracts depending on the amount of circulation and the discrepancy between nearshore water level and the water table of the beach. The zone of saturation remained saturated under all conditions of tide and swash.

Common to these studies of beach ground water are two critical shortcomings: (1) the data were collected neither at sufficiently high resolution nor at sufficiently high frequency to allow for an insight into the actual physical mechanisms and (2) because the pertinent first-order parameters were not measured simultaneously, the real-time interaction between input waves, swash, groundwater, and beach changes could not be established. Furthermore, results from these studies were mostly qualitative.

SWASH AND INPUT WAVES

One of the fundamental interrelations of a beach system occurs between swash and groundwater; however, to understand this coupling, it is necessary to understand and characterize the natural swash process.

A normal isolated shorebreaking wave can reasonably be approximated by a bore configuration (LeMehaute, 1965). As a consequence of this similarity, work of Shen and Meyer (1962) provides a useful description of the mechanism of wave break at a shoreline, followed by the uprush and backwash of swash. Their analytical treatment, which evaluates the nonlinear long wave equations governed by conditions appropriate to a beach environment, produces predictions of some characteristics of an undisturbed swash mass as it moves up the active beach face.

Immediately prior to shorebreak, i.e. which is likened to initiation of bore collapse, the front of the wave is vertical. Following this initial condition there is a sequential disintegration of the bore configuration as the leading edge of swash moves up the beach face (Fig. 1). As the front edge moves upslope, the angle between the swash surface and the sand surface decreases. This has been described as the progressive thinning of the swash. Eventually, the leading edge reaches a maximum run-up and begins backwash. During

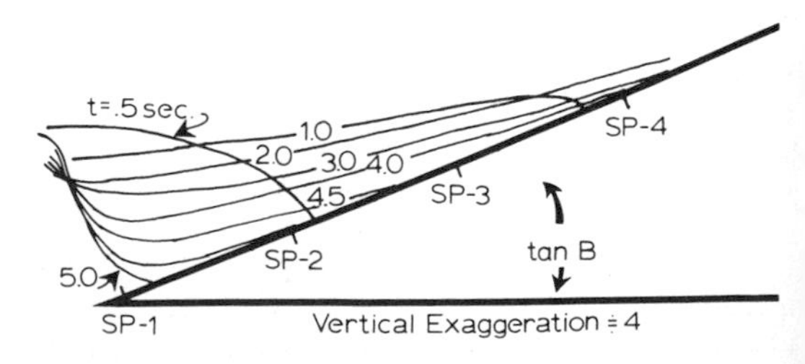

FIG. 1.—Swash geometry. Initial steep faced front (bore configuration) deteriorates by progressive thinning throughout the entire swash cycle. As a result, uprush and backwash have distinctly different flows.

this backswash phase, a swash mass continues to thin as the leading edge of the water mass moves down the beach face. It is important to realize that backwash flow is definitely not the reverse of uprush. There are distinctly different internal velocity fields during these two different phases of a swash cycle.

Observed and predicted histories of swash depth at several across beach locations are given in Figures 2 and 3. There is fair qualitative agreement between those predicted by Shen and Meyers (1962) as shown in Figure 2 and those observed by Waddell (1973) and shown in Figure 3. In Figure 3 the secondary depth maximum at approximately $T = 3$ secs, could be a secondary "retrogressive" bore as predicted by Shen and Meyer. This bore originates at midslope and propagates downslope while facing upslope.

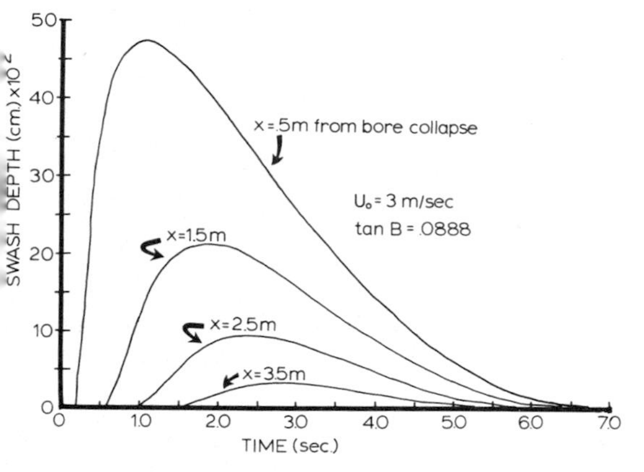

FIG. 2.—Predicted history of undisturbed swash depth. An initial rapid rise in water depth is followed by a more gradual decrease. Note that duration of inundation and maximum and average water depth decrease upslope.

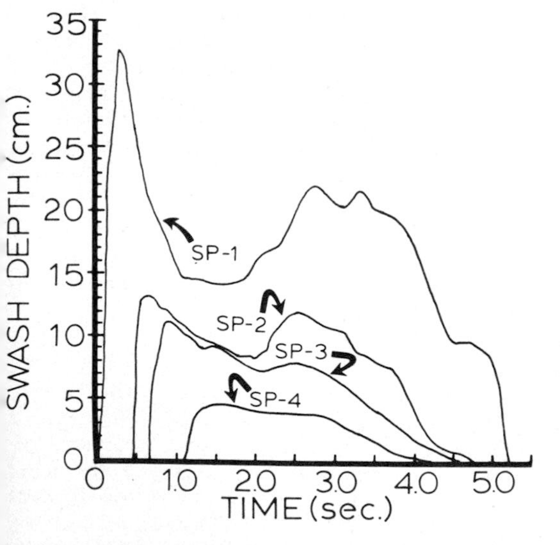

FIG. 3.—Observed history of swash depth. Qualitiative similarities with prediction are good. The secondary depth maximum following the initial rise to maximum water depth may be associated with a retrogressive bore. It is initiated at midslope and sequential moves downslope.

As given by Shen and Meyer (1962) and supported by my unpublished laboratory data, the path of the leading edge of an undisturbed swash is parabolic and approximated by

$$Xs(t) = \mu_0 t - (1/2g \tan\beta)\, t^2 \qquad (1)$$

where $Xs(t)$ = location of leading edge relative to initial shoreline, $x = 0$; μ_0 = initial horizontal velocity of leading edge; g = acceleration due to gravity, $\tan\beta$ = slope of the swash slope, t = time. Note that the path of the leading edge is uniquely and completely determined by the initial energy conditions and the beach face slope. It should be noted that this development does not consider frictional effects and in many situations friction or apparent friction can be significant (Miller, 1968; Meyer, 1970).

As a result of the above equation (1) and an approximation of μ_0 in terms of breaker height, it is possible to determine swash duration, Ts, as a function of breaker height, Hb, for various beach slopes (Fig. 4),

$$Ts = \frac{6\, Hb^{1/2}}{g^{1/2} \tan\beta.} \qquad (2)$$

In the evaluation of the governing equations which result in equation (2), it was assumed that only one swash mass acted on the beach face at any given time. Thus Figure 4 and equation (2) are only valid when collisions between successive swashes are not occurring, i.e., when the period of incoming waves is less than the duration of swash. Examination of Figure 4 indicates that for the range of beach slopes normally found, swash duration exceeds input wave period, and hence, interaction between swashes occurs. This is true even for many laboratory generated (monochromatic) wave fields. Consequently, any examination of processes or mechanisms active on a subaerial beach should incorporate a consideration of collisions between swashes (Kemp, 1960; Kemp and Plinston, 1968).

For a natural wave field which has a joint distribution of wave height and period, one of the most obvious consequences of interaction between swashes is that the upslope movement of selected swashes is curtailed or completely inhibited. This results in a distinctly higher energy

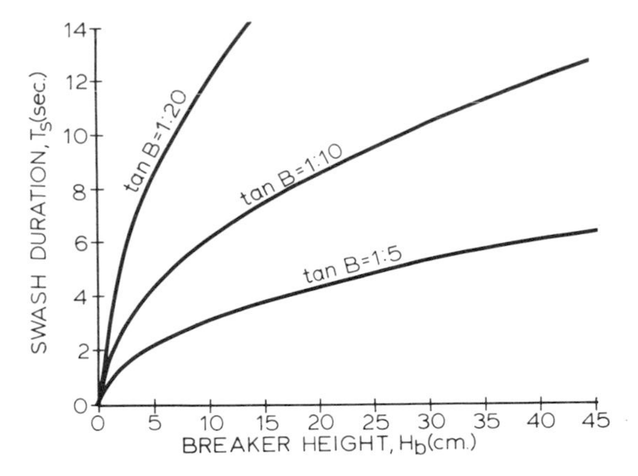

FIG. 4.—Graph of Equation 2 for various beach slopes. Examination indicates that only waves having long periods combined with low amplitudes can avoid producing collisions between swash masses.

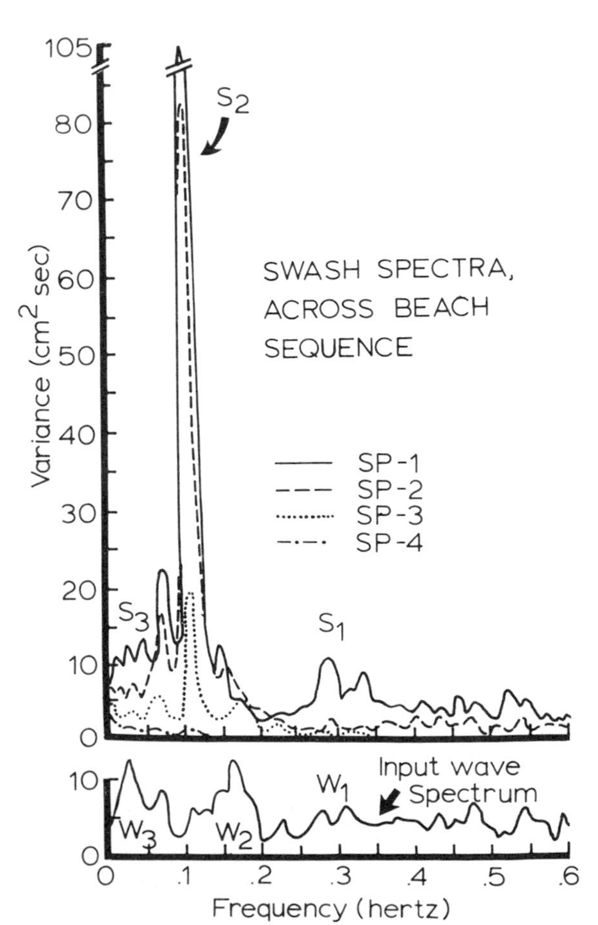

FIG. 5.—Variance spectra for input waves and associated swash. A low frequency shift of variance occurs across the shorebreak as the result of collisions between swashes.

environment at the base of the slope in addition to a distinct low frequency shift between average input wave frequency and the frequency associated with the average period between successive swashes. In terms of a systems function acting on a random input, shorebreak and interaction do not produce a high frequency attenuation function (linear lowpass filter) as suggested by Emery and Gale (1951) but, in fact, produce a nonlinear transfer of energy to lower frequencies. This becomes more obvious when examining the variance spectra for input waves and swash (Fig. 5). There are virtually no components of the incoming wave field having periods corresponding to that of the period between successive swashes.

Irrebarren and Nogales (1949) determined that significant incoming wave energy was reflected from a shoreline where the beach slope was greater than

$$\tan\beta = \frac{5.65\, Hb^{1/2}}{g^{1/2}\, T_{\text{input}}} \tag{3}$$

If the beach slope is less than that given by equation (3) a much reduced reflection of wave energy occurs. Because imperfect reflection of input wave energy can produce partial standing waves in the nearshore (Ippen, 1966), this critical beach slope, in fact, segregates those conditions producing significant nearshore standing waves from those which produce insignificant standing waves.

It was suggested by Irrebarren and Nogales (1949) that separation of these different reflection environments results from variation in the intensity of shorebreak. Similarities between equations (2) and (3) suggest that significant shorebreak coincides with initiation of interaction. Apparently, occurrence and intensity of swash-to-swash interaction can influence the generation of nearshore standing waves (Waddell, 1973). As will be shown, these nearshore standing waves can have a significant effect on the water table in the beach deposit.

GROUNDWATER

Fluctuations of the water table in the vicinity of the beach face have been identified over a wide range of frequencies. Tidal frequency response of groundwater has received the most intense discussion. Recently, several investigators (Dominck and others, 1971; Harrison and others, 1971) have developed one-dimensional computer simulations of the response of the water table in a beach matrix. These have been checked with field data and shown to be rather accurate. Waddell (1973) used continuously recording electronic sensors placed in a series of wells adjacent to

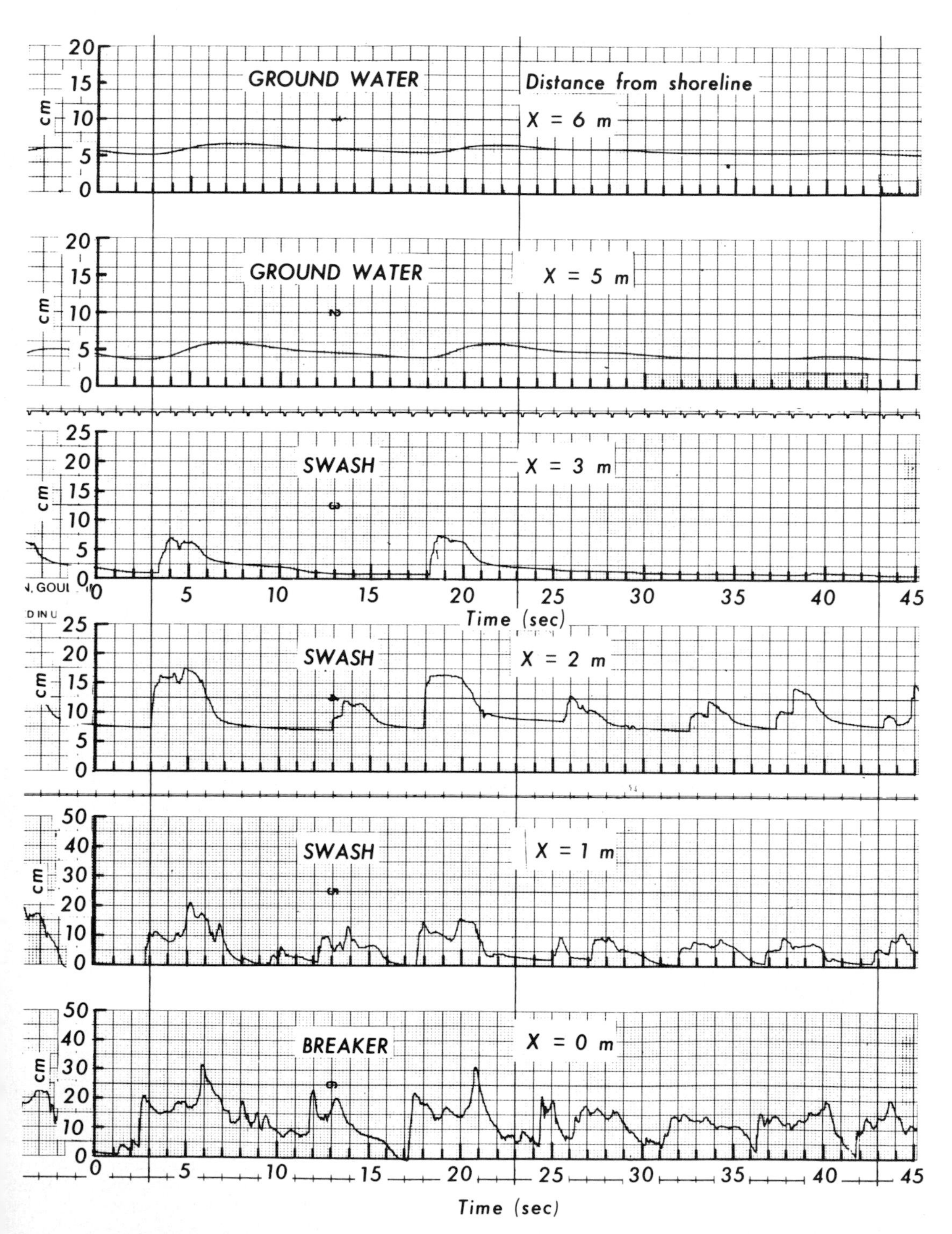

FIG. 6.—Examples of time series of water levels of breakers, swash and beach watertable. These data show groundwater responding to pressure from a swash while it is still at the base of the beachface.

the beach face in order to measure high frequency fluctuations of groundwater. These measurements support many of the statements of Emery and Gale (1951) regarding response of the water table to excitation by individual swashes.

Evidence of the response of a beach water table to individual swashes is seen in Figure 6 which shows simultaneous measurements of groundwater and swash depth. For these records, the water table intersects the beach face between swash probes #3 and #4 ($X = 2M$ and $X = 3M$) which were located two and three meters respectively upslope of the break point. At (Fig. 6) $t = 2.5$ sec, the leading edge of swash is at the base of the swash slope. At the same time, the water table in groundwater well #1 (*GW* #1) reaches a minimum and begins to increase. This increase continues until the swash is in the backwash phase.

While the leading edge of swash is downslope of the line of water table-beach-slope intersection, any increase in water table must be produced by mass flux through the saturated lower beach. After the leading edge reaches the water table, it is no longer possible to partition between water influx through the saturated beach face and infiltration into the nonsaturated beach face.

Variance spectra resulting from analysis of simultaneous time-series of nearshore water level, swash depth, and water table elevation provide further evidence of mass flux through the saturated beach surface (Fig. 7). There are two primary peaks in the input wave spectra: *W*2 which was associated with input swell, and *W*3 which was produced by a standing wave in the nearshore. The swash spectrum exhibits only one significant peak, *S*2, which is associated with the period between successive swashes. Note that the input wave spectrum has virtually no variance in the well defined frequency band of the swash peak. Also, the swash spectrum, obtained from measurements taken half way up the beach face (2 meters upslope from break point), contains virtually no variance at the frequencies which contain the nearshore standing waves. This suggests that these standing waves did not cause corresponding oscillations in the location of the swash zone. Instead, the influence of these standing waves was limited to the very lower portion of the beach face. If the influence of these longer period waves was limited to the base of the beach face yet was still able to induce water table oscillations, then it is apparent that this coupling and associated mass flux was transmitted through the saturated portion of the beach matrix.

Transmission of pressure through the beach matrix can be influenced by the sediment textures and fabric of the beach deposit. If there is a well defined coarse step deposit, then pressure could be transmitted more readily than in a homogeneous fine textured deposit. While this dependence on structure and texture can be important, the pattern of groundwater response is not totally dependent on it because such sequences of interaction between swash and groundwater have been observed in beaches having no apparent vertical structure and beaches having no well defined aquifers.

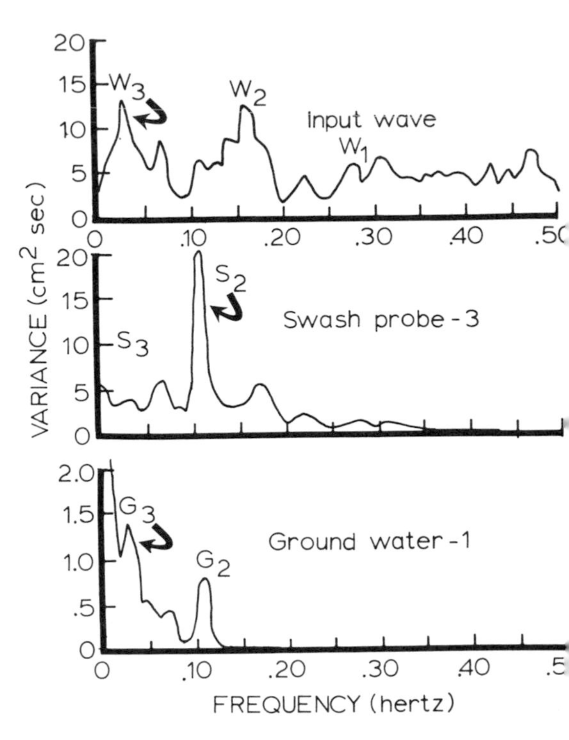

FIG. 7.—Spectra of input waves, swash and groundwater. Two peaks in groundwater spectrum correspond exactly to peak swash frequency ($f \cong 1$ hertz) and nearshore standing wave frequency ($f \cong 0.025$ hertz).

Examination of a transfer function between input wave spectrum and groundwater level spectrum shows that the beach matrix acts as a low pass filter (Fig. 8). In this particular example, despite input swell and standing waves having similar root mean square heights, influence of swell was completely attenuated at a location just landward of the swash slope (5 meters inland of the breakpoint) while standing waves could still produce significant water table fluctuations. Because groundwater oscillations attenuate with distance, the location of measurement of groundwater can influence the specific filter applied to the forcing function; however, all such frequency response functions should have general characteristics of a low pass filter which tends to accentuate longer period fluctuations.

In previously mentioned numerical modeling of

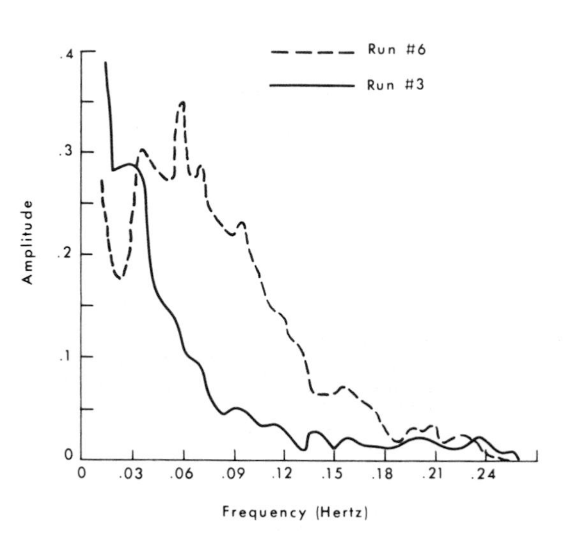

FIG. 8.—Representation of low pass filter characteristics of the beach matrix. During Run #6, the break point had migrated slightly further landward, thus the groundwater well was in closer proximity to the generating forces.

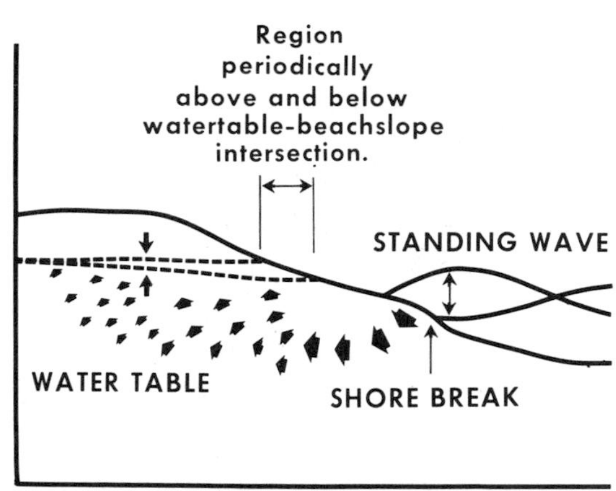

FIG. 9.—That portion of the upper half of a beach face which can be sequentially nonsaturated and saturated is illustrated.

tide induced water table movement, some discrepancies existed between predicted and observed elevations. These simulations however, considered only tidal period forcing functions, while field measurements, against which the models were evaluated, were taken at 10 or 15 minute intervals. As a result, observed elevations would include oscillations at a frequency not considered in the numerical model. Consequently, what appear as errors in prediction may in fact be the result of lack of resolution in the frequency domain. Specific differences which support this likelihood are that discrepancies between prediction and observation decrease away from the beach face, and that at each location the magnitude of this difference is of the same order as could be induced by nearshore standing waves.

All changes in groundwater level can cause a corresponding shift in the location of the line of intersection between the water table and swash slope. As the water table lowers, this line of intersection moves seaward and when it rises, the line of intersection moves landward. As a result of these water table fluctuations, there is a zone on the beach face which is sequentially saturated and unsaturated (Fig. 9).

A portion of a swash mass, when acting on a nonsaturated beach surface, infiltrates into the sand deposit which decreases the energy of the swash (Duncan, 1964; Nelson and Miller, 1974). This mass loss does not have to be large in order to affect the capacity for sediment transport by swash. During the entire cycle of swash, water depth normally decreases following the maximum (Fig. 3). Thus, near the location of maximum run-up, the swash depth is relatively shallow. Consequently, even a small loss of water due to infiltration can significantly decrease potential energy which becomes available to transport sediment downslope during backwash. Nelson and Miller (1974) found that losses due to infiltration become more critical as the waves become smaller or the beach slope lower.

A reduced mass of swash during backwash would tend to encourage deposition on the nonsaturated portion of the beach. If the beach face is saturated however, then mass loss would not be as great and the tendency for deposition would be reduced. Because of fluctuations of the water table, there can be a portion of the beach face which is alternately saturated and unsaturated which in turn, would suggest that this portion of the beach could alternately experience conditions encouraging deposition and discouraging deposition. If this is true, then measurement of sand levels on the swash slope should reveal a sequential rise and fall in elevations of the sand surface at frequencies similar to those for the water table fluctuations.

BEACH ELEVATIONS

Figure 10 shows a 6-hour record of sand level elevations which occurred at two locations, one meter apart, on the upper half of the swash slope. These records were made by electronically measuring exposed sand surfaces between successive swashes (Waddell, 1973). Obvious fluctuations of

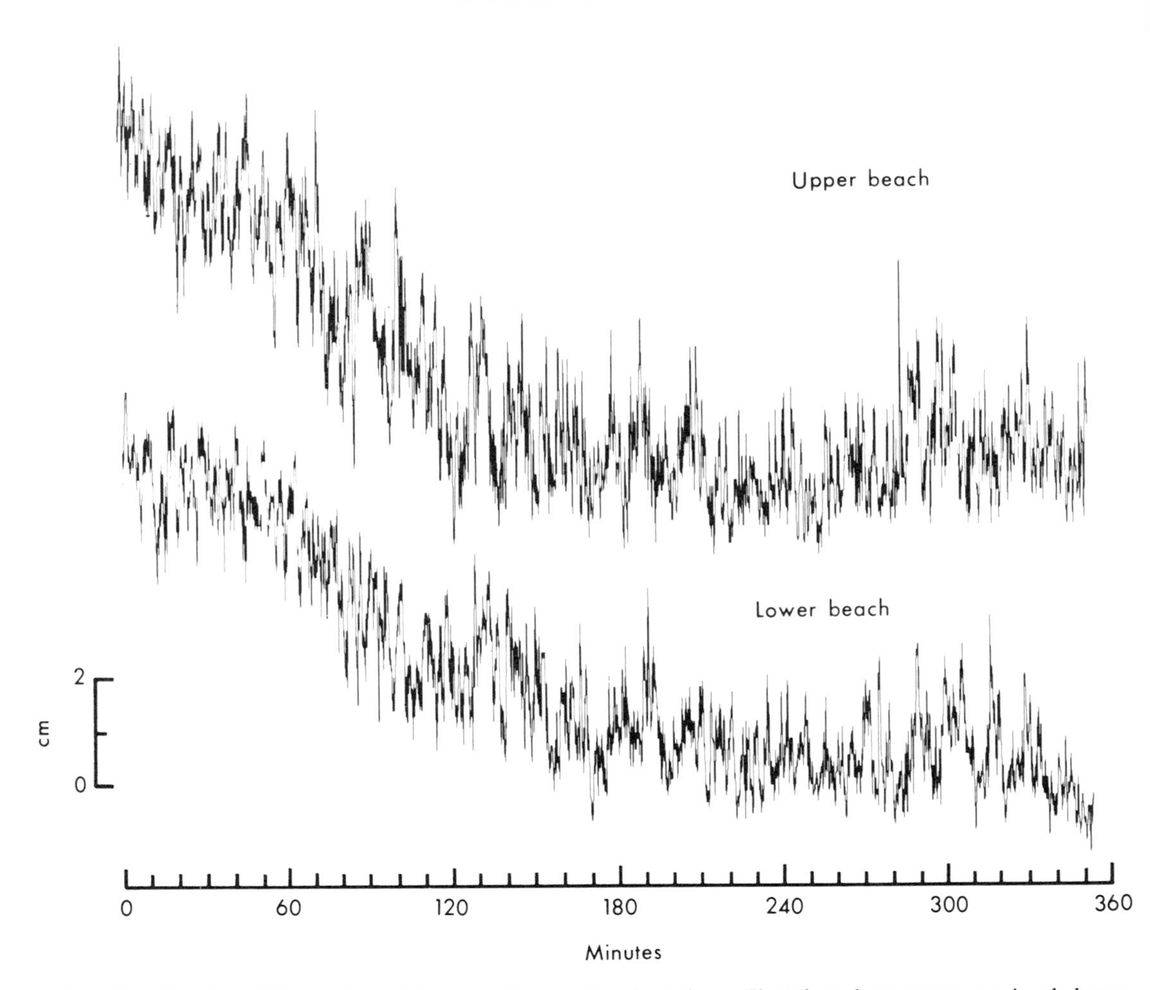

FIG. 10.—Sand level fluctuations at lower and upper beach stations. First three hours were erosional; latter three hours were approximately stable.

sand level occurred even during the 3-hour period of net erosion.

Visual examination of original time-series data indicates that the fluctuations in sand level were the result of multiple swashes causing deposition and subsequent erosion. Generally, these changes were composed of a gradual and sequential increase in sand level followed by a gradual decrease. The deposition and subsequent erosion was rather symmetrical about the time of maximum elevation. This is not to suggest that large changes can not be produced by an individual swash mass. On occasion, single swash elevation changes as large as 5 cm occurred. When such a large discontinuous change occurred, it was usually erosional.

Spectral analysis of these time-series indicates that virtually all sand level variation occurred at periods greater than 40 seconds, and thus, there was no evidence of a significant contribution of variance at the swash frequency (Fig. 11). Cross spectral analysis between the upper and lower beach stations indicates that at all frequencies beach elevation changes on the lower beach consistantly lagged behind elevation changes on the upper beach.

An explanation for these results was suggested by Waddell (1973) and subsequently supported, in part, by the laboratory work of Nelson and Miller (1974). Sand moves upslope as suspended load which Nelson and Miller found to be heavily concentrated in the highly turbulent leading portion of the uprush. Due to loss of capacity as a result of infiltration, a portion of this material remains on the unsaturated upper beach. The net accumulation seen in Figure 9 can be the result of repeated swash action. As groundwater level rises, a previously unsaturated portion of the slope

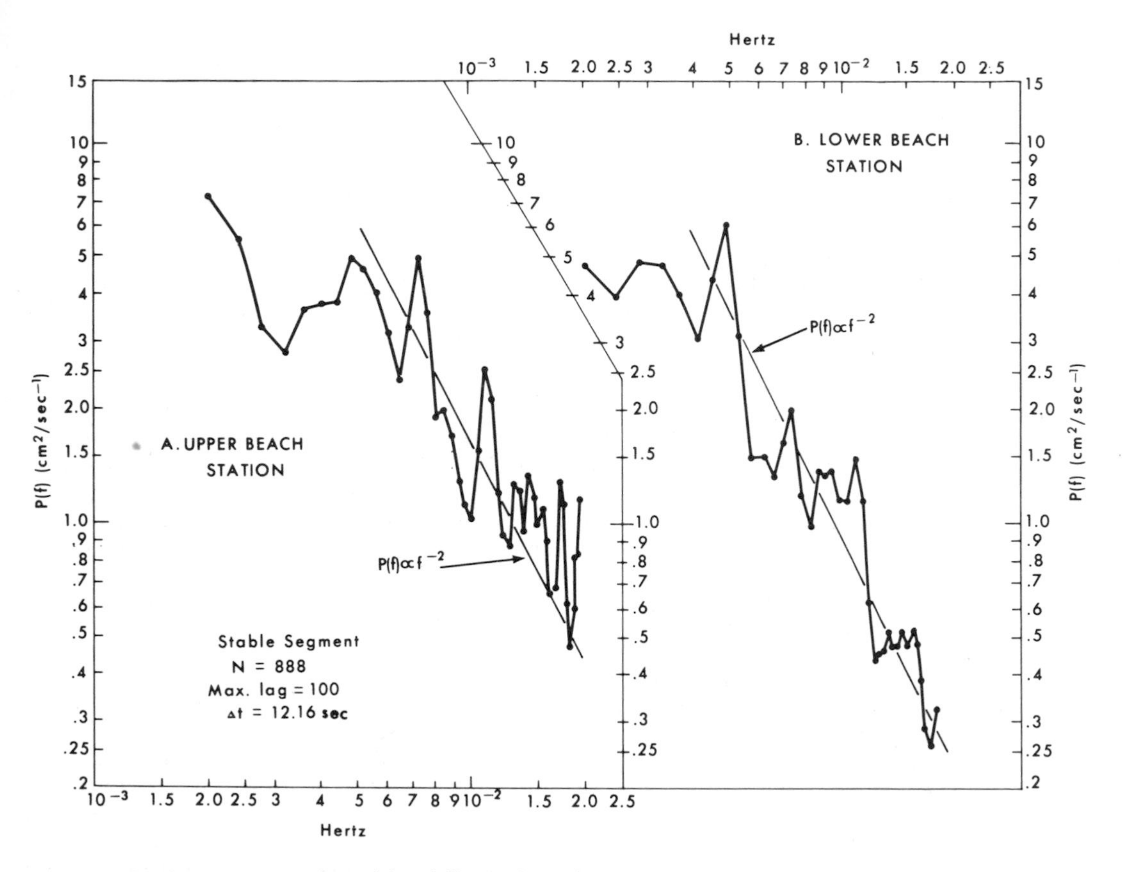

FIG. 11.—Variance spectra of sand level fluctuations during non-erosional segments. Straight line ($P \propto f^{-2}$) is the predicted slope of the spectrum using the development of Hino (1968) for sandwaves in unidirectional flow.

becomes saturated resulting in reduced local water loss, and hence, increased sediment transporting capacity. The undirectural backwash sequentially removes sediment and moves it downslope as bedload.

Another possible explanation for this sequential deposition and erosion is suggested by some of the conclusions of Nelson and Miller (1974). They suggest that low energy swashes tend to deposit while higher energy swash tends to have no net deposition. When this relationship is viewed in the context of swash interaction, it can readily be seen that for the upper beach face, swashes having intense interaction will generally be of lower energy while noninteracting swashes will be more likely to have characteristics of higher energy. As a result of this dependence, the pattern of swash-to-swash interaction can possibly induce depositional and erosional environments on the upper portion of the swash slope. It is possible to incorporate groundwater oscillations into this model because interaction inhibits energy reflection which does not encourage generation of standing waves and hence would encourage a lower water table. Minimal interaction encourages energy reflection which can create standing waves and hence tend to cause the water table to oscillate.

Obviously, any study using detailed measurement of sand levels on the active beach face as a parameter must consider these higher frequency variations. Survey accuracy of 0.5 cm is of limited use if measurements are taken at 10 minute intervals during which time elevation may vary 3 cm or more.

SUMMARY

Identifying some of the fundamental processes which are active on the subaerial beach requires laboratory and field measurements which permit resolution of high frequency interactions. Until recently such measurements were not available; however, several studies have identified some of the salient characteristics of swash and related phenomena (Shen and Meyer, 1962; Miller, 1968; Waddell, 1973; Nelson and Miller, 1974).

The fundamental mode of swash, i.e. an undisturbed swash cycle, is not the expected condition on most laboratory and natural beaches. Because of the dependence of swash duration on breaker height, collisions between swashes are to be expected. Evidence suggests that generation of standing waves in the nearshore may be intimately linked to the occurrence and intensity of interaction because increasing interaction decreases energy reflection.

These standing waves generate significant variations of the groundwater level in the vicinity of the active beach face. Although individual swashes do cause water table fluctuations, dampening of these impulses by the beach matrix limits higher frequency fluctuations to the immediate vicinity of the swash slope. Because of the low pass filter characteristics of the beach deposit, the water table variations induced by standing waves are of larger magnitude and are transmitted further into the beach.

Rapid changes of beach elevations on the upper half of the beach face have been measured. These variations occur such that a group of swashes produce a regular and sequential pattern of deposition and subsequent erosion. This pattern is produced by the collected influence of many swashes and is generally a regular pattern with large discontinuous changes in elevation being the exception.

To explain these regular elevation changes, a model which is consistent with other experimental data is suggested. Oscillations in groundwater level create a zone which is periodically saturated or nonsaturated. When nonsaturated, infiltration occurs which tends to encourage local deposition. When saturated, infiltration does not occur, deposition is not encouraged, and previously deposited material can be eroded. Material which is deposited is moved upslope as suspended load. Material which is subsequently eroded is moved downslope as bed load. In this arrangement, deposition and erosion would be the result of action by several swashes. It is possible that this downslope movement occurs as low amplitude sand waves which move successively further downslope with each eroding swash.

With the development and adoption of accurate and continuously recording instruments it is now becoming possible to realistically attempt to identify those mechanisms which result in observed beach morphologies. With this ability it may now be possible to learn in detail not only what is occurring, but, why many things occur which have long been recognized as characteristic of a beach. Only with this ability will beach systems truly be understood.

ACKNOWLEDGEMENTS

Much of the field work which forms the foundation for this paper was supported by the Geography Programs, Office of Naval Research, under contract N 00014-69-0211-0003, Project NR 388-022 with the Coastal Studies Institute, Louisiana State University.

REFERENCES

Dominick, T. F., Wilkins, Bert, Jr., Roberts, H. H., and Ho, C. L., 1971, A study of beach groundwater hydrology and chemistry: Louisiana State Univ., Coastal Stud. Inst. Tech. Rep. 152, 96 p.

Duncan, J. R., 1964, The effects of water table and tide cycle on swash-backwash, sediment distribution, and beach profile development: Marine Geol., v. 2, p. 186-197.

Emery, K. O., and Foster, J. F., 1948, Water tables in marine beaches: Jour. Marine Res., v. 7, p. 644.

—— and Gale, J. F., 1951, Swash and swash marks: Am. Geophys. Union Trans., v. 32, p. 31-36.

Grant, U. S., 1948, Influence of watertable on beach aggradation and degradation, Jour. Marine Res., v. 7, p. 655-660.

Harrison, Wyman, 1969, Empirical equations for foreshore changes over a tidal cycle: Marine Geol., v. 7, p. 529-551.

——, Booj, J. D., Fang, C. S., Fausak, L. E., and Wang, S. N., 1971, Investigation of the watertable in a tidal beach: Virginia Inst. Marine Sci., Spec. Sci. Rep. 60, 168 p.

Hino, M., 1968, Equilibrium range spectra of sandwaves formed by flowing water: Jour. Fluid Mechanics, v. 134, p. 565-573.

Ippen, A. T., 1966, Estuary and coastline hydrodynamics: McGraw-Hill Book Co., New York, 744 p.

Irrebarren, M. R., and Nogales, M. C., 1949, Protection of ports: section II: Internat. Assoc. Navigation, 17th Congr., Commun. 4, Ocean Navigation, p. 31-82.

Kemp, P. H., 1960, Relationships between wave action and beach profile characteristics: Am. Soc. Civil Engineers, Proc. 7th Conf. on Coastal Eng., v. 1, p. 262.

—— and Plinston, D. T., 1968, Beaches produced by waves of low phase difference: Am. Soc. Civil Engineers Proc., Jour. Hydraulics Div., v. 94, p. 1183-1194.

Lemehaute, Bernard, 1965, On non-saturated breakers and the wave run-up: Am. Soc. Civil Engineers, Proc. 8th Conf. on Coastal Eng., p. 77-92.

Longuet-Higgins, M. S., and Parkin, D. W., 1962, Sea waves and beach cusps: Geog. Jour., v. 128 p. 194-201.

Machemehl, J. L., and Herbich, J. B., 1970, Effects of slope roughness on wave run-up on composite slopes: Texas A & M Univ., Coastal and Ocean Eng. Rep. 129, 243 p.

Mayer, R. E., 1970, Notes on wave run-up: Jour. Geophys. Res., v. 75, p. 687–690.

Miller, R. L., 1968, Experimental determination of run-up of undular and fully developed bores: Jour. Geophys. Res., v. 73, p. 4497–4510.

Nelson, C. L., and Miller, R. L., 1974, The interaction of fluid and sediment on the foreshore: Univ. Chicago, Dep. Geophys. Sci., Fluid Dynamics and Sed. Trans. Lab. Tech. Rep. 15, 175 p.

Pollack, L. W., and Humman, W. D., 1971, Cyclic changes in interstitial water content, atmospheric exposure, and temperature in a marine beach: Limnology and Oceanography, v. 16, p. 522–535.

Shen, M. C., and Meyer, R. E., 1962, Climb of a bore on a beach, Part 3, Run-up: Jour. Fluid Mechanics, v. 16, p. 113–125.

Waddell, Evans, 1973, Dynamics of swash and its implications to beach response: Louisiana State Univ., Coastal Stud. Inst. Tech. Rep. 139, 49 p.

WAVE-FORMED SEDIMENTARY STRUCTURES—A CONCEPTUAL MODEL

H. EDWARD CLIFTON
U.S. Geological Survey, Menlo Park, California 94025

ABSTRACT

Sedimentary structures produced by shoaling waves can be integrated into a conceptual model based on four parameters: maximum bottom orbital velocity (u_m), velocity asymmetry (Δu_m), median grain size (D), and wave period (T). The model can be visualized as a series of block diagrams with axes of u_m, Δu_m and D, each at a constant value of T. The parameters can be related to physical aspects of the environment by applicable wave theories. The structural relations observed under unidirectional flow form one boundary to this 4-dimensional model under conditions of $\Delta u_m = u_m$ and $T = \infty$. The interrelations of wave-formed structures show increasing similarity to those of the flow regime concept as this boundary is approached.

The model comprises four main fields: (1) no sediment movement, (2) symmetric bedforms, (3) asymmetric bedforms and (4) flat bed (sheet flow). Threshold velocities required to initiate grain movement depend directly on both grain size and wave period (Komar and Miller, 1973). The critical velocity at which a sheet flow develops depends on grain size (Dingler, 1974). The transition between symmetric and asymmetric bedforms occurs at values of velocity asymmetry between 1 and 5 cm/sec.

Within the symmetric field, bedform size is gradational. Three kinds of ripples can be defined, depending on the relations of ripple wave length to D and u_m at a given wave period. "Orbital" ripples form under short-period waves; their wave length depends directly on the length of the orbital diameter of the oscillatory current. "Suborbital" ripples form under longer period waves; their wave length increases with increasing grain size but decreases with increasing orbital diameter. "Anorbital" ripples form where orbital diameter is very large; their wave length depends directly on grain size and is independent of orbital diameter. Each type of ripple can be recognized in ancient deposits by the ratio of the ripple wave length to the square root of the median grain size.

Asymmetric bedforms can presently be described only in qualitative terms. A consistently observed shoreward progression suggests that the following sequence occurs with increasing velocity and velocity asymmetry: long crested asymmetric ripples—irregular asymmetric ripples—cross ripples—lunate megaripples—flat bed. The abrupt increase in bedform size from ripples to lunate megaripples resembles the conversion of ripples to "dunes" in unidirectional flow. The size relations of asymmetric ripples seems not to follow those indicated for their symmetric counterparts.

Internal structures produced by shoaling waves include planar parallel lamination, medium-scale foreset bedding, and ripple foreset bedding. Planar parallel lamination can be produced by sheet flow over a flat bed or by migrating long-creasted ripples under conditions of slowly accumulating sediment. Medium-scale cross-bedding occurs in the medium to coarse sand fraction; it is not likely to develop in very fine sand as the result of oscillatory currents alone. Ripple foreset-bedding may be the most common structure in the upper part of the sediment column but has a relatively low preservation potential.

The conceptual model has its greatest application to geologic interpretation in a generalized form showing differences between waves of long, intermediate, and short period. Individual lines passing through the three-dimensional diagram toward increasing u_m, Δu_m, and (generally) D can be used to explain the shoreward progression of sedimentary structures on any particular coast. Because wave-generated unidirectional currents introduce considerable complexity landward from the breaker zone, geologic interpretation based on this model can be made most confidently seaward from the breakers.

INTRODUCTION

During the past decade, geologists have become increasingly aware of the advantages of applying hydrodynamic principles to the interpretation of sedimentary structures. The flow regime concept has received broad acceptance as the basis for interpreting the structures produced by unidirectional currents, because it provides a means of integrating a set of basic parameters and of identifying the stability fields of different structures in terms of the integrated parameters.

Despite the evident success of this approach, it has not been applied to a second major category of current structures, those produced by oscillatory currents under shoaling waves. This lack of application almost certainly relates to the greater complexity of wave-generated currents and the difficulty of reproducing these currents on a full scale in the laboratory. Oceanic waves create bottom currents that may flow with great intensity in one direction, then abruptly reverse themselves and flow with similar intensity in the opposite direction. Rapid accelerations and variable pressure fields add factors not often encountered in unidirectional flow. The bottom slope may be unrelated to the current. Finally, the shoaling waves may themselves locally generate unidirectional currents that complicate the pattern of their oscillatory currents.

Such complexity raises questions as to whether an integrating framework is possible for wave-generated sedimentary structures; questions that are reinforced by the disparity of description of the structures reported from different coasts. Nonetheless, detailed observations made over the past several years on the swell-swept coast of southern Oregon, on the relatively calm southeastern coast of Spain, and on protected sandy shores of Willapa Bay, Washington, indicate that the pattern of structural development within these diverse environments is surprisingly consistent. These observations, complemented by recent contributions by Komar and Miller (1973, 1974), Komar (1974a), and Dingler (1974), form the basis for the conceptual model for the origin and interrelation of wave-formed sedimentary structures that is presented here.

BASIS FOR THE MODEL

The sedimentary structures produced by waves result from the interaction of diverse variables. These include fluid factors, such as density and viscosity; flow factors, such as existing undirectional currents; wave factors, particularly the height, period and direction of approach of the waves and their variability; bottom configuration factors, such as water depth, local slope, and overall slope profile; and sediment factors that include grain diameter, sorting, density and grain shape. To reduce the number of independent variables, the basic variables could at this stage be combined to form nondimensional variables, or some variables could be assumed constant and the others left in dimensional form; the latter approach is followed here. Thus, it will be assumed initially that the fluid is of constant density (1.00 g/cm^3) and kinematic viscosity (0.01 cm^2/sec) and that no unidirectional currents exist. Further assumptions are: the waves are of uniform height and period, and their direction of approach is normal to a straight shoreline; local slope is low enough to be neglected; and the overall slope follows a simple, continuously shallowing profile. Further, the sediment is assumed to consist of well-sorted, equidimensional quartz grains, the density of which equals 2.65 g/cm^3.

Under these assumptions, the relation reduces to: bed configuration = f(wave height [H], wave period [T], water depth [h], and grain diameter [D]). Even as simplified, however, the interaction of H, T, and h is so complex that a causal relation to bedform development is difficult to define. It is therefore useful to employ integrating parameters that describe the actual oscillatory currents. The parameters considered here are the maximum velocity of the oscillatory current immediately above the boundary layer, the length of oscillatory orbit of water movement, and the velocity asymmetry of the oscillatory current. Other factors, such as acceleration, may be important to bedform development, but their effect is not sufficiently understood to warrant inclusion at this time. Although shear stress might provide a more generally valid parameter than the velocity measures, velocity is more easily grasped, can be measured more readily, and relates directly to wave and water depth combinations. Moreover, the velocity parameters probably provide an adequate approximation for a conceptual model.

As a wave shoals, it imparts to the water particles an elliptical motion which at the sea floor is essentially flat. These particles reach a maximum velocity (u_m) during passage of the wave crest (the direction of wave propagation is presumed to be positive, and u_m is therefore a positive vector quantity). Wave theory provides a basis for estimating this maximum velocity. For example, waves that are small in amplitude relative to wave length and water depth generate, by linear theory:

$$u_m = \frac{\pi H}{T \sinh kh} \tag{1}$$

where $k = 2\pi/L$ and L = wave length. Combinations of higher waves and/or shallower water depths may require the use of other wave theories. A summary of the methods for calculating u_m and the other parameters is provided in the Appendix.

The maximum velocity is related to the diameter of the orbital motion (d_o) by $u_m = \pi d_o/T$. Thus, values for any two of the variables determine the third. This relation provides an option of utilizing either d_o or T as a second parameter of the conceptual model. Although d_o is useful for developing parts of the model, the wave period is a more primary factor. It is more readily measured and has geologic significance in its relation to basin size and energy input.

An important element of any reversing flow, whether purely oscillatory (motion whereby fluid particles return to their original position) or partly translatory (motion resulting in some net translation of fluid particles) is the asymmetry of that flow. Asymmetry can be viewed in different ways. Probably the most significant aspect regarding bedform development is the difference in maximum velocity, in the two directions; almost certainly it is the most readily measured or calculated aspect. This difference in maximum velocity (Δu_m) is here defined as the sum of the maximum velocity under the wave crest in the direction of wave propagation (taken as a positive vector)

TABLE 1.——VELOCITY ASYMMETRY AND NATURE OF FLOW FOR VARIOUS TYPES AND COMBINATIONS OF OSCILLATORY AND TRANSLATORY FLOW

	OSCILLATORY FLOW $u_m^{\rightarrow} > 0,\ u_m^{\leftarrow} < 0$				PURELY TRANSLATORY FLOW $u_m^{\rightarrow} = u_m^{\leftarrow} = 0$, $\bar{u} > 0$
	DEGREE OF TRANSLATORY COMPONENT				
Nature of oscillatory flow	None $\bar{u} = 0$	Relatively small $\bar{u} + u_m^{\leftarrow} < 0$	Relatively large $\bar{u} + u_m^{\leftarrow} = 0$	Relatively large $\bar{u} + u_m^{\leftarrow} > 0$	
$u_m^{\rightarrow} + u_m^{\leftarrow} = 0$	symmetrical oscillatory flow $\Delta u_m = 0$	Asymmetric reversing translatory flow $\Delta u_m = 2\bar{u}$	Pulsing discontinuous unidirectional flow	Pulsing continuous unidirectional flow	Steady unidirectional flow
$u_m^{\rightarrow} + u_m^{\leftarrow} > 0$	asymmetrical oscillatory flow $\Delta u_m = u_m^{\rightarrow} + u_m^{\leftarrow}$	$\Delta' u_m = u_m^{\rightarrow} + u_m^{\leftarrow} + 2\bar{u}$			
$u_m^{\rightarrow} + u_m^{\leftarrow} = u_m^{\rightarrow}$	limit of purely oscillatory flow $\Delta u_m = u_m^{\rightarrow}$				

The possibility exists for an oscillatory flow with a maximum near-bottom orbital velocity ($u_m^{\rightarrow}$, in the text referred to as u_m) in the direction of wave propogation such that $u_m^{\rightarrow} > 0$, and a maximum near-bottom orbital velocity ($u_m^{\leftarrow}$) in the opposite direction such that $u_m^{\leftarrow} \leq 0$. The possibility exists for translatory flow with a velocity $\bar{u}$ in the direction of wave propagation such that $\bar{u} \geq 0$ (a flow in the opposite direction would produce similar relations but would reverse the sign of the vector). Velocity asymmetry refers to difference in maximum velocites in reversing flow and is defined by $\Delta u_m = (u_m^{\rightarrow} + \bar{u}) + (u_m^{\leftarrow} + \bar{u})$. Here Δu_m is shown only for reversing flow, although the concept can be applied to the various types of unidirectional flow as well.

plus the maximum velocity under the trough of the wave (which for reversing flow is a negative vector). This measure of velocity asymmetry applies to both totally oscillatory and partly translatory reversing flow (Table 1). Shoaling waves impart some net translation to the fluid particles, but seaward from the breaker zone this translatory motion contributes relatively little to the asymmetry of the flow. Therefore this paper is primarily concerned with asymmetry developed by purely oscillatory motion.

The asymmetry results from several factors. As a wave begins to shoal it becomes asymmetric about a horizontal plane; the crest elevates and steepens and the trough shallows and becomes flatter. Neglecting water mass transport, the volume of water carried forward under the crest of such a wave must equal the volume of water carried backward under the trough. Because the steepened crest passes a given point in a shorter time than the broader trough, the velocity of the oscillatory current under the crest of the wave will exceed that under the trough. Any net forward translation or mass transport of water as the wave passes must add to the velocity asymmetry. In very shallow water, as just seaward from the breaker zone, the landward slope of the wave becomes steep relative to the seaward slope, a deformation that further complicates the asymmetry of the orbital velocities (Adeyemo, 1970).

Velocity asymmetry (Δu_m) can be quantitatively estimated from the Stokes second-order wave equations whereby

$$\Delta u_m = \frac{14.8\, H^2}{LT \sinh^4 kh}. \quad (2)$$

This approximation loses its validity where h/L values decrease below about 0.1 (Adeyemo, 1970). An alternate approach (Dingler, 1974) is to attribute asymmetry in the turbulent boundary layer to water-mass transport as calculated by Longuet-Higgins (1953):

$$\Delta u_m = 2\left(5/4 \frac{u_m^{2} T}{L}\right). \quad (3)$$

This function applies specifically to movement over a hydrodynamically smooth bed; over a rippled bed, mass transport may decrease substantially (Bagnold, 1947). In situ observation indicates that mass transport contributes relatively little to the total asymmetry as a wave approaches the breaker zone.

Table 1 indicates the nature of velocity asymmetry for different combinations of oscillatory and translatory flow.

NATURE OF THE MODEL

Under the stated assumptions and considerations, the nature of wave-generated structures resolves to:

$$\text{bed configuration} = f(u_m, \Delta u_m, T, D). \quad (4)$$

A model based on these four parameters is by definition four-dimensional, but it can be readily visualized by considering three of the parameters in three-dimensional diagrams at constant values of the fourth. In this way, the structures can be examined in their relation to u_m, Δu_m, and D for specific wave periods (Fig. 1).

The natural limits of the chosen parameters are an important part of the model. The maximum stable wave height for a given wave period and water depth imposes such a limit. Two constraints exist: wave height generally does not exceed about $0.142\ L \tanh kh$ in deep water (Miche, 1944) nor $0.78\ h$ in shallow water (Iverson, 1951). A wave higher than either of these values tends to be unstable and will break. The approximate maximum stable height for waves of certain periods at any depth is shown in Figure 2.

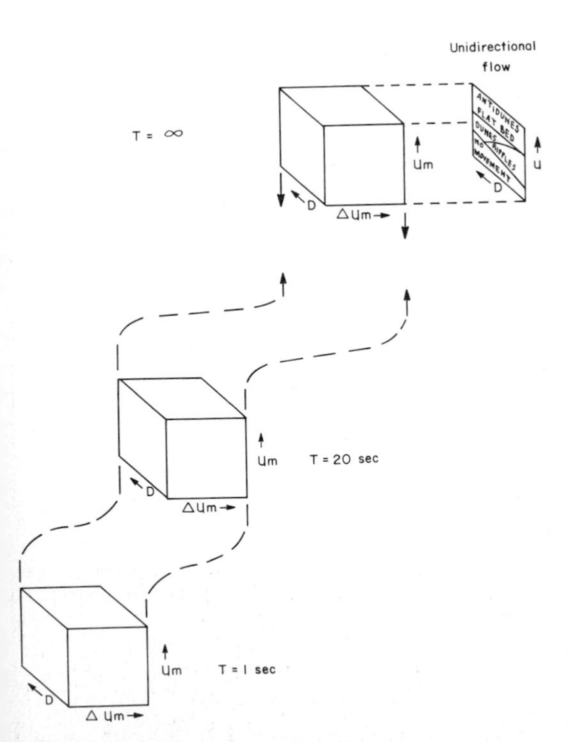

FIG. 1.—Schematic showing nature of conceptual model and relation to unidirectional flow ($\Delta u_m = u_m$ and T = infinity).

The maximum bottom orbital velocity associated with these maximum stable wave heights is shown in Figure 3. Both linear theory and an approximation based on solitary waves (Inman, 1963) are used to calculate u_m; either approach produces a curve that reaches a limit to u_m just seaward of the breaker zone. The greater of these

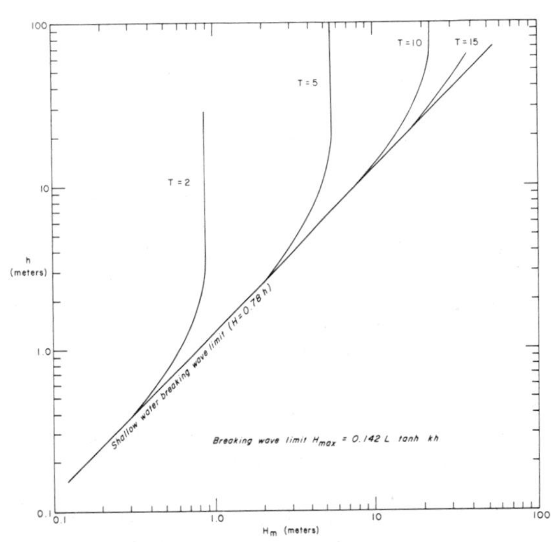

FIG. 2.—Maximum stable wave height (H) as a function of water depth (h) for waves of selected period (T).

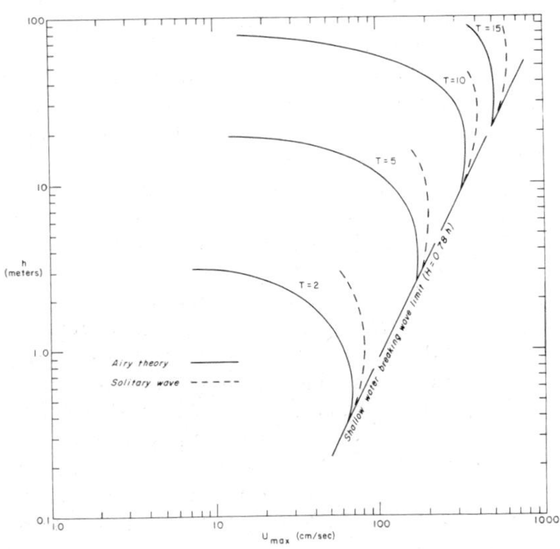

FIG. 3.—Maximum orbital velocity (u_m) associated with maximum stable-wave heights in Figure 2 as a function of water depth for waves of selected period. Calculations based on linear (Airy) and solitary wave theory (see Appendix).

values (the one based on solitary wave theory) is used here as the general limit to bottom orbital velocity for a wave of that period. For example, a 5-second wave would not be expected to generate a u_m value greater than about 2 meters/sec.

In addition, a more restricted limit to u_m can be established for certain values of Δu_m. The Stokes second-order equation for Δu_m indicates the combination of h and H that will produce a specific Δu_m for a given wave period. A curve for u_m drawn for these same combinations of h and H reaches a maximum at the approximate deep water breaking limit ($H = 0.142\ L_o$, where L_o is deep water wave length). Because the Stokes second-order equation is not valid for all ranges of Δu_m, this approach is somewhat restricted. It does, however, provide for certain theoretical limits to u_m and will be used as such in a following section.

The natural limits of Δu_m are less well defined, primarily because wave theory is less well developed for conditions where strong velocity asymmetry exists. A conceptual limit exists for purely oscillatory flow (Table 1) at $\Delta u_m = u_m$. At this point the orbital current becomes unidirectional, a condition that could be met only by superimposing a translatory flow.

The limits to grain diameter and wave period are drawn more arbitrarily. The model is developed here only for sand-sized particles (0.062-2.000 mm), although it could be extended, if desired, to a broader range of grain size. Wave period in the natural environment rarely exceeds 20 seconds (Bascom, 1964) and waves of periods less than one second are inconsequential in their effects on the sea floor.

A limit of $T = \infty$ is of special conceptual interest because when combined with the limit $\Delta u_m = u_m$, it approaches conditions of steady unidirectional flow (Fig. 1). Under these limits, a planar boundary with axes of u_m and D exists on which the structural relations are essentially those described by the unidirectional flow regime concept. This consideration is important for two reasons. First, it establishes that unidirectional flow forms a boundary to the more complicated system required to describe structures produced by oscillatory currents. Second, similarities such as a transition from rippled bed to flat bed at higher current velocities occur in both unidirectional and oscillatory flow. These similarities become increasingly pronounced with larger values of Δu_m and T, or as the unidirectional boundary is approached within the system.

RELATIONS WITHIN THE MODEL

Thresholds.—The model contains two important thresholds: the velocity required to initiate grain movement and the velocity at which a rippled bed converts to a flat bed (sheet flow). The velocity at both thresholds depends directly on grain size.

The initiation of grain movement under oscillatory flow has recently been examined by Komar and Miller (1973; 1974) and determined experimentally for progressive waves by Dingler (1974). Both studies note the effect of wave period on threshold velocity; as T increases, greater velocities are required to initiate grain movement. Komar and Miller indicate that this threshold depends on whether flow in the bottom boundary layer is laminar (grain size <0.5 mm) or turbulent (grain size >0.5 mm), whereas Dingler found the threshold relation to remain the same for grain sizes up to 1.0 mm. The threshold velocities for assumed conditions of fluid and grain densities are graphed on Figure 4 for 2-, 5-, 10- and 15-second waves.

The transition from rippled to flat bed has been studied by Dingler (1974), who used field experiments to establish criteria for the threshold velocity. He defines a relation that, under the stated assumptions, reduces to $u_{m(\text{transition})} = (3.88\ D \times 10^5)^{1/2}$ where D is in cm and u_m in cm/sec. This definition of the transition velocity was made on the basis of rather long (7-10 sec) waves. Future studies may show that the tran-

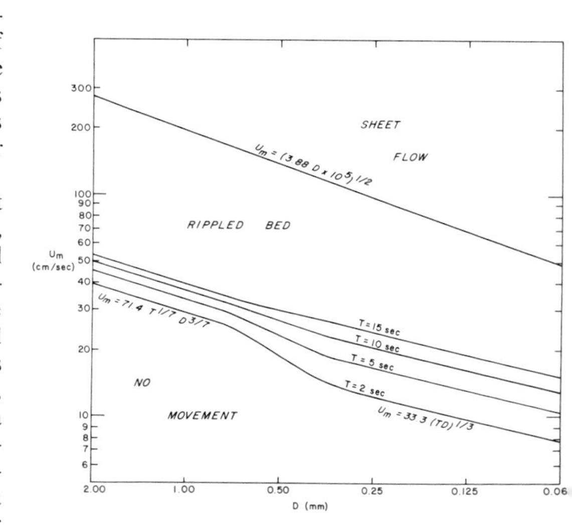

FIG. 4.—Threshold velocities for initiation of grain movement (Komar and Miller, 1974) and conversion of rippled to flat bed (Dingler, 1974) for uniform quartz sand (density = 2.65) in water (density = 1.00, viscosity = 0.01 cm^{2}sec^{-1}). Threshold for grain movement becomes progressively less accurate for 2-second waves at the coarsest grain size. Experimental studies by Dingler (1974) indicate a set of curves with equation $u_m = (2744\ TD)^{1/2}$ (in centimeters and seconds) for initiation of movement of grains smaller than 1 mm in diameter.

sition velocity, like the grain-movement threshold velocity, depends in part on wave period.

In the absence of data to the contrary, the velocities required to initiate grain movement and to produce a flat bed are assumed to be independent of the asymmetry of the oscillatory currents. Conceivably, however, future studies could show that such factors as increased acceleration associated with asymmetric flow may alter the value of the threshold velocities.

Symmetry of structures.—Wave ripples (i.e. oscillation ripples) may be either symmetric or asymmetric, as demonstrated by Evans (1941) and as noted in nearly every subsequent careful study of wave-formed ripples. Unfortunately there is a persistent misconception that wave ripples are, by definition, symmetrical (see Glossary of Geologic Terms, Gary and others, 1972).

The degree of bedform asymmetry almost certainly depends on the asymmetry of the generating oscillatory current (Δu_m). The value of Δu_m required to produce asymmetric bedforms constitutes a critical boundary within a model of wave-formed structures. Although data are meager, the different sets of pertinent field observations and experimental studies consistently point to a transition from symmetric to asymmetric structures at a rather narrow range of Δu_m values.

One set of pertinent data comes from a comparison of observations on the coasts of southern Oregon, southeastern Spain and in Willapa Bay, Washington. In each place, the depth at which the symmetric-asymmetric transition occurs can be related to the characteristics of a "typical" wave at the time the observations were made (Table 2). The combinations of water depths and wave characteristics lie within the field of applicability of the Stokes second-order equation (U.S. Army Coastal Engineering Research Center, 1973; fig. 2-7), and the location of the transition is sufficiently seaward from the surf zone that slope asymmetry of the wave (Adeyemo, 1970) should not be critical. Using Stokes equation for Δu_m, the transition from symmetric to asymmetric bedforms occurs in each case at Δu_m values that range approximately between 1 and 5 cm/sec (Fig. 5). The Longuet-Higgins mass transport velocity equation yields surprisingly similar values (Fig. 5).

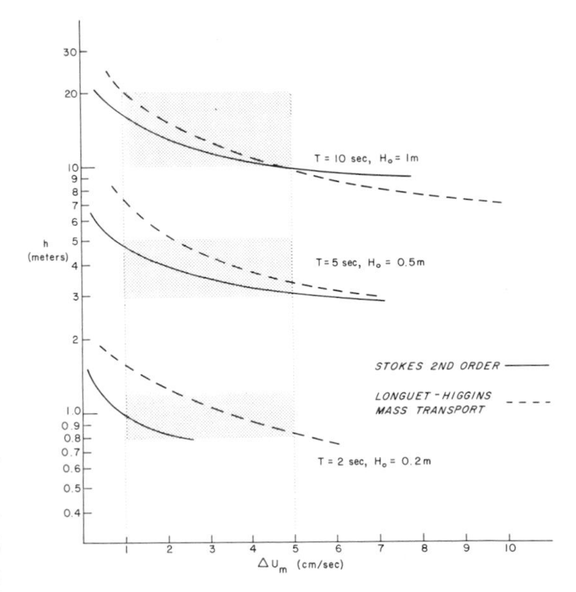

FIG. 5.—Velocity asymmetry associated with typical waves from the coast of Oregon ($T = 10$ sec, $H_o = 1.0$ m), southeastern Spain ($T = 5$ sec, $H_o = 0.5$ m) and Willapa Bay, Washington ($T = 2$ sec, $H_o = 0.2$ m). Calculations based on Stokes' second-order equation and Longuet-Higgins mass transport velocity superimposed on oscillatory flow following Airy Theory. Stippled patterns indicate general water depth of transition between symmetric and asymmetric structures for each location.

Additional data come from the few studies where enough wave and ripple data are given to determine (1) the applicability of the Stokes second-order equation, (2) Δu_m based on this equation, and (3) the ripple asymmetry in quantitative terms. Both Inman (1957) and Tanner (1971) supply such information; evaluation of their data (Fig. 6) supports the concept that the transition from symmetric to asymmetric ripples lies within a Δu_m range of 1-5 cm/sec.

This interpretation is reinforced by experimental studies of the effects of unidirectional flow superimposed on symmetric oscillatory currents by Inman and Bowen (1963) and Harms (1969). Inman and Bowen found that a superimposed current of a few centimeters per second produced

TABLE 2.—DEPTHS OF TRANSITION, SYMMETRIC-ASYMMETRIC RIPPLES

LOCATION	"TYPICAL WAVE" T(sec)	$H_o(m)$	TRANSITION DEPTH
Southern Oregon	10	1.0	10–20 m
Southeastern Spain	5	0.5	3–5 m
Willapa Bay, Washington	2	0.2	1 m

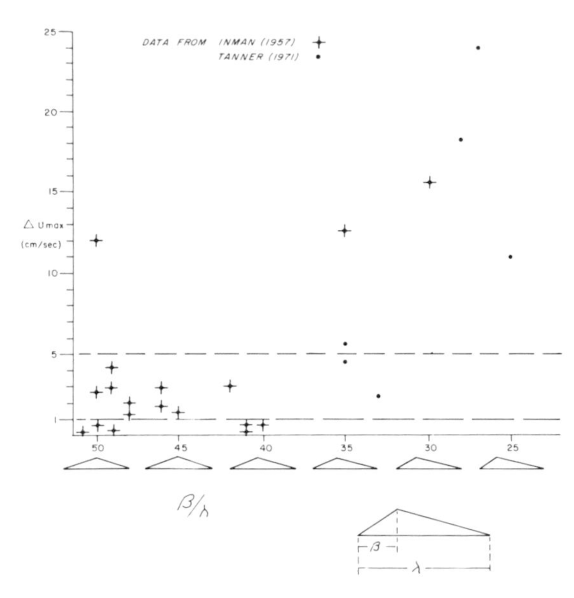

FIG. 6.—Calculated velocity asymmetry (Δu_m) and associated ripple asymmetry. Data from Inman (1957) and Tanner (1971).

asymmetry in the ripples. Although Harms' experiment did not extend to such low velocities, his slowest superimposed current (19 cm/sec) produced distinctly asymmetric ripples.

The foregoing observations indicate that symmetric bedforms occur where Δu_m is less than 1 cm/sec and asymmetric bedforms result where Δu_m exceeds 5 cm/sec. This analysis implies that asymmetric ripples may exist at any point in the marine environment where a unidirectional current has a velocity component normal to the ripple crest that exceeds a few centimeters per second. Although such a current might by itself be too small to initiate grain movement, it could, where superimposed on oscillatory flow, significantly affect bedform geometry or sediment transport (Komar, 1974b).

The distinction between symmetric and asymmetric ripples is important in ways beyond mere classification. The two classes show basic differences in crestal form, ripple size relative to grain size and flow velocity, and tendency to migrate. Active asymmetric ripples generally migrate in the direction of their slip face, whereas active symmetric ripples need not change their position (Evans, 1941). Symmetric ripples formed in one meter of water at high tide in Willapa Bay do not shift position during the fall of the tide. Time lapse photography of symmetric ripples off southern California (Cook, 1969; Cook and Gorsline, 1972) showed no net ripple migration over periods as long as 48 hours. The rate of ripple migration and attendant production of new internal structure bears on the question of structural preservation, as discussed in the section on internal structures.

The transition from symmetric to asymmetric bedforms also has interpretive significance. Curves for wave heights and water depths that will produce $\Delta u_m = 1.0$ and 5.0 cm/sec for waves of 2-, 5-, 10-, and 15-seconds, for example, show that asymmetric ripples in ancient lake deposits, where wave period is assumed to be 2 seconds, must have formed in water shallower than 2 meters. In contrast, under open coastal conditions similar to the present Pacific coast, the presence of symmetric ripple marks indicates water depths of at least 5 meters (based on $T = 10$ sec and significant deep-water wave height exceeding 0.5 m).

The relation of Figure 7 can be combined with the threshold velocities of Figure 4 and the maximum wave heights indicated in Figure 2 to produce stability fields for certain types of ripples. Figure 8 shows the stability fields for symmetric ripples in 0.125 mm sand formed by 2-second waves and 5-second waves. Combinations of wave heights and water depths that occur beneath the stability field will generate asymmetric bedforms. At combinations above the top of the stability field, the sand will not move.

Symmetric structures.—Symmetric ripples are characterized by a form whereby the crest occupies a central position between two equally deep troughs (β/λ between 0.40 and 0.60, Fig.

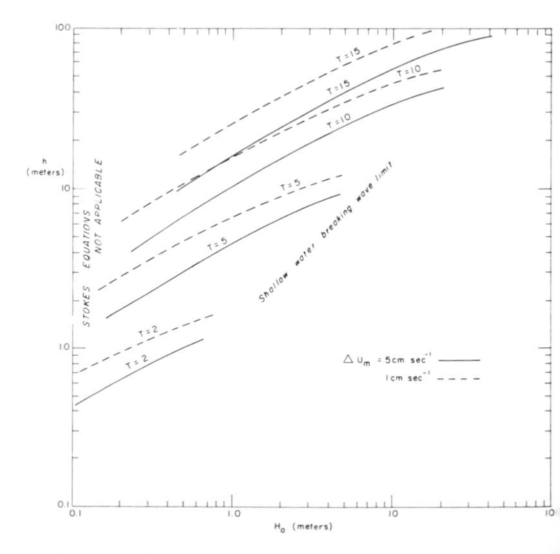

FIG. 7.—Combinations of deep-water wave height (H_o) and water depth to produce $\Delta u_m = 1.0$ cm/sec and 5.0 cm/sec for waves of selected period, representing minimum energy conditions for asymmetric ripples and maximum energy conditions for symmetric ripples.

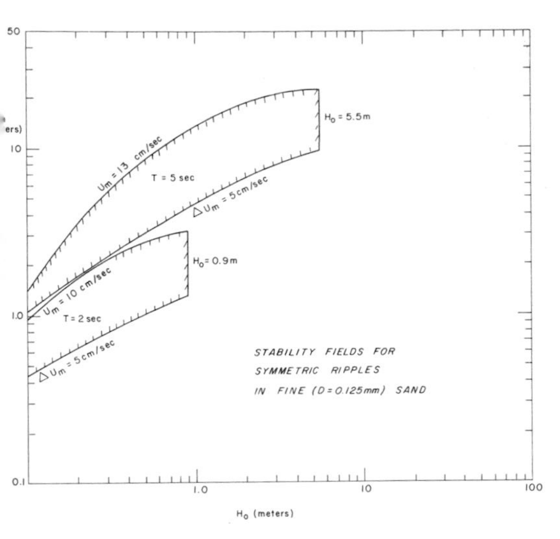

FIG. 8.—Stability fields in terms of deep-water wave ıeight and water depth for symmetric ripples in 0.125 ɲm quartz sand under waves of 2 seconds and 5 seconds. At values above top of stability field, sand does not move; at values below base of stability field, asymmetric ripples would be expected.

)). The crests are generally long and relatively straight, bifurcating sporadically.

The size of the ripples depends on two factors: orbital diameter and sediment grain-diameter. The basic relations were identified experimentally by Bagnold (1946) and in the field by Inman (1957). They have recently been examined in more detail by Komar (1974a) and Dingler (1974). All these studies note a relation at smaller orbital diameters wherein ripple wave length closely approaches the length of the orbital diameter (ripple wave length [λ] = 0.8 d_o, according to Komar, 1974a). At larger orbital diameters, this relation ceases to exist and ripple size decreases until it appears to approach a value that is stable for a particular grain size.

Dingler (1974) demonstrates these relations by plotting λ/D against d_o/D. His illustrations reinforce the concept of two distinct ripple types separated by an intermediate variety; they also ndicate that λ/D depends slightly on grain size even at large values of d_o/D. A similar plot, Figure 9, has $\lambda/D^{1/2}$, the parameter used by Bagnold (1946), as the ordinate rather than λ/D. This dimensional parameter introduces the usual problems of unit definition but substantially reduces the dependence on grain size at large values of d_o/D.

Three types of relation between $\lambda/D^{1/2}$ and d_o/D are evident from Figure 9. At d_o/D values less than about 2000 (1000 for finer sand), ripple spacing is independent of grain size and relates directly to the length of the orbital diameter. Such ripples are here termed "orbital ripples." At d_o/D values between 2000 and 5000 (1000 and 5000 for finer grains), ripple wave length may be inversely related to orbital diameter, but also depends on grain size. Such ripples are here termed "suborbital ripples." At d_o/D values greater than 5000, ripple spacing is largely independent of orbital diameter and is dependent on grain size in the approximate relation: $\lambda = 60D^{1/2}$, where λ and D are both given in centimeters. These ripples are here termed "anorbital ripples." This class probably includes the structures described by Inman (1957) as reversing ripples, the symmetry of which changes with the passage of each wave crest and trough. The d_o/D values associated with the reversing ripples of Inman range from 6500 to 13,000. The concept of ripple symmetry in its relation to Δu_m does not apply to reversing ripples, whose form at any one time will depend on the direction of the most recent strong orbital motion. Although no data exist regarding the $\lambda/D^{1/2} - d_o/D$ relations for coarse sand at d_o/D values greater than 5000, the data for finer sand span a size range of 2.5ϕ without showing detectable trends that relate to grain size. Figure 9 will therefore be used as a basis for inferring the sizes of symmetric ripples for all sand sizes.

Figure 9 is useful for interpreting ancient symmetric ripples. Ripple wave length and grain size are readily measured parameters. If $\lambda/D^{1/2}$ exceeds 100 cm$^{1/2}$, d_o/D probably lies within the range of 500 to 5000. A $\lambda/D^{1/2}$ value between 50 and 100 cm$^{1/2}$ implies one of two situations: reversing ripples preserved in a symmetric position (d_o/D greater than 5000) or orbital ripples (d_o/D = 500–1000). If the deposit contains sufficient ripples, it should not be difficult to ascertain whether ripple size depends on grain size (anorbital ripples) or is independent of grain size (orbital ripples). If $\lambda/D^{1/2}$ is less than 50 cm$^{1/2}$, d_o/D is probably less than 500, a condition that is likely to occur only with short-period (2–4 second) waves in shallow water (Komar, 1974a).

Assume, for example, that an ancient ripple with a wave length of 50 cm occurs in sand of mean diameter of 0.500 mm. For such an example, $\lambda/D^{1/2}$ = 224 cm$^{1/2}$, implying a d_o/D ratio between 500 and 5000, and, accordingly, d_o between 25 and 250 cm. Using the relation $d_o = H/\sinh kh$, the combination of values for H, T, and h required to produce d_o = 25 and d_o = 250 cm can be readily calculated following the procedure outlined in the Appendix. Figure 10 shows the combinations of water depth and wave height for waves of 2-, 5- and 10-second periods that satisfy these requirements. It can be seen that such a ripple would not form in water deeper than 1.5

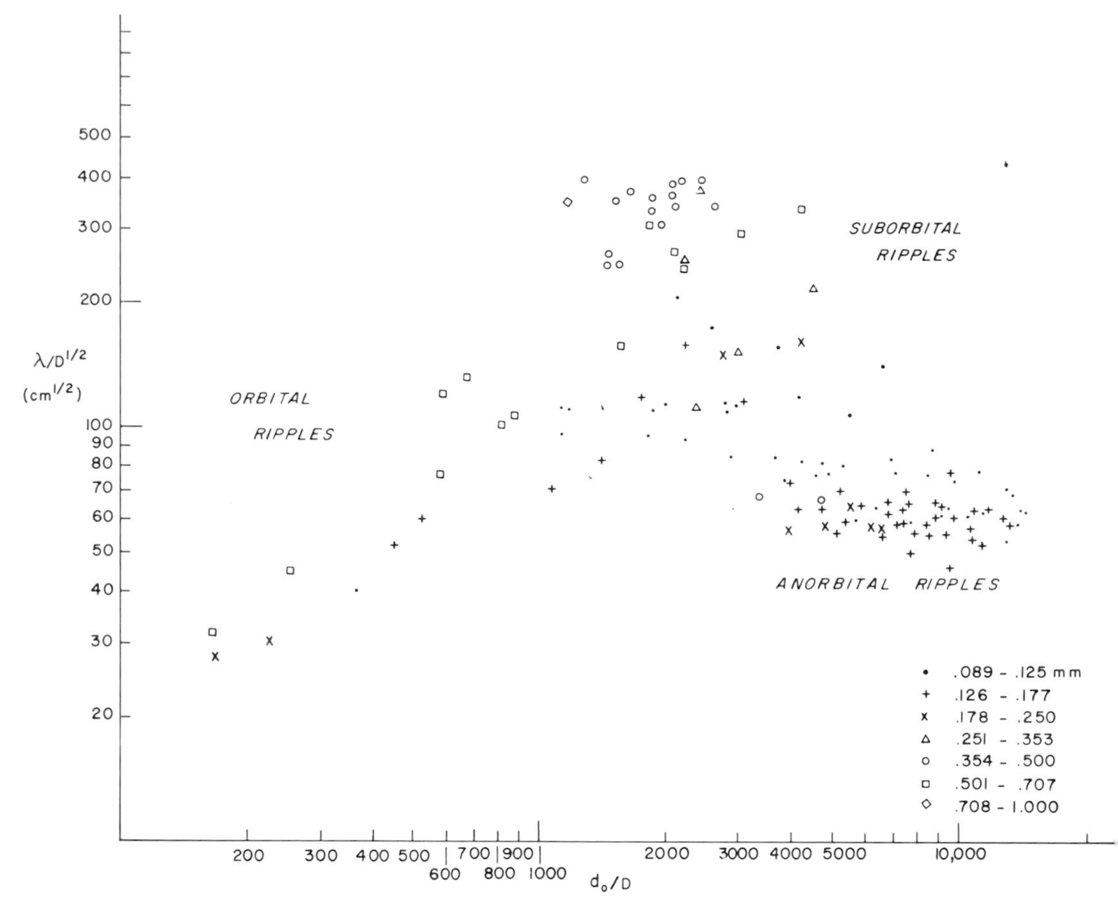

FIG. 9.—Relation between ratio of ripple wave length to square root of grain diameter ($\lambda/D^{1/2}$) and ratio of orbital diameter to grain diameter (d_o/D) for symmetric and reversing ripples. Data from Inman (1957) and Dingler (1974).

m if wave period were 2 seconds or less. Such a ripple is not likely to form at a depth of 30 meters unless wave period exceeds 5 seconds, but it could result at that depth from a moderately high (between 0.45 and 4.5 meters) 10-second wave. Figure 9 suggests that a $\lambda/D^{1/2}$ value of 224 $cm^{1/2}$ is probably associated with a d_o/D value between 1000 and 5000. This more restricted interpretation would, for example, imply a d_o between 50 and 250 cm. Such an assumption further restricts the possible combinations of H, T and h. It would, for example, indicate that the ripple mark could not have been formed by a 2-second wave.

The relations shown on Figure 9 provide a basis for predicting the size of symmetric ripples for a given grain size and orbital velocity. Because $u_m = \pi d_o/T$, a scale of d_o can be shown on any plot of u_m in relation to D at constant wave period (Fig. 11). On such a plot, it is possible to graph curves for $d_o/D = 5000$ and $d_o/D = 2000$. At values of d_o/D greater than 5000, ripple wave length can be approximated by the equation $\lambda = 60D^{1/2}$. At d_o/D values less than 2000 (less than 1000 for the finest sand size), the ripple wave length can be approximated as $\lambda = 0.8\ d_o$ (Komar, 1974a). At values of d_o/D between 2000 and 5000, the contours of equal ripple wave length are drawn schematically. Figure 11 predicts the size of symmetric ripples for 5-second waves. For smaller ($T = 2$ sec) waves, most of the symmetrical ripples are orbital; suborbital ripples exist only at higher velocities for the finer grain sizes (Fig. 12). For waves of longer period ($T = 10$–15 sec), the finer grain sizes produce anorbital ripples, and orbital ripples exist only for the coarsest sizes at velocities just above threshold (Figs. 13, 14). The relations shown in Figures 11–14 are consistent with Komar's (1974a) conclusions that in fine sand, ripple spacing increases shoreward

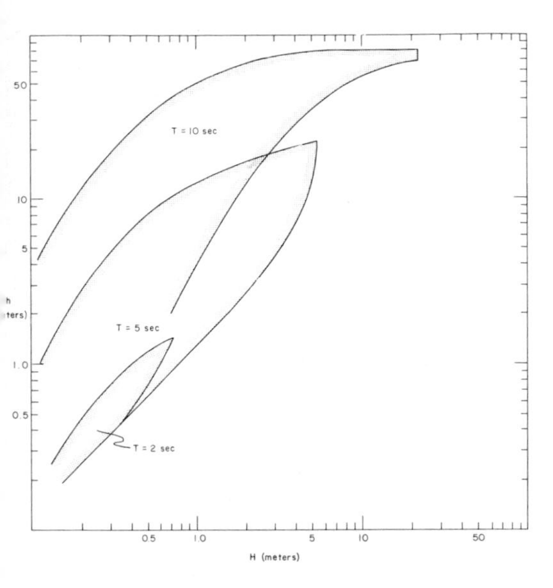

FIG. 10.—Combinations of wave height and water depth that produce orbital diameters between 25 and 250 cm under waves of 2, 5, 10 seconds (grain diameter = 0.05 cm, d_o/D between 500 and 5000).

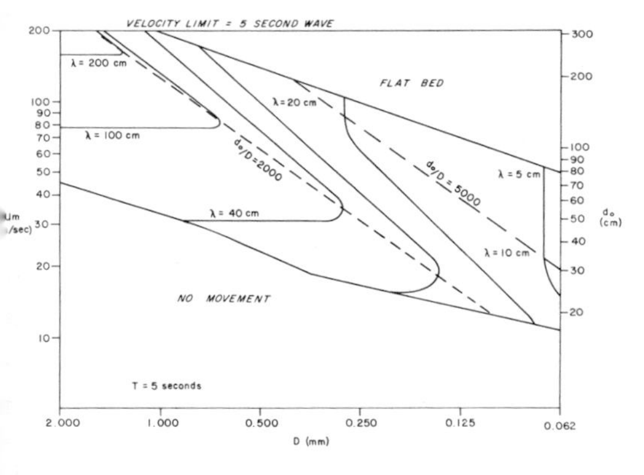

FIG. 11.—Sedimentary structures, produced by symmetrical oscillatory currents generated by 5-second waves, as a function of maximum orbital velocity and grain size. At d_o/D values less than 2000, orbital ripples develop ($\lambda \approx 0.8\ d_o$). At d_o/D values between 2000 and 5000, suborbital ripples occur, and at d_o/D values greater than 5000, the ripple form becomes anorbital ($\lambda \approx 60\ D^{1/2}$). "Velocity limit" is maximum possible orbital velocity for a 5-second wave (Fig. 3). General form of the diagram resembles that shown schematically by Allen (1970, fig. 5.8) for wave-formed ripples.

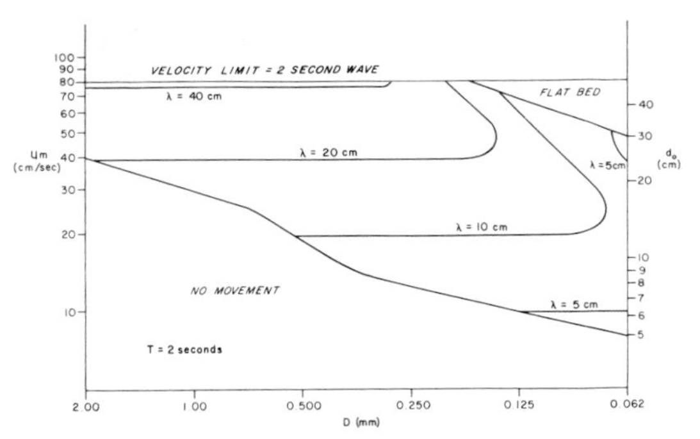

FIG. 12.—Sedimentary structures produced by symmetric oscillatory currents generated by 2-second waves as a function of maximum orbital velocity and grain size.

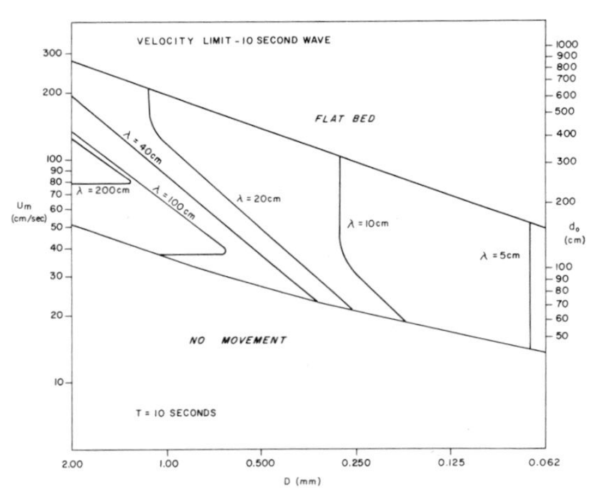

FIG. 13.—Sedimentary structures produced by symmetric oscillatory currents generated by 10-second waves as a function of maximum orbital velocity and grain size.

under short period (2 sec) waves but tends to decrease shoreward under longer period waves.

The size of the structures within the symmetric field is gradational. Abrupt increase in bedform size, such as that which occurs between "ripples" and "dunes" in unidirectional flow, has not been observed, nor is it predicted by the model. The gradation in bedform dimension raises a problem of terminology: should a distinction be made between ripples and megaripples on the basis of some arbitrary size? Should, for example, structures of wave length greater than, say, one meter be given a different name than those that are smaller? In the absence of a natural break in size, my preference is to refer to all symmetric structures as "ripples" regardless of the dimensions. Such is not the case when dealing with asymmetric structures.

The ratio of ripple height to ripple wave length has been the subject of considerable discussion. Dingler (1974) observes that this ratio remains at a constant, approximately 0.15, until just prior to the occurrence of sheet flow (flat bed). At

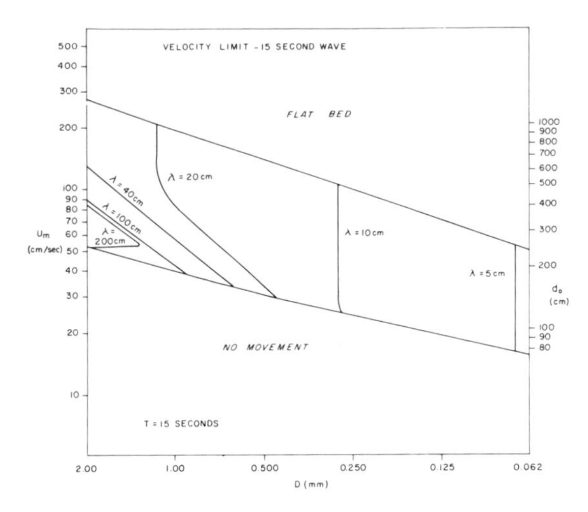

FIG. 14.—Sedimentary structures produced by symmetric oscillatory currents generated by 15-second waves as a function of maximum orbital velocity and grain size.

that point, the ripple steepness decreases rapidly.

Asymmetric structures.—Structures in the asymmetric field must at present be considered only in qualitative terms. The application of wave theory becomes rather uncertain under the conditions which produce marked asymmetry of oscillatory currents. Moreover, virtually no systematic measurements of current velocities exis that can be related to bedform developments, an no full-scale experiments have been devised tha approach the flow intensities of asymmetric oceanic oscillatory currents.

Observations made off the coasts of souther Oregon and southeastern Spain and in Willap Bay, Washington, indicate that asymmetric struc tures develop in a consistent pattern (Fig. 15) at least for medium- to coarse-grained sand. I each case a transformation begins in deeper wate from symmetric ripples through a similar serie of asymmetric structures in an onshore directio (toward increasing u_m and Δu_m). The asymmetri structures associated with less intense orbita velocity and smaller velocity asymmetry occu in all the environments, whereas the asymmetri structures associated with more intense flow occu only under longer period waves, such as thos on the Pacific coast.

Structural transformation in the outer part o the system is gradational, as the transition betwee symmetric and asymmetric structures occurs ove a distance of at least several meters. Th seawardmost asymmetric ripples are much lik their symmetric counterparts; ripple crests ar relatively long and straight, and bifurcate sporadi cally. Landward the crest length shortens an the ripples become increasingly irregular. Suc a transformation consistently occurs under cond

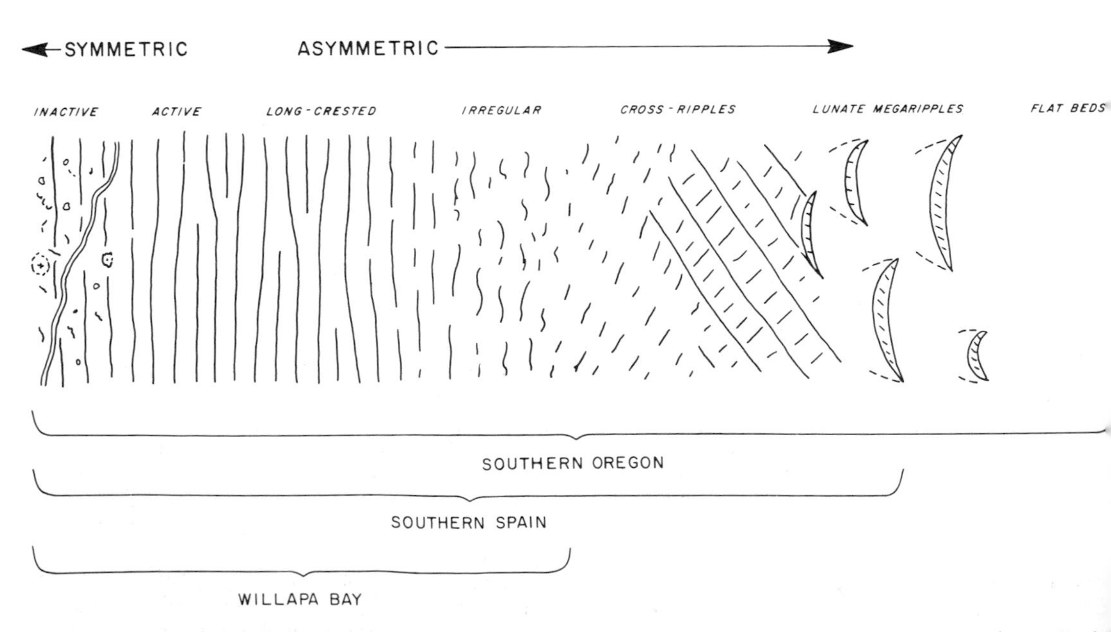

FIG. 15.—Sequence of structures commonly observed off the coast of southern Oregon, southeastern Spai and Willapa Bay, Washington. Brackets below diagram indicate the range of structures observed in eac environment.

ons of increasing energy in unidirectional flow Allen, 1968, p. 93; Reineck and Singh, 1973, p. 1).

If the flow intensity increases, the randomly paced irregular ripples may become aligned en chelon oblique to the oscillatory current (Fig. 5). With increasing velocity and orbital asymmetry, such patterns grade into "cross-ripples." These structures, described briefly by Clifton and others (1971), consist of two sets of ripples, both oriented oblique to the oscillatory current. One set tends to be long crested and the other set tends to be composed of shorter ripples that occupy the troughs of the longer set. This ripple pattern resembles that of interference ripples, but the ripples respond only to the oscillatory current, which approximately bisects the angle between the two sets. Sand transport occurs mostly during the landward component of the oscillatory current which is markedly stronger than the seaward component) in the form of helical vortices directed landward from the points of ripple intersection. Both ripple sets migrate in an onshore direction, and the ripples in the shorter set have been observed to migrate more rapidly than the associated long-crested ripples. Directional relations suggest that ancient examples of cross-ripples occur in rocks of Ordovician age in southwestern Scotland (Kelling, 1958).

The origin of cross-ripples is uncertain. Their occurrence is not detectibly related to extraneous currents nor to waves approaching from different directions. Perhaps such ripples form in response to the rapid acceleration of the oscillatory currents in this regime. The observed helical flow that occurs behind the crest of oblique ripples provides a means of water transport in the vortex downcurrent from the ripple that is absent in normal transverse ripples.

Where present, cross-ripples appear to be structurally transitional between small, asymmetric, irregular ripples and lunate megaripples (Clifton and others, 1971). The lunate megaripples are large-scale bedforms one to several meters in wave length and as much as one meter high. These structures are common in medium- to coarse-grained sand under conditions of intense asymmetric flow generated by long period waves. They are common in 2 to 4 meters of water off the coast of southern Oregon and were observed twice in somewhat shallower water off the coast of southeastern Spain. The value of u_m and Δu_m required to produce lunate megaripples is uncertain. Calculations from typical wave conditions under which they were observed off the coasts of Oregon and Spain indicate that these structures form under velocities of the order of a meter per second and velocity asymmetry of at least 20–30 cm per second.

Off the coast of Oregon, a flat bed generally develops just seaward of and under the breaker zone. Although an active flat bed was not observed on the coast of Spain, parallel flat bedding in cores from the nearshore attest to its presence under storm-wave conditions.

The pattern of asymmetric structures produced in fine sand resembles that in medium sand only at relatively low values of u_m and Δu_m. The fine sand that occurs off the Scripps Institution of Oceanography pier in La Jolla, California, is worked into symmetric ripples in relatively deep water. These ripples transform shoreward to reversing ripples and thence to flat bed (Dingler, 1974, and personal observation). Lunate megaripples are conspicuously absent. The wave regime and bottom configuration here are quite similar to those off the coast of southern Oregon where lunate megaripples predominate in medium to coarse sand seaward of the breaker zone.

Lunate megaripples have not been observed in fine gravel. The predominant bedform of such material consists of large straight-crested asymmetric ripples a meter or more in length and 10 to 20 cm high. At a few places on the Oregon coast, concentrations of fine gravel in large straight-crested asymmetric ripples occur locally within a field of medium-grained sand shaped into lunate megaripples.

The foregoing discussion illustrates the effect of grain size on bedform development in the asymmetric field. Even at very low orbital diameters, the size of structures in the asymmetric field relative to grain size and flow velocity differs from those in the symmetric field. Tanner (1971) conducted a number of experiments on asymmetric ripple development in sand of diverse sizes under small, short-period waves. His results are summarized in Figure 16, which resembles Figure 9 only in that $\lambda/D^{1/2}$ values less than 50 $cm^{1/2}$ are mostly associated with d_o/D values less than 500. Tanner's data clearly indicate that the wave length of asymmetric ripples formed under small orbital diameters, unlike symmetric ripples, depends in part on grain size.

The abrupt transition from small asymmetric wave-generated ripples to lunate megaripples strongly resembles the transition from ripples to "dunes" in the lower flow regime of unidirectional flow. Clifton and others (1971) note the similarity of onshore structural transformation off the Oregon coast to that which occurs with increasing stream power in unidirectional flow. The absence of lunate megaripples in fine sand under oscillatory flow is consistent with relations described for unidirectional flow. "Dunes" of the lower flow regime form in fine sand under a relatively narrow range of flow conditions (Simons and others, 1965) and marked inconstancy of flow

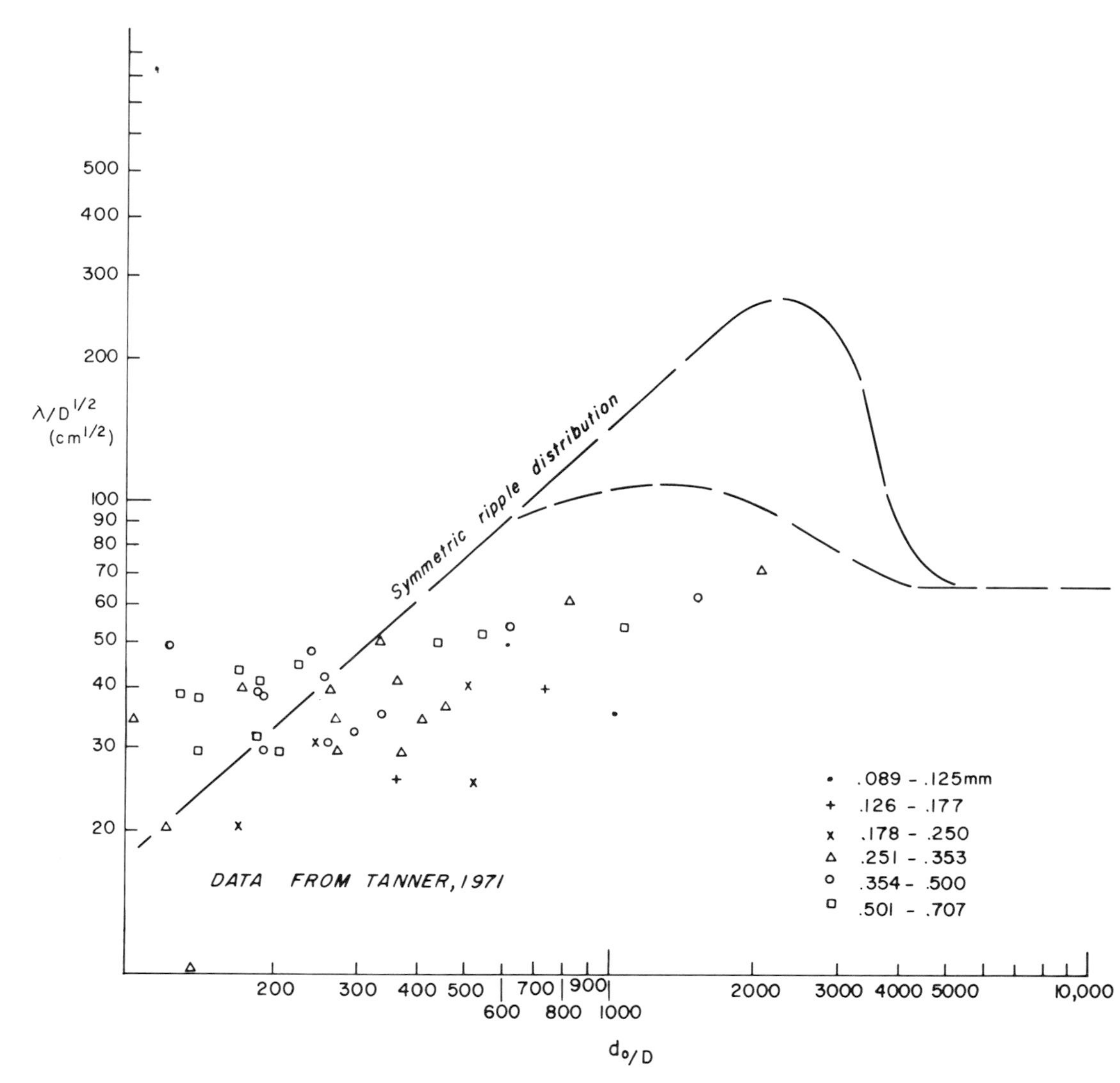

FIG. 16.—Nature of $\lambda/D^{1/2} - d_o/D$ relationships for asymmetric ripples described by Tanner (1971). Note difference between distribution of these ripples and the symmetric ripples shown in Figure 9.

would probably inhibit their development. The variability of wave-generated currents may thereby preclude the development of lunate megaripples in fine sand.

Figures 17–20 schematically relate the structures described above to u_m, Δu_m, and D, for 2-second, 5-second, 10-second, and 15-second waves. These illustrations form the basis for extending the model to real conditions.

APPLICATION OF THE MODEL TO THE NATURAL ENVIRONMENT

The initial simplifying assumptions include two that are particularly incompatible with conditions in the natural environment. The assumption that the waves are of uniform height, period and direction of approach is generally invalid and the assumption of no superimposed currents is unjustified for part of the profile swept by shoaling waves. These inconsistencies must be resolved before the conceptual model can be considered as generally applicable to the natural system.

Effects of wave variability.—The surface of the ocean rarely, if ever, consists of monochromatic waves. Rather, it is traversed by waves of diverse height, period, and direction in such complication that prediction of bedforms may seem hopeless. A "typical wave" may be described, but this generally is only a gross approximation of the actual surface.

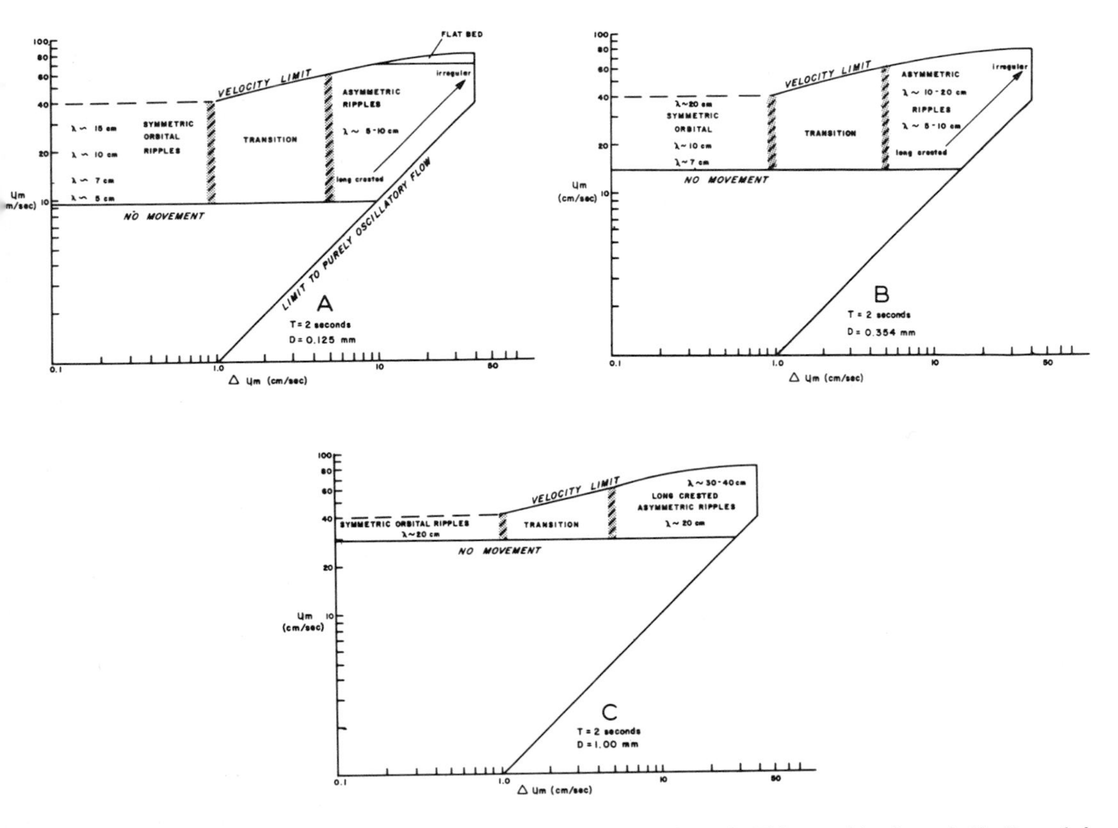

FIG. 17.—Structural relations for specific grain sizes for 2-second waves. The velocity limits to this diagram represent an estimate of the overall velocity limit (Fig. 5) and the maximum possible velocity associated with $\Delta u_m = 0.01$ m/sec and 0.05 m/sec. The velocity limits associated with $\Delta u_m = 0.1$ and 0.05 m/sec are drawn from Figure 7 which shows the combinations of wave height and water depth required to produce these values of Δu_m for waves of 2, 5, 10 and 15 seconds. A plot of the maximum velocity associated with these combinations indicates that the value of u_m associated with either value of Δu_m increases with increasing water depth. The limiting u_m is therefore defined by the intersection of the curves shown in Figure 7 with the curves for the maximum stable waves shown in Figure 2. This intersection establishes the wave height-water depth combination associated with the velocity limit related to $\Delta u_m = 0.01$ and 0.05 m/sec for the wave periods indicated. This combination of H, T, and h lies within the field of applicability of the Stokes' 3rd and 4th order equation (U.S. Army Coastal Engineering Research Center, 1973, fig. 2-7). Stokes' 2nd-order equations used to approximate velocity limits associated with $\Delta u_m = 0.01$ and 0.05 m/sec in Figures 17 through 20. Velocity limit associated with Δu_m values less than 0.01 m/sec schematic.

Overall velocity limit presumed to be associated with highest velocity asymmetries. Inman (1957) notes that velocity under crests of waves near breaker zone is about two-thirds of the theoretical velocity of solitary waves and velocity under the trough resembles one-third the theoretical velocity. Accordingly, Δu_m in this situation approximately equals one-half u_m. Therefore on Figures 17-20, the limit to Δu_m is defined as one-half overall velocity limit determined from Inman's approximation).

Wave variability affects the conceptual model in several ways. Changes in wave height or direction of approach alter the values of u_m and Δu_m, but not the position of structural boundaries shown on Figure 17 through 20. In contrast, variations in wave period change not only u_m and Δu_m, but also their basic influence on structural development. Wave variability may also inhibit the development of some of the larger structures, particularly the lunate megaripples, which appear to require a certain stability of currents in order to develop. These bedforms tend to be either absent or poorly developed on the Oregon coast when the sea surface is highly complicated, and the lunate megaripple facies is best developed when the waves are regular (Clifton and others, 1971). As previously noted, wave variability may account for the absence of lunate

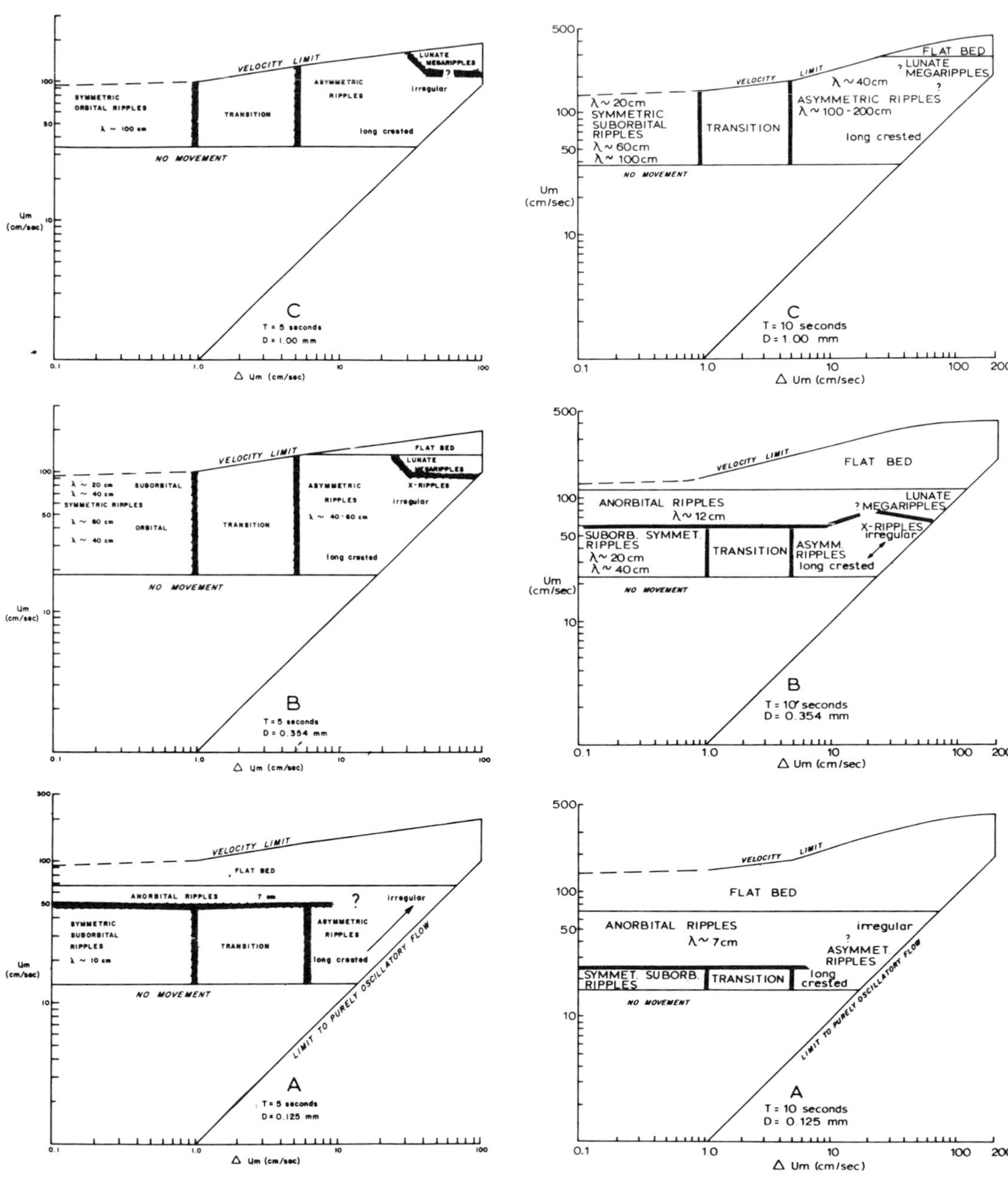

FIG. 18.—Structural relations for specific grain sizes for 5-second waves. As anorbital ripples are likely to be reversing, their symmetry is not delineated on this or subsequent figures.

FIG. 19.—Structural relations for specific grain sizes for 10-second waves.

megaripples in fine sand.

The model can be applied to conditions of variable waves by expanding the linear boundaries drawn for monochromatic waves to indeterminate zones. Although the transitions between structures can no longer be related to specific values of the model's parameters, general relations can still be predicted. It is in these terms that the model has its greatest application to geologic interpretation.

Figures 21-23 indicate the general relations of structures formed under the influence of short-, intermediate-, and long-period waves, respectively. For any grain size, structural changes

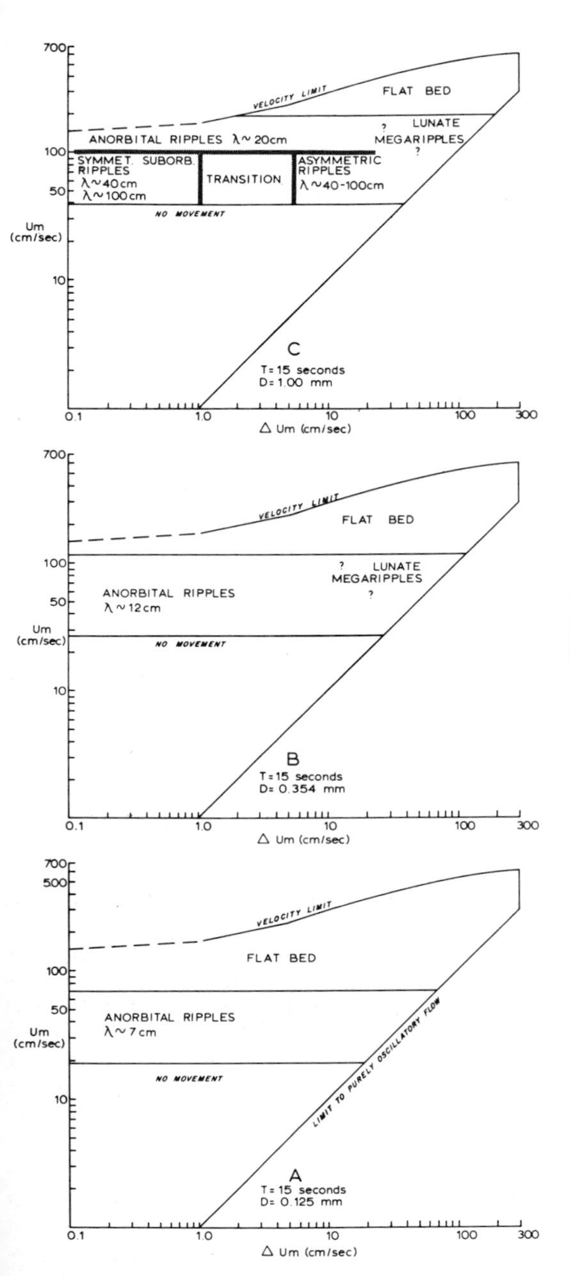

FIG. 20.—Structural relations for specific grain sizes for 15-second waves.

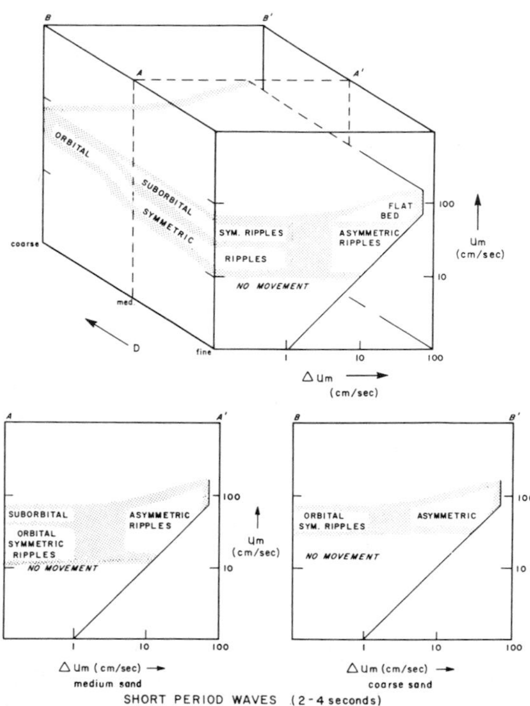

FIG. 21.—Conceptual model of sedimentary structures produced by short-period (2-4 seconds) waves. Stippled pattern indicates area of boundary uncertainty. For estimates of bedform sizes, see Figures 17 and 18.

associated with increases in u_m and Δu_m imply shallower depths and/or larger waves. Features associated with the highest velocity values are probably produced only under extreme wave conditions.

In lakes and enclosed bays where wave period is short (2-4 seconds), ripples may form in water depths to 25 m. Symmetric ripples may be orbital or suborbital; in coarser sand, the orbital type predominates. Asymmetric ripples will occur at shallow depths near the shoreline, but lunate megaripples would be uncommon even in medium or coarse sand. A flat bed will form only in fine sand and only in shallow water under extreme wave conditions (Fig. 21).

Intermediate-period waves (5-8 seconds) predominate at the margins of enclosed or semi-enclosed seas (e.g., Mediterranean) or on an open ocean coast where longer period swell is absent (e.g., the Atlantic coast of the United States). Such waves may ripple the bottom to depths of nearly 100 meters. Suborbital ripples dominate the symmetric ripple field except for the coarsest sand fraction. Anorbital ripples would be expected at higher velocities in fine- to medium-grained sand. A flat bed would be expected in shallow water under the larger waves for all grain sizes but the coarsest sand (Fig. 22).

Coasts dominated by long-period (10-15 second) swell such as the Pacific coast of the United States may have ripples generated by surface waves to depths exceeding 300 m (Komar and others, 1972). Symmetric suborbital ripples will form in relatively deep water. Symmetric orbital ripples are likely

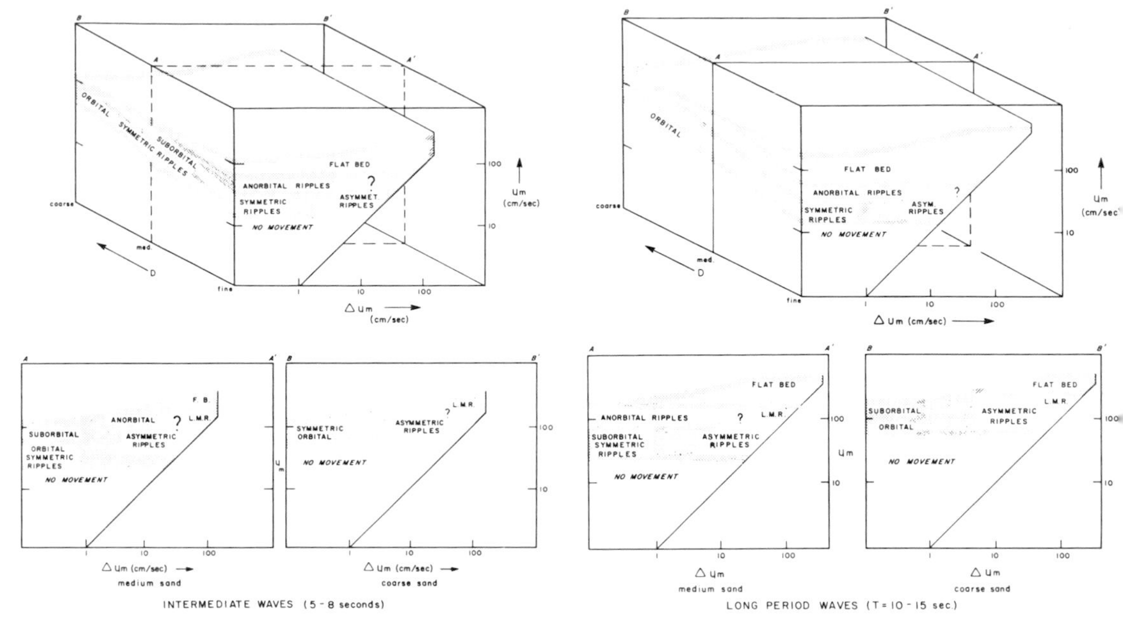

FIG. 22.—Conceptual model of sedimentary structures produced by intermediate-period (5-8 seconds) waves. For estimates of bedform sizes, see Figures 18 and 19 (LMR = lunate megaripples, FB = flat bed).

FIG. 23.—Conceptual model of sedimentary structures produced by long-period (10-15 seconds) waves. For estimates of bedform sizes, see Figures 19 and 20 (LMR = lunate megaripples).

only in coarse to very coarse sand, whereas in fine sand, anorbital ripples dominate the entire field of ripple development. In medium and coarse sand, the asymmetric field is characterized by the transition to lunate megaripples as shown in Figure 15. A flat bed can develop at higher velocities for all sand sizes, but will form in coarse sand only in shallow water under extreme conditions (Fig. 23).

The maximum velocity, velocity asymmetry, and, in general, sediment grain size increase in a shoreward direction. The progression of structures on any particular coast can be represented by a line passing through the appropriate three-dimensional diagram. Such a line would, in general, extend from the lower left front corner of the diagram toward the upper rear right corner. Its exact pathway depends on the actual conditions and grain sizes encountered in the shoreward progression.

Effects of unidirectional currents.—Discussion of wave-formed sedimentary structures would not be complete without some discussion of the effects of unidirectional currents that are generated by shoaling waves. Although currents do not necessarily alter structural relations within the model, they can drastically affect their interpretation in terms of wave characteristics and water depth. Two general types of wave-generated currents may exist: return flow and longshore currents. Each type may complicate the form or the orientation of structures produced by oscillating currents.

Return flow refers to a seaward transport of water in response to a hydraulic head generated on the beach by wave action or other factors. This flow can theoretically exist as a general seaward drift, although it is more commonly channelized into relatively narrow, seaward-flowing rip currents (Bowen, 1969). General seaward flow would be most significant very close to shore, and probably accounts for seaward-facing bedforms observed at the base of gently-sloping beaches (Clifton and others, 1971). Cook and Gorsline (1972) suggest that such a general flow may produce a seaward asymmetry to the maximum velocity of the oscillatory current somewhat farther offshore. Rip currents are well-documented phenomena (Shepard and Inman, 1951; Bowen, 1969; Bowen and Inman, 1969; Cook, 1970). Seaward-facing megaripples are common structures in the channels occupied by such currents (Clifton and others, 1972; Davidson-Arnott and Greenwood, 1974).

Longshore currents flow parallel to the coast in response to waves obliquely incident to the coast. They are strongest inshore from the breaker zone and are best developed in the troughs of longshore bars if such exist. Longshore currents commonly become rip currents at breaks in such bars.

The currents generated by shoaling waves are most intense inside the breaker zone. The complexity of internal structures observed in cores taken within the breaker zone and shoreward of it on the Oregon coast (Clifton and others, 1971) reflects the complexity of processes close to the shoreline. Use of the conceptual model to interpret wave characteristics and water depths is therefore best restricted to the environment seaward from the zone of breaking waves.

INTERNAL STRUCTURES

The internal structures produced by shoaling waves generally are more important for geologic interpretation of outcrops or cores than the bedforms themselves, which require rather special conditions for their preservation. Vertical sequences of structures or structural associations may provide much information regarding the depositional environment and the nature of basin infilling. The internal structures most likely to be produced by shoaling waves include more or less planar parallel lamination, medium-scale foreset bedding and ripple foreset bedding.

Planar parallel lamination.—Parallel, more or less horizontal, lamination may develop in shallow marine sands in two very different ways. It may be generated by sheet flow in the oscillatory equivalent of the upper flow regime or by the migration of ripple forms accompanied by a very slow rate of sediment accumulation (Newton, 1968).

Sheet flow produces parallel lamination at relatively high orbital velocities. As the internal structure associated with a flat bed, this type of lamination requires less velocity to form in fine sand than in coarser sand. Under waves of short or intermediate periods, this structure forms only as a result of intense wave activity close to the shoreline. It may develop in deeper water under the influence of large, long-period waves.

Planar lamination produced by ripple migration is essentially a form of very gently climbing ripple stratification. It is most likely to occur under long-crested ripples at depths which depend on wave height and period. Waves of all periods may generate this structure, but intermediate- to long-period waves are required to produce it in deeper water.

Distinction between the two types of lamination may be difficult. Parallel lamination produced by sheet flow may show evidence of shear sorting (Inman and others, 1966) of components of different size, density or shape, although gently climbing ripples may also produce internal sorting within laminae (Hunter, 1974). Planar concentrations of mica flakes similar to those produced experimentally by sheet flow are associated with flat bed deposits in the southern Spanish nearshore; such concentrations may provide a useful criterion for sheet flow in deposits that contain abundant mica. Small concavo-convex shells in sheet flow deposits tend to lie convex-up (Fig. 24). Bioturbation may be relatively undeveloped in sheet flow deposits, owing to the intense reworking by the waves, unless the burrowing infauna is particularly prolific. Planar lamination produced by sheet flow is likely to be associated with medium-scale foreset bedding.

Planar lamination produced by ripple migration may show poorly defined climbing ripple foresets. Small concavo-convex shells are likely to lie predominantly concave-up, owing to rotation as they pass over ripple crests (Clifton and Boggs, 1970). Bioturbation may be intense, and ripple foresets and shell lag deposits are probable associated structures.

Medium-scale foreset bedding.—Foreset units more than 6 cm thick occur in medium- to coarse-

FIG. 24.—X-ray radiograph of box core taken on rippled sand at a depth of 1.7 meters off southeastern coast of Spain after a local storm. Lower part of core shows planar bedding produced by sheet flow during storm (note convex-up small pelecypod shells). Landward-dipping ripple foreset bedding at top of core formed after storm ceased. Light spots, shadows of pebbles within sand. Width of core, 15 cm.

grained sand as the result of the migration of lunate megaripples or, if the sedimentation rate is rapid, of long-crested ripples. Lunate megaripples are likely to produce trough cross-bedding (Clifton and others, 1971), whereas long-crested ripple migration would be expected to produce more tabular units. The foresets generally will be oriented in the direction of wave propagation (more or less onshore), except where longshore or rip currents on other intense unidirectional flow prevails along the bottom. Cross-bedding that persistently dips in an onshore direction implies a degree of orbital asymmetry generally restricted to relatively shallow water.

The apparent absence of lunate megaripples in fine sand plus the size restrictions that seem to exist for symmetric ripples lead to a useful inference. The maximum $\lambda / D^{1/2}$ value shown in Figure 9 for sand in the 0.125-0.088 mm size range is slightly more than 200 $cm^{1/2}$. If a limiting value of 250 $cm^{1/2}$ is assumed, the maximum wave length for symmetric ripples in fine sand is approximately 30 cm. Using the general ratio between ripple height and wave length of 0.15 (Dingler, 1974), the maximum height of symmetrical ripples is around 4 cm. It therefore seems reasonable to conclude that larger ripples (including lunate megaripples) are not likely to be produced in the finest sand solely by oscillatory currents. Medium-scale cross-bedding (units more than 6 cm thick) in very fine sand would thus imply the presence of unidirectional currents.

Ripple foreset bedding.—Ripple foreset bedding is most likely to be produced by irregular asymmetric ripples, which rework the sediment to the depth of the deepest ripple trough as they migrate. It will also result from any small ripple migration during rapid sediment accumulation. The foresets within asymmetric ripples generally dip onshore, in the direction of wave propagation, although unidirectional currents may impose other orientations. Foreset bedding within symmetric ripples also commonly dips onshore (Newton, 1968, and personal observation), although Reineck and Singh (1973), in describing the internal structure of symmetric ripples, note that a chevronlike internal structure can occur.

Preservation.—Sedimentary structures preserved in the stratigraphic record may not reflect the relative abundance of bedforms at the sediment surface. Two factors, abnormally large waves and faunal reworking, contribute to this disparity. Both processes can rework the sediment below the normal sediment surface to a depth not generally disturbed by the prevailing waves.

Most coasts are subject to intense but relatively infrequent storms or abnormally large swell. The shoreward sequence of structures produced under these conditions may resemble the normal sequence but generally reflects larger orbital velocities, and possibly velocity asymmetry. The large waves will rework the sediment most deeply in shallow water; the thickness of the reworked layer should decrease in progressively deeper water until at some depth where waves normally have no effect on the bottom, only the sediment surface is disturbed.

The structures produced by abnormally large waves will differ, depending on the size of the waves, the grain size and the water depth. In relatively shallow water, passing storm waves will likely generate a flat bed in fine sand, or a flat bed or megaripples in medium to coarse sand. Although structures produced under normal conditions may predominate near the sediment surface, the lower part of the active sediment column is likely to consist of structures produced by large waves (Fig. 24). Such structures, therefore have a relatively higher preservation potential.

Faunal mixing, or bioturbation, commonly extends to a depth of some centimeters beneath the sediment surface (Rhoads, 1967; Clifton and Hunter, 1973). Physical processes that do not extend below this depth are not likely to be permanently recorded, particularly if the rate of bioturbation exceeds the rate of reworking by physical processes (Fig. 25). The nature and rate of bioturbation depends strongly on sediment texture. Faunal mixing seems to be most intense in the presence of a well-developed infauna, which in turn is inversely related to sand grain size (Clifton and Hunter, 1973). Accordingly, structures produced in fine sand, such as small ripples and ripple foreset bedding, are more likely to be destroyed than structures formed in coarser sediment (large ripples and cross-bedding). On the southeastern coast of Spain, large ripples in coarse sand are well preserved at a water depth of 12 meters, whereas small ripples in fine sand at the same water depth a short distance away are partly destroyed and have a much disrupted internal structure.

The degree of bedform asymmetry may influence structural preservation. The internal structures produced by symmetric bedforms, inherently have a lower preservation potential than those in their asymmetric counterparts, because symmetric bedforms need not migrate. Unless wave conditions change to the point of forcing symmetric ripples to migrate, no new internal structure develops. One can, indeed, conceive of a field of active, persistent but immobile, symmetric ripples in which internal structure has been totally destroyed by bioturbation.

Consideration of the effects of storms and large swell together with the long-range effect of bioturbation permits prediction of the general nature of the preserved deposit. In shallow water, the effects of the largest waves—flat bedding and

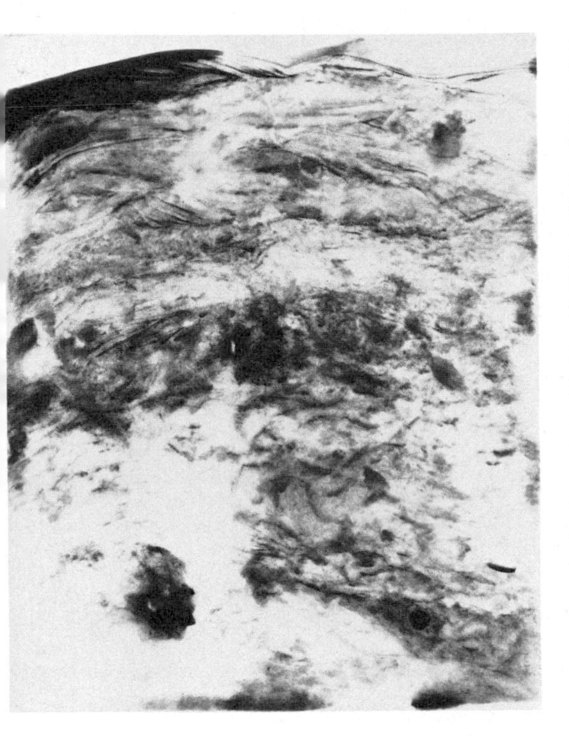

FIG. 25.—X-ray radiograph of box core taken on rippled sand at a depth of 3.8 meters directly offshore from core shown in Figure 24. Lower part of core is thoroughly bioturbated; ripple foreset bedding at top of core dips onshore. Width of core, 15 cm.

cross-bedding—will dominate the stratigraphic record. In deeper water, bioturbation will obliterate most internal structure. Parallel, nearly flat bedding produced by ripple migration may occur, but ripple foreset bedding should be relatively uncommon even though ripples predominate at the sediment surface most of the time. Examination of a number of shallow marine deposits—Pleistocene terrace deposits in Oregon and California and in southern Spain, as well as middle Miocene shoreline deposits in the Caliente Range of the southern California Coast Ranges—verifies that ripple foreset bedding is distinctly uncommon relative to flat bedding, cross-bedding and bioturbated sediment.

CONCLUSIONS

A combination of maximum orbital velocity, velocity asymmetry, wave period, and grain size appears to provide a basis for a conceptual model of wave-formed sedimentary structures. The values of the integrating parameters, velocity and velocity asymmetry, depend on a combination of wave height, wave period, and water depth; analysis of sedimentary structures can therefore provide insight into this combination. If water depth can be assumed or inferred, the wave characteristics can be estimated, and vice versa. The hydrodynamic consideration of unidirectional flow can be incorporated into the model under the special case of total asymmetry and infinitely long oscillation.

Much work is needed to verify the relations within the model. In particular, the asymmetry of oscillatory currents and its relation to wave characteristics needs to be better understood. In situ measurements of oscillatory current velocity and its relation to existing bedforms will provide a much better basis for interpreting structures, particularly those in the asymmetric field. Studies of such effects as acceleration on bedform development will much improve our understanding of wave-formed structures.

All boundaries within the model as it presently exists must be considered as approximations. It would be unrealistic to use the model in its present form to predict precise relations between structures and physical parameters of the depositional environment. Future research must further resolve these boundaries before the model can, with confidence, be applied quantitatively.

Application of the model for interpreting ancient deposits requires care. It specifically does not consider the effects of unidirectional currents, and the possiblity of such currents cannot be disregarded. The model may, however, be useful for distinguishing between purely wave formed structures and those influenced by a unidirectional current.

The model incorporates observations of structures from a wide range of environments into an integrated framework. It provides a basis for the hydrodynamic interpretation of these structures. And as a conceptual framework, it allows the geologist to better interpret the effects of waves in ancient depositional environments.

ACKNOWLEDGEMENTS

Many people participated in useful discussions of the material presented herein; I particularly want to thank R. E. Hunter, J. R. Dingler, R. L. Phillips and A. H. Sallenger, Jr. for their help. In addition, R. L. Street and J. B. Southard read the manuscript and made many constructive comments.

REFERENCES

ADEYEMO, M. D., 1970, Velocity fields in the wave breaker zone: Am. Soc. Civil Engineers, Proc. 12th Conf. on Coastal Eng., p. 435–460.

ALLEN, J. R. L., 1968, Current ripples; their relation to patterns of water and sediment motion: North-Holland Pub. Co., Amsterdam, 433 p.

——, 1970, Physical processes of sedimentation: American Elsevier, New York, 248 p.
BAGNOLD, R. A., 1946, Motion of waves in shallow water, interactions between waves and sand bottoms: Royal Soc. London Proc., Ser. A, v. 187, p. 1-18.
——, 1947, Sand movement by waves; some small-scale experiments with sand of very low density: Excerpt Jour. Inst. Civil Engineers, v. 27, p. 447-469.
BASCOM, W. A., 1964, Waves and beaches—The dynamics of the ocean surface: Anchor Books, Doubleday and Co., Inc., New York, 267 p.
BOWEN, A. J., 1969, Rip currents, 1. Theoretical investigations: Jour. Geophys. Res., v. 74, p. 5467-5478.
—— AND INMAN, D. L., 1969, Rip currents, 2. Laboratory and field observations: Jour. Geophys. Res., v. 74, p. 5479-5490.
CLIFTON, H. E., AND BOGGS, S. JR., 1970, Concave-up pelecypod (*Psephidia*) shells in shallow marine sand, Elk River Beds, southwestern Oregon: Jour. Sed. Petrology, v. 40, p. 888-897.
—— AND HUNTER, R. E., 1973, Bioturbational rates and effects in carbonate sand, St. John, U.S. Virgin Islands: Jour. Geology, v. 81, p. 253-268.
——, ——, AND PHILLIPS, R. L., 1971, Depositional structures and processes in the non-barred high energy nearshore: Jour. Sed. Petrology, v. 41, p. 651-670.
——, —— AND ——, 1972, Depositional models from a high-energy coast [abs.]: Am. Assoc. Petroleum Geologists Bull., v. 56, p. 609.
COOK, D. O., 1969, Sand transport by shoaling waves: Ph. D. Dissertation, Univ. Southern California, 148 p.
——, 1970, Occurrence and geologic work of rip currents in southern California: Marine Geol., v. 9, p. 173-186.
—— AND GORSLINE, D. S., 1972, Field observations of sand transport by shoaling waves: Marine Geol., v. 13, p. 31-56.
DAVIDSON-ARNOTT, R. G. D., AND GREENWOOD, B., 1974, Bedforms and structures associated with bar topography in the shallow-water wave environment, Kouchibouguac Bay, New Brunswick: Jour. Sed. Petrology, v. 44, p. 698-704.
DINGLER, J. R., 1974, Wave-formed ripples in nearshore sands: Ph. D. Dissertation, Univ. California, San Diego, 136 p.
EVANS, O. F., 1941, The classification of wave-formed ripple marks: Jour. Sed. Petrology, v. 11, p. 37-41.
GARY, M., MCAFEE, R., JR., AND WOLF, C. L. (EDS.), 1972, Glossary of Geology: Am. Geol. Inst., 805 p.
HARMS, J. C., 1969, Hydraulic significance of some sand ripples: Geol. Soc. America Bull., v. 80, p. 363-396.
HUNTER, R. E., 1974, Types of eolian strata and pseudostrata [abs.]: Am. Assoc. Petroleum Geologists and Soc. Econ. Mineralogists and Paleontologists Ann. Meeting Abstracts, v. 1, p. 47-48.
INMAN, D. L., 1957, Wave-generated ripples in nearshore sands: U.S. Army Corps of Engineers, Beach Erosion Board Tech. Memo. 100, 66 p.
——, 1963, Ocean waves and associated currents: *In* F. P. Shepard, Submarine geology: Harper and Row Pub., New York, p. 49-81.
—— AND BOWEN, A. J., 1963, Flume experiments on sand transport by waves and currents: Am. Soc. Civil Engineers, Proc. 8th Conf. on Coastal Eng., p. 137-150.
——, EWING, D. C., AND CORLISS, J. B., 1966, Coastal sand dunes of Guerro Negro, Baja California, Mexico: Geol. Soc. America Bull., v. 77, p. 787-802.
IVERSON, H. W., 1952, Studies of wave transformation in shoaling water, including breaking: Natl. Bur. Standards Circ. 521, p. 9-32.
KELLING, G., 1958, Ripple-mark in the Rhinns of Galloway: Edinburgh Geol. Soc. Trans., v. 17, p. 117-132.
KOMAR, P. D., 1974a, Oscillatory ripple marks and the evaluation of ancient wave conditions and environments: Jour. Sed. Petrology, v. 44, p. 169-180.
——, 1974b, Transport of cohesionless sediment on continental shelves: *In* D. J. Stanley and D. J. P. Swift (eds.), The NEW concepts of continental margin sedimentation—sediment transport and its application to environmental management, short course lecture notes: Am. Geol. Inst., p. 215-267.
—— AND MILLER, M. C., 1973, The threshold of sediment movement under oscillatory water waves: Jour. Sed. Petrology, v. 43, p. 1101-1110.
—— AND ——, 1974, Sediment threshold under oscillatory waves: Am. Soc. Civil Engineers, Proc. 14th Conf on Coastal Eng., p. 756-775.
——, NEUDECK, R. H., AND KULM, L. D., 1972, Observations and significance of deep-water oscillatory ripple marks on the Oregon continental shelf: *In* D. Swift, D. Duane, and O. Pilkey (eds.), Shelf sediment transport: Dowden, Hutchinson and Ross, Inc., Stroudsburg, Pennsylvania, p. 601-619.
LEMEHAUTE, B., DIVOKY, D., AND LIN, A., 1969, Shallow water waves; a comparison of theories and experiments: Am. Soc. Civil Engineers, Proc. 11th Conf. on Coastal Eng., p. 86-96.
LONGUET-HIGGINS, M. S., 1953, Mass transport in water waves: Royal Soc. London Philos. Trans., Ser. A., v. 245, p. 535-581.
MADSEN, O. S., 1974, Wave climate of the continental margin (its mathematical description): *In* D. J. Stanley and D. J. P. Swift (eds.), The NEW concepts of continental margin sedimentation—sediment transport and its application to environmental management, short course lecture notes: Am. Geol. Inst., p. 42-108.
MICHE, R., 1944, Undulatory movements of the sea in constant and decreasing depth: Ann. de Ponts et Chaussecs, May-June, July-August, p. 25-78, 131-164, 270-292, 369-406.

NEWTON, R. S., 1968, Internal structure of wave-formed ripple marks in the nearshore zone: Sedimentology, v. 11, p. 275-292.

REINECK, H. E., AND SINGH, I. B., 1973, Depositional sedimentary environments: Springer-Verlag, New York, 439 p.

RHOADS, D. C., 1967, Biogenic reworking of intertidal and subtidal sediments in Barnstable Harbor and Buzzards Bay, Massachusetts: Jour. Geology, v. 75, p. 461-476.

SHEPARD, F. P., AND INMAN, D. L., 1951, Nearshore circulation: Univ. California, Council on Wave Res., Proc., 1st Conf. on Coastal Eng., p. 50-59.

SIMONS, D. B., RICHARDSON, E. V., AND NORDIN, C. F., JR., 1965, Sedimentary structures generated by flow in alluvial channels: *In* G. V. Middleton (ed.), Primary sedimentary structures and their hydrodynamic interpretation: Soc. Econ. Paleontologists and Mineralogists, Spec. Pub. 12, p. 34-52.

TANNER, W. F., 1971, Numerical estimates of ancient waves, water depth and fetch: Sedimentology, v. 16, p. 71-88.

U.S. ARMY COASTAL ENGINEERING RESEARCH CENTER, 1973, Shore protection manual: U.S. Army Coastal Eng. Res. Cent., 3 volumes.

WIEGEL, R. L., 1954, Gravity waves, tables of functions: Univ. California, Berkeley, Eng. Foundation, Council on Wave Res., 30 p.

APPENDIX

Calculation of estimates of wave conditions and waters depths associated with wave-formed structures.—The conceptual model described herein is based on integrating parameters: maximum velocity and velocity asymmetry of the oscillatory current at the bottom. It therefore does not directly relate sedimentary structures to factors such as wave height and water depth. The integrating parameters used, however, are controlled by wave characteristics and water depth. By using calculations based on wave theory, it is possible to estimate the combinations of wave height, wave period, and water depth associated with a particular value or range of values of u_m or Δu_m.

This appendix briefly summarizes the technique for making such calculations. For derivations of basic wave equations, the interested reader is referred to the excellent summaries provided by Inman (1963), Madsen (1974) or Volume I of the Shore Protection Manual of the U.S. Army Coastal Engineering Research Center (1973). Simplifying assumptions of "deep water" or "shallow water" conditions are not made; the equations, though somewhat more complex, are still readily solved, particularly by using a calculator with hyperbolic functions.

The basis for the equations is outlined in Volume I of the Shore Protection Manual cited above. A wave in deep water (h greater than $L/2$) can be considered to have a specific deep-water wave length ($L_o = 1.56\ T^2$ meters). As the wave shoals, the wave length decreases and the wave height (H) first decreases slightly, then near the breaker zone, it increases. The values at depth (h) can readily be determined from tables computed by Wiegel (1954) and reprinted in Volume III of the Shore Protection Manual. For any value of L_o/h, corresponding values of L/h, $2\pi h/L$ (or kh), and H/H_o are given that provide a ready estimate of L and H or H_o (deep water wave height).

From the relations shown in the model, one can determine an approximate value (or range of values) of u_m or Δu_m associated with a specific structure or with a boundary between structural fields. He can then calculate various combinations of wave height, wave period and water depth that would produce this value (or range of values). Accordingly, most of the equations below are expressed in terms of H; the equation can be solved for different values of T and h, and a graphic plot of the results will show the combination of conditions required to produce a certain u_m or Δu_m (Fig. 7).

The predicted relation of maximum velocity to wave characteristics and water depth differs depending on the wave theory used. Figure 2-7 of the Shore Protection Manual Vol. I (1973) indicates the conditions under which different theories are applicable. In combinations of relatively deep water and relatively small waves, the linear theory applies whereby

$$H = \frac{T\,u_m \sinh kh}{\pi} \tag{5}$$

with all values considered to be in meters, seconds or meters/second. This equation provides a reasonable approximation, even in shallow water (LeMehaute and others, 1969). In water depths and wave combinations where the Stokes second-order theory is applicable (at $HL^2/2h^3$ less than 13.16 according to Madsen, 1974)

$$T = \frac{L \cosh kh}{4.9H}\left[u_m - \frac{3}{4}\left(\frac{\pi H}{L}\right)^2 \frac{(1.56\,L \tanh kh)^{1/2}}{\sinh^4 kh}\right] \tag{6}$$

If values are assumed for both u_m and Δu_m, the same relation can be expressed:

$$H = \frac{T \sinh kh}{\pi}\left(u_m - \frac{\Delta u_m}{2}\right) \tag{7}$$

In very shallow water ($h < 2.5H$), the empirical relation described by Inman (1963) relating u_m to solitary waves may be useful:

$$u_m = 1/3\ H/h\ [g(h + H)]^{1/2} \tag{8}$$

Velocity asymmetry is more difficult to estimate quantitatively. Under conditions where the Stokes second-order equation is applicable,

$$H = (0.0675\ \Delta u_m\ T\ L\ \sinh^4 kh)^{1/2} \tag{9}$$

This equation is invalid near the surf zone where use of the more complex cnoidal wave theory may be required for reasonable approximations of Δu_m (Adeyemo, 1970). The empical relation of Inman (1963) using solitary wave theory indicates that near the breaker zone Δu_m is approximately $1/2 u_m$, a relation somewhat supported by data of Adeyemo (1970, figs. 2, 3).

Another parameter of oscillatory currents that is geologically useful is the orbital diameter (d_o), a term valuable for estimating environmental characteristics on the basis of the wave length and grain size of symmetric ripples. The expression

$$H = d_o \sinh kh \tag{10}$$

is useful for determining the combination of wave heights and water depths associated with critical values of d_o indicated in Figure 9.

FACIES RELATIONSHIPS ON A BARRED COAST, KOUCHIBOUGUAC BAY, NEW BRUNSWICK, CANADA

ROBIN G. D. DAVIDSON-ARNOTT AND BRIAN GREENWOOD
Scarborough College, University of Toronto, Canada

ABSTRACT

Bedforms and sedimentary structures formed by wave oscillatory and wave-generated unidirectional currents in modern nearshore environments provide a useful analogue for the interpretation of ancient sediments. This paper presents: (1) a basic facies model for a modern, barred nearshore environment, and (2) an examination of the application of the flow regime concept to flow and bedform generation in the nearshore area. Work was carried out in Kouchibouguac Bay, New Brunswick, Canada in an area characterized by two marine bar systems: (1) an inner system of bars up to 1 m in height and breached at intervals by rip channels and (2) a single continuous, crescentic outer bar system up to 2.5 m high, located 200 to 300 m offshore. Fetch restrictions limit waves to a maximum of 2.5 m high and 7 to 8 second period during northeasterly storm winds. Bedforms were observed using SCUBA, but only in the inner system during high energy conditions. Sedimentary structures were studied in resin peels made from 90 box cores taken from the nearshore zone.

Four subfacies associated with each bar and trough, and extending parallel to the shoreline were identified: (1) a seaward slope facies characterized by sets of small-scale oscillation ripple cross-lamination and units of seaward dipping plane bed. Considerable bioturbation may be present particularly in the outer system; (2) a bar crest facies characterized by units of sub-horizontal plane bedding and medium-scale cross-bedded units produced by migration of lunate megaripples. Cross-bedding dip may be both landward and seaward; (3) a landward slope facies characterized in the outer system by oscillation ripple cross-lamination and units of low-angle, landward-dipping plane bed, and in the inner system by high-angle landward dipping, medium-scale cross-bedding up to 1 m in thickness produced on an avalanche slope by bar migration; (4) a trough facies characterized by poorly preserved oscillation and current ripple cross-lamination, dipping roughly parallel to shore. A distinct organic component produces a dark coloration. A fifth subfacies, the rip channel facies, is found only in the inner system associated with distinct rip channels. It is characterized by units of medium-scale cross-bedding dipping seaward and produced by megaripple migration under the influence of rip currents.

Comparison of the sequence with that of Clifton and others (1971) and consideration of flow conditions in the nearshore area suggests the existence of three distinct flow regime sequences: (1) a symmetric oscillatory flow regime controlled by near-symmetric oscillatory currents with a bedform sequence, no movement-ripples-plane bed; (2) an asymmetric oscillatory flow regime controlled by highly asymmetric oscillatory and translational flow with a bedform sequence, no movement—ripples—lunate megaripples—plane bed; (3) a unidirectional flow regime controlled by unidirectional longshore and rip currents with a bedform sequence, no movement—ripples—megaripples—plane bed.

INTRODUCTION

The transport of sediment, whether in water, air or some other fluid, is usually accompanied by some deformation of the sediment-water interface. It has long been recognized that these deformations or bedforms exhibit regular variations in shape in response to differences in flow conditions and flow velocities. If the sedimentary structures produced by these bedforms are subsequently preserved, they can be used to characterize the depositional environment. Bedforms and structures have been studied from a wide range of depositional environments, including rivers, dunes, turbidity currents and coastal areas. There is a considerable body of literature dealing with field, laboratory and theoretical studies, particularly concerning bedforms generated by unidirectional currents flowing over cohesionless material. There are, however, comparatively few studies of bedforms produced under oscillatory flow conditions, particularly those associated with breaking waves in the nearshore area.

Most studies of bedforms and structures produced by wave action have been carried out either seaward of the area where rapid wave transformation begins, or on the beach face. Early work in the lower shoreface was carried out by Evans (1941) and a detailed study was made by Inman (1957). Other notable works include those of Tanner (1959, 1963, 1971), Davis (1965), Risk (1967), Newton (1968) and Cook (1969). The structures found in the swash zone and beach face have been described in even greater detail, probably because of their accessibility at low tide. Among the earliest work was that of Thompson (1937) and subsequent studies have included those of Emery and Stevenson (1950), McKee (1957), Andrews and van der Lingen (1969), Hayes (1969) and Clifton (1969). Studies of berm and intertidal bars (ridge and runnel) include those of Thompson (1937), Davis and others (1972) and Reineck (1963).

Studies of bedforms and structures in the inner nearshore area, however, have been very limited, the notable exceptions being those of Clifton and others (1971) on the high energy Oregon coast, Hunter and others (1972) on the Gulf coast and Reineck and Singh (1971, 1973) on the Mediterranean coast. Risk (1967) studied bedforms associated with bar and trough topography in Lake Huron but the observations were made only under low energy conditions and did not include examination of structures. A search of the literature fails to reveal any other significant field studies of bedforms and structures associated with waves breaking either on plane or barred coasts.

The nearshore, shoreface marine environment is complex, being dominated by the processes of waves, tides and secondary currents associated with breaking waves. The primary objective of this study is to provide information on the types of bedforms and resulting sedimentary structures present within the barred nearshore system in Kouchibouguac Bay, New Brunswick, Canada, and on the size characteristics of the sediments themselves. Because direct observations on the manner by which sediment is transported during high energy conditions is difficult, the study of sedimentary structures can provide: (1) useful information on bed conditions in the different zones; (2) relationships to changing energy levels and (3) the direction of sediment transport. By taking cores from all zones within the bar systems, together with direct observations of bedforms, it was anticipated that a general model could be formulated relating bedforms and sedimentary structures to wave and current processes and topographic features within the systems. Such a model complements studies on the form and movement of the bars and provides additional information on the nature of the wave and current processes which control their formation and movement. Finally, the model should be useful in the study of bedforms in other shallow water areas and in the study of ancient sediments.

Two major aims of this paper are: (1) to establish the basic facies characteristics of a modern nearshore barred environment as an analogue for future paleo-environmental reconstruction and (2) to examine the concept of flow regime and bedform generation under oscillatory flow conditions.

LOCATION AND DESCRIPTION OF STUDY AREA

The study was carried out in Kouchibouguac Bay, which is located at the western end of the Northumberland Strait on the New Brunswick coast of Canada (Fig. 1). A series of barrier islands and barrier spits 29 km long extends in a gentle arc from Pt. Sapin in the north to Richibucto Cape in the south. The barrier islands are separated by three permanent inlets located opposite the tidal estuaries of rivers draining into the bay. The general characteristics of the barrier island system have been described by Davidson-Arnott (1971) and Bryant and McCann (1973).

Detailed work was carried out on one of the barrier islands, North Richibucto Beach, primarily in the two study areas shown in Figure 1 and in a third area located midway between them. North Richibucto Beach is characterized by the presence of two distinct bar systems in the nearshore area: (1) an outer system consisting of a single bar, crescentic in form, which is continuous for most of the length of the barrier island; and (2) an inner system of one, two or, occasionally, three discontinuous bars which are frequently crescentic in form (see Fig. 2) although straight and transverse forms also occur. An example of the morphology of the bar systems is shown in Figure 3.

The outer bar along North Richibucto Beach ranges in height from 1.5 m to 2.5 m and its crest is located 200 m to 300 m offshore in 2.5 m to 3.5 m of water. It is separated from the inner system by a distinct deep trough. The inner bars range in height from 0.5 m to 1.25 m and extend up to 150 m offshore. The shoal areas of the innermost bars may be exposed for short periods at spring low tides but otherwise the bars are submerged. Both bar systems are permanent features of the nearshore bathymetry. The bars maintain their form and position over periods of several months with only gradual changes and are not destroyed during storms. The morphology of the bars and their response to wave and current processes have been described in detail by Greenwood and Davidson-Arnott (1975).

The average beach slope off North Richibucto Beach is 0.5% from 0 to 10 m and 0.1% from 10 to 20 m. The beach and nearshore area is composed of sand-sized sediments (to a depth of 10 m) but some relict Pleistocene gravels and bedrock are found in deeper water (Kranck, 1967). Size characteristics of sediments associated with the inner and outer systems are given in Table 1.

During the summer months the predominant winds blow offshore from the southwest (Fig. 1). Important wave generation occurs under the influence of winds blowing from the northeast (the direction of maximum fetch) during the passage of storms which occur on the average two to three times a month. Mean wave height during the storms ranges from 1.0 to 2.0 m with periods of 4 to 8 seconds. Because of the effects of refraction by a large shoal area extending in a northwesterly direction from Prince Edward Island, and by further refraction at the entrance to the Strait, waves generated in the Gulf of St. Lawrence with periods greater than 8 seconds

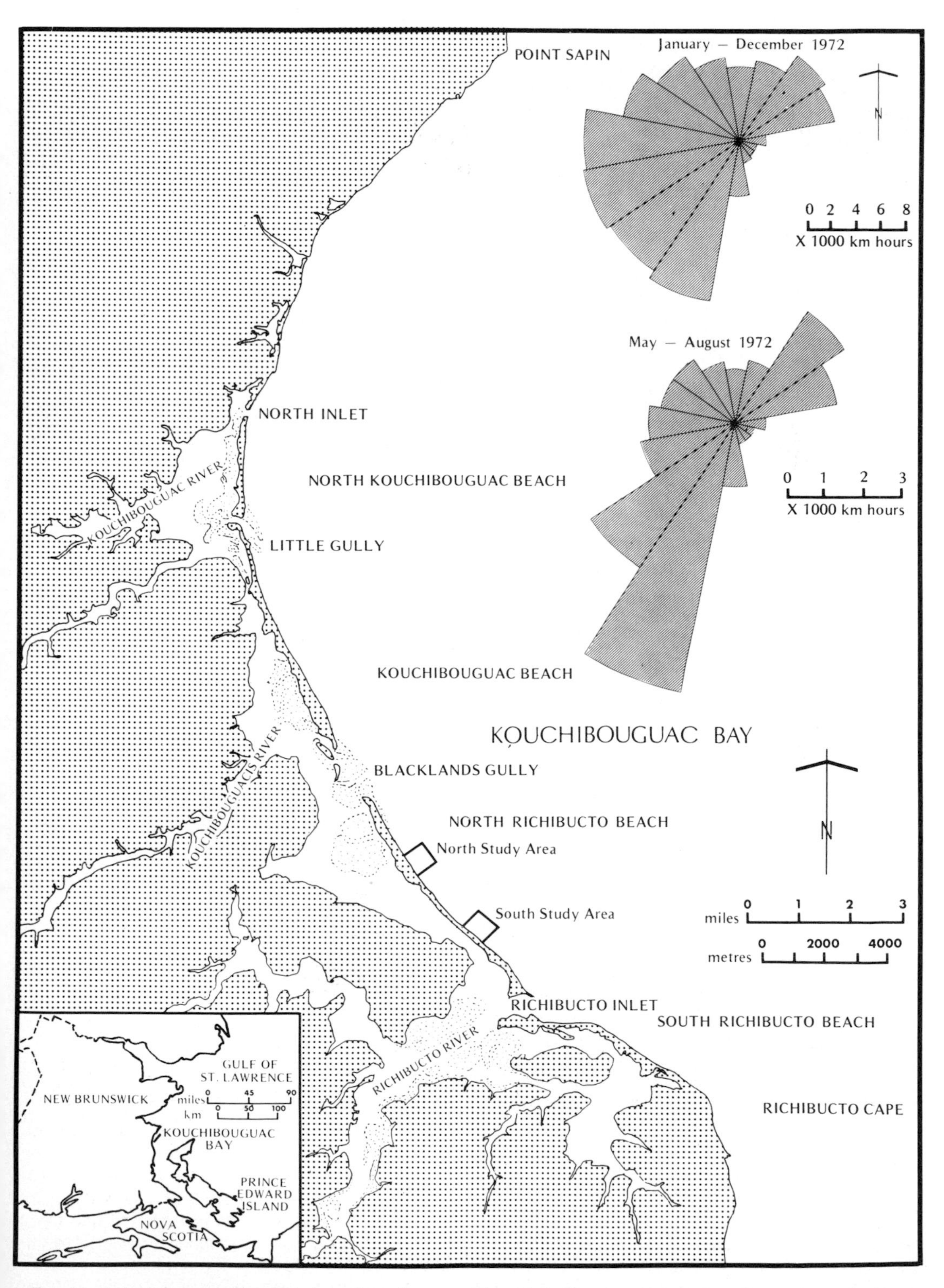

FIG. 1.—Index map showing location of study area. The wind diagrams summarize the product of average wind speed and duration for each direction.

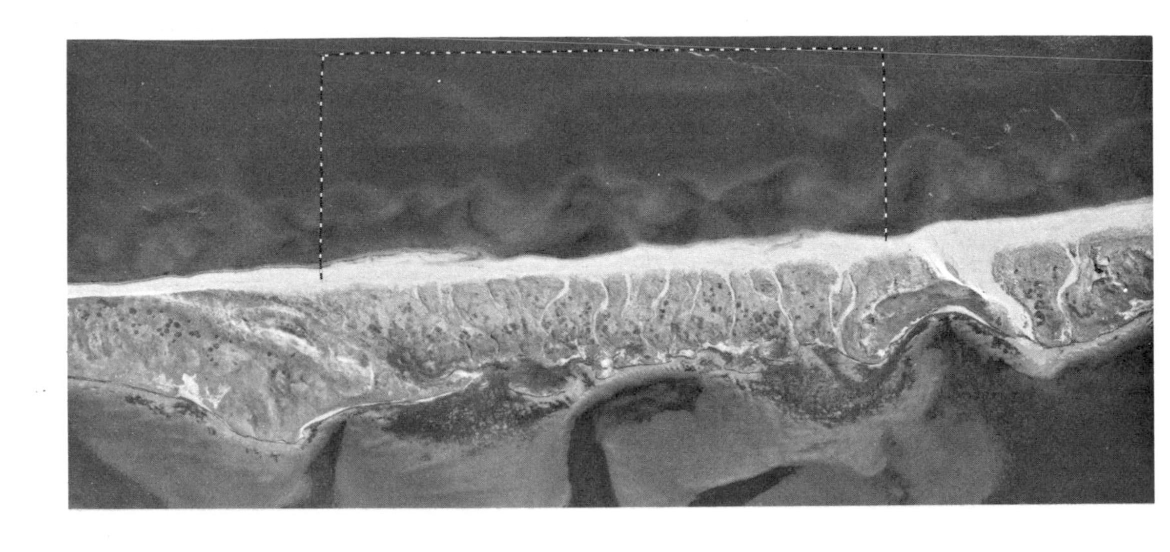

FIG. 2.—Aerial photograph, taken July 18, 1973, illustrating inner and outer bar systems. The area outlined is illustrated in detail in Figure 3.

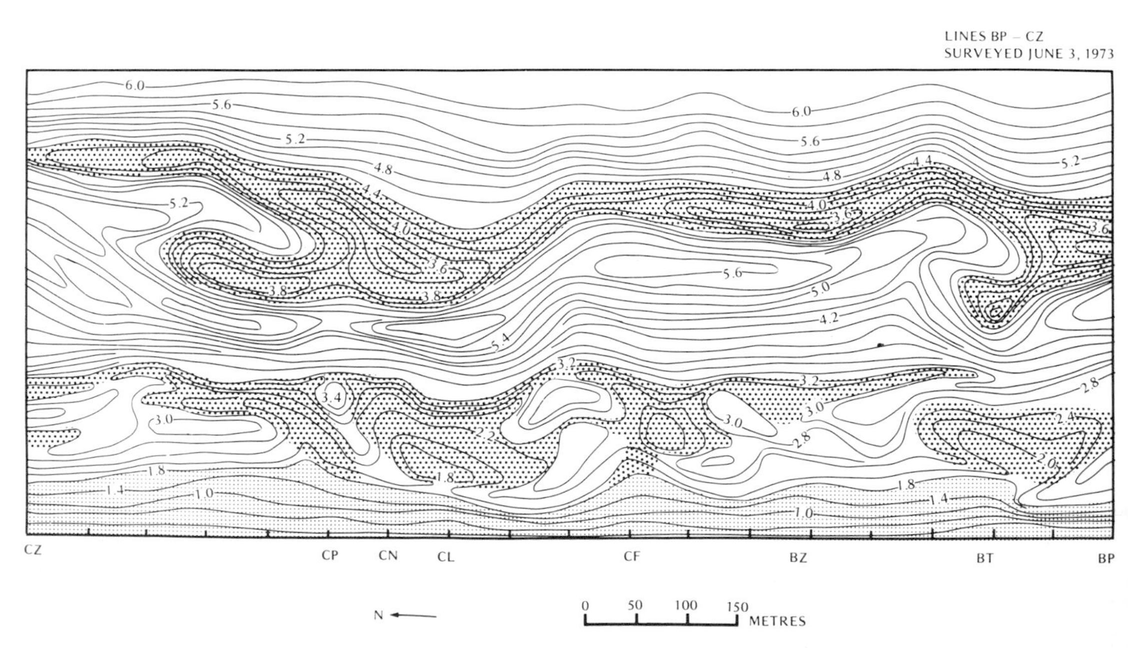

FIG. 3.—Map of the North Study Area illustrating inner and outer bar systems. Survey carried out by echo sounding along lines 60 m apart, and by levelling in the inner system along lines 30 m apart. Depths are in meters below a fixed datum on the beach (0.5 m above high high water) and the contour interval is 20 cm. Light stipple denotes beach above low low water; heavy stipple outlines bar form.

TABLE 1.—MEAN AND RANGE OF SIZE-FREQUENCY STATISTICS OF SEDIMENTS FROM THE INNER AND OUTER BAR SYSTEMS

Location	Number of Samples	mϕ min.	mϕ av.	mϕ max.	sϕ min.	sϕ av.	sϕ max.
Inner bar-trough system	103	1.61 (0.33 mm)	2.25 (0.21 mm)	2.57 (0.17 mm)	0.24	0.39	0.69
Outer bar-trough system	62	0.83 (0.56 mm)	2.12 (0.23 mm)	2.83 (0.14 mm)	0.36	0.53	1.10

Note: mϕ = phi mean; sϕ = phi standard deviation: computed using the method of moments.

do not affect the Bay (McCann and Bryant, 1972). Waves from the easterly and southeasterly directions are of limited importance because of the low frequency of winds from this direction and the limited fetch length (maximum 35 km). Significant wave heights and periods for waves recorded over a two month interval during the 1973 field season are given in Table 2.

During storms the short, steep waves break as spilling breakers on the outer bar, reform in the trough and repeat the process in the inner system. Rip cell circulation patterns are well-defined in the inner system, where longshore currents flow in the troughs, and rip currents flow seaward not only in distinct channels but also across the center of crescentic bars. Measured current velocities in the trough frequently range from 20 to 40 cm/sec and rip current velocities may exceed 75 cm/sec. Current velocities in the outer system are less well documented but seem to be considerably lower than in the inner system. Maximum tidal range is 1.25 m and tidal currents seaward of the outer system rarely exceed 15 cm/sec.

FIELD METHODS

Observations and measurements of bedforms were carried out by diving, with and without SCUBA. However, during storms these were limited to the inner system because the difficulties of operating a small boat under such conditions and poor underwater visibility made diving impossible.

Sedimentary structures were studied in resin peels made from cores taken with a box corer (modified after Klovan, 1964) with all coring being carried out under low-wave conditions. The corer is 45 cm high, 30 cm wide and 20 cm deep. Cores were obtained in shallow water by wading and in deep water by a SCUBA diver working from a small boat. In order to preserve the surface bedforms, an open metal frame 5 cm high was placed over the selected area and dyed sand spread over the surface (Newton, 1968).

Peels were obtained using the method of Burger and others (1969). After removing the peels, individual units within the core were sampled for size analysis. Approximately 90 cores were obtained from the area extending from the beach face to the seaward side of the outer bar. These form the basis for identification of the facies discussed in this paper.

TABLE 2.—FREQUENCY DISTRIBUTION OF SIGNIFICANT WAVE HEIGHT AND PERIOD FOR A 2 MONTH INTERVAL, SUMMER 1973

Height (cm)	Period (seconds) 1-1.9	2-2.9	3-3.9	4-4.9	5-5.9	6-6.9
20-39	36	168	79	32	6	2
40-59	0	27	37	38	8	6
60-79	0	5	21	23	13	3
80-99	0	0	4	5	18	8
100-119	0	0	0	2	5	8
120-139	0	0	0	0	2	1
140-159	0	0	0	0	2	1
160-179	0	0	0	0	4	1
180-199	0	0	0	0	1	0

FACIES MODEL FOR BARRED SHORELINES

Although the entire nearshore zone from the point at which waves first begin to shoal to the point of final wave collapse may constitute one environment associated with the zone of wave transformation, it is evident that distinct subenvironments can be identified associated with the varying bar topography. Each of these subenvironments has distinct bedforms and sequence of sedimentary structures; five subfacies can be documented on the basis of sediment characteristics and are related closely to varying topography, wave and current characteristics. These facies have been termed in accord with the subenvironments they reflect: (1) seaward slope; (2) bar crest; (3) landward slope; (4) trough; and (5) rip channel. The first four recur where more than one bar is present while the last is confined to the inner system of bars, where distinct rip channels periodically breach the system. Each subfacies is characterized by a distinct assemblage of sedimentary structures which reflects shifts in bedform type with changes in water depth and wave transformation.

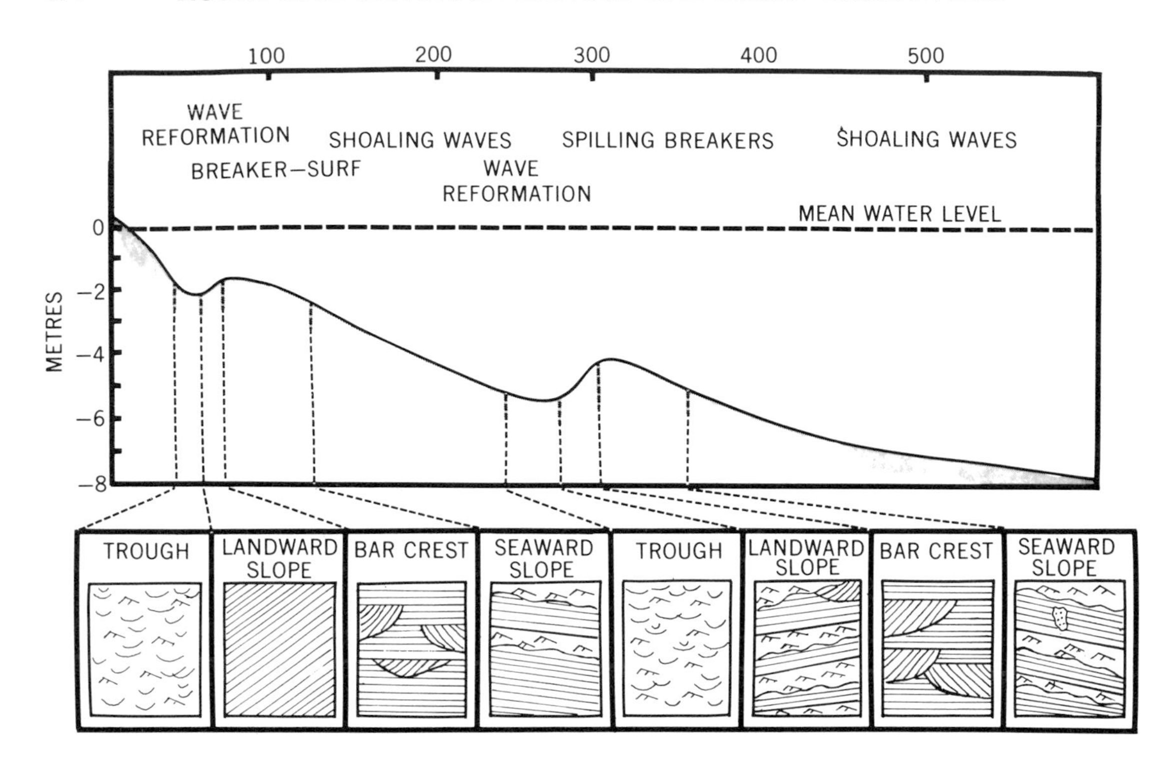

FIG. 4.—Facies model of nearshore barred topography illustrating characteristic sedimentary structures and wave transformation zones.

Bedforms on the seaward slope and crest of the bar are primarily controlled by wave oscillatory and translatory currents and those in the trough and rip channels by unidirectional longshore and rip currents. The landward slope comes under the influence of both types of currents, particularly in the outer system and near the center of crescentic inner bars.

Bedforms within the zones vary with differing wave characteristics. The zones on the bar crest and seaward slope in particular, shift their boundaries with variations in wave height and tidal stage. Most of the preserved sedimentary structures originate under high wave conditions when sediment transport rates are highest and strong currents are generated in the troughs.

The major sequence of structures and associated wave characteristics is illustrated in Figure 4. The following discussion of the subfacies will attempt to define the relationships more closely.

SEAWARD SLOPE FACIES

Bedforms on the seaward slope of the bar are generated by shoaling waves and consist of two principal types—ripples and plane bed. Under low current velocities at the bed, ripples form with crests aligned roughly perpendicular to the direction of predominant wave advance. As current velocities increase, either toward the bar crest or with increasing wave height or period, the ripples are washed out to form a plane bed.

The seaward slope facies of both the inner and the outer system is characterized, therefore, by interbedded sets of small scale ripple cross-lamination which dips predominantly landward, and by seaward dipping, low angle, plane bedding (Fig. 5a, b and f). These may form composite bedsets (Campbell, 1967) of "plane-to-ripple" bedding.

In general, plane bed sets are more common than sets of ripple lamination in the seaward slope facies. This reflects not only the fairly shallow water depths, and thus relatively high current velocities, but also the low rates of sediment transport in ripples formed by oscillatory conditions. As a result only thin units of ripple lamination are preserved. Toward the bar crest the units of plane bed increase in importance and they may occupy almost the whole of the core (Fig. 6c). Where plane beds overly ripples, the contact is usually a sharp seaward dipping erosion surface whereas the transition from plane bed to ripples is usually more irregular. The thickness of both units tends to decrease seaward, again reflecting a decrease in the rate of sediment transport in deeper water.

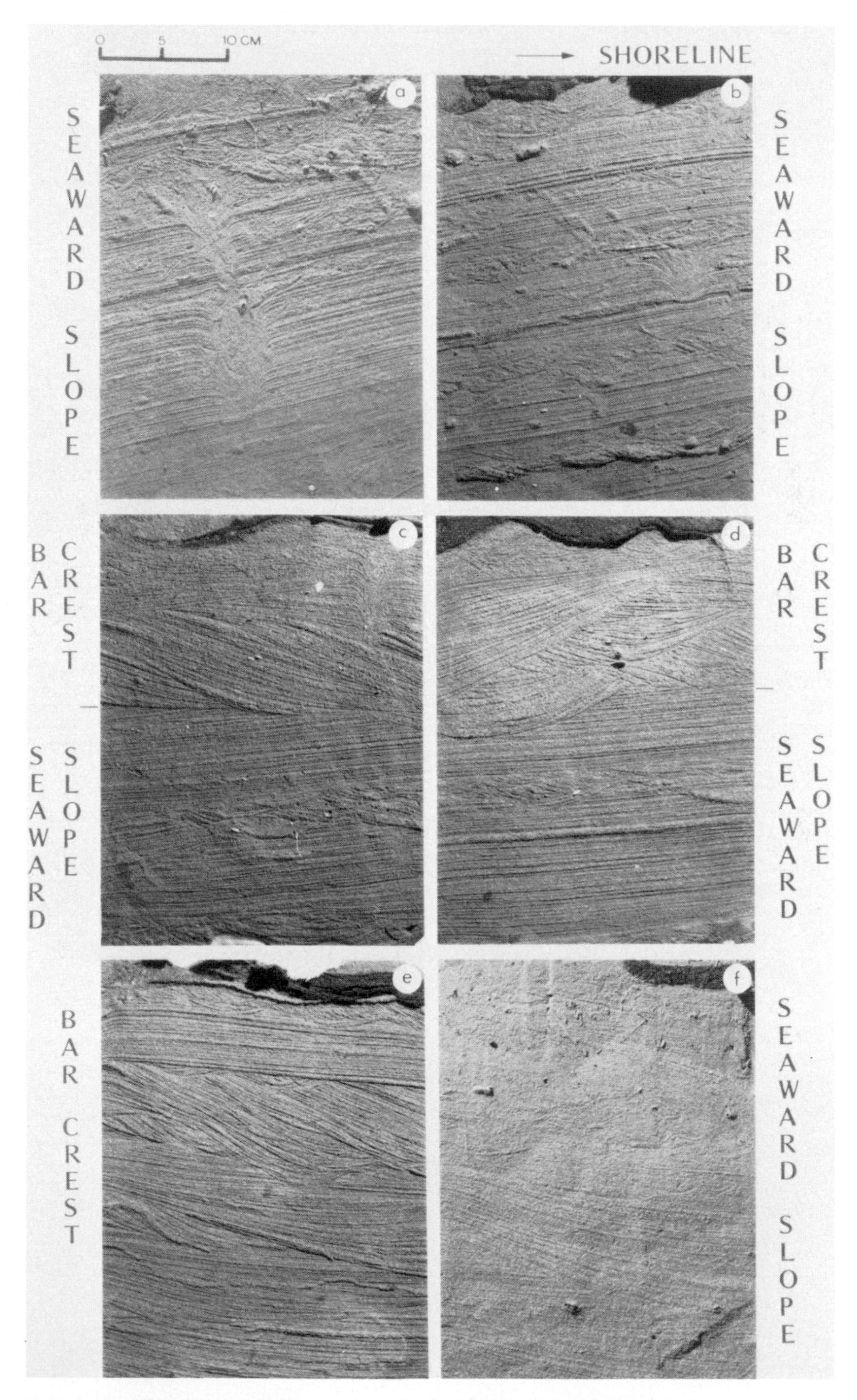

Fig. 5.—Box cores from outer bar illustrating characteristics of sedimentary structures: *a–e*, sequence of cores along one profile across seaward slope (a,b) and bar crest (c,d,e); *f*, seaward slope.

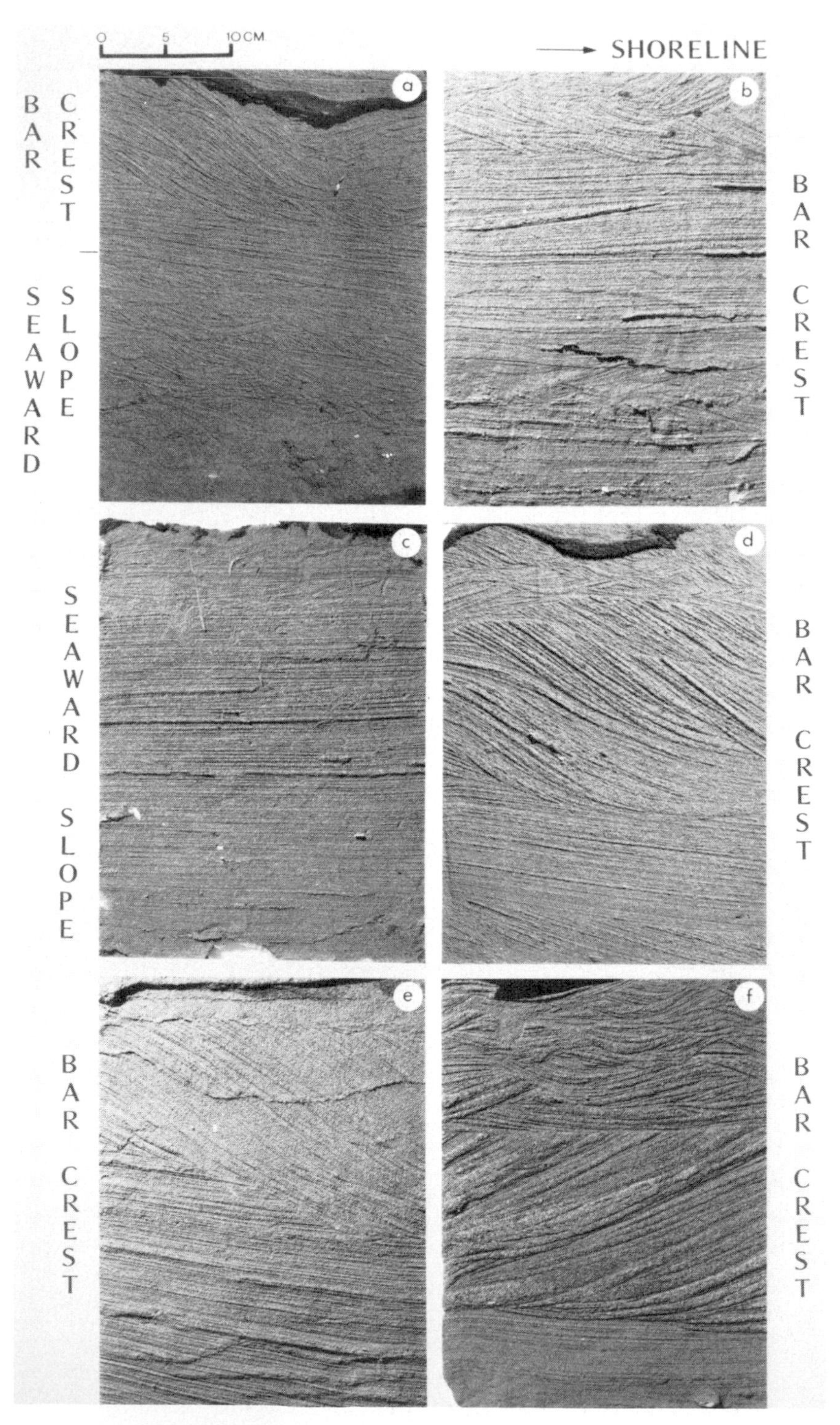

FIG. 6.—Box cores: *a*, seaward slope, inner bar; note that the apparent landward slope of the plane bedding is due to the deviation of the corer from the vertical; *b*, seaward slope, bar crest transition, inner bar; *c*, seaward slope, outer bar; *d*, crest, inner bar; *e*, crest, inner bar; *f*, crest, outer bar.

Sediments on the seaward slope in the outer system are finer than those in the inner system and grain size decreases seaward into deeper water in both systems. Sedimentary structures in the outer system are frequently disturbed by the activities of various marine organisms producing bioturbation structures (Figs. 5a–c, f and 6c) typical of the inner shoreface zone (Howard, 1972). Bioturbation decreases toward the bar crest because of greater and more frequent physical reworking. Bioturbation is rare in the inner system.

BAR CREST FACIES

During high wave conditions the bar crest is an area of shoaling and breaking waves. On the outer bar and most of the inner bars, waves form spilling rather than plunging breakers and the wave form is not completely destroyed across the bar. A true surf zone develops only on the shallowest inner bar and shoal areas. Bedform generation is controlled primarily by waves breaking on the bar, but is also influenced by the interaction of waves with currents flowing seaward across the crest.

The principal bedforms are plane beds, and lunate megaripples similar to those described by Clifton and others (1971). Small-scale oscillation ripples are generated during periods of low wave activity but, because of the low rates of sediment transport, sedimentary structures produced by their migration are usually obliterated during subsequent periods of high wave conditions, particularly in the inner system. The lunate megaripples develop both landward and seaward dipping slip faces on either side of the scour hollow (Fig. 6a). In areas of wave domination the megaripples migrate landward, producing landward dipping cross-stratified units. However, where a seaward flowing unidirectional current is superimposed on the wave orbital currents, the net migration of the lunate megaripples may be seaward. This produces seaward dipping cross-stratification. It occurs primarily in the center of crescentic bars, both in the inner and outer systems, where seaward flowing currents move across the bar crest, but where no topographically defined channel exists.

In the outer system, and on the deeper bars of the inner system, lunate megaripples develop over the entire bar crest. On the shoal areas and broad inner bars, however, a surf zone may develop which causes lunate megaripples to form primarily on the seaward side of the crest and plane bed predominates on the landward side.

The bar crest facies is therefore characterized by interbedded sets of subhorizontal plane beds, developed in somewhat coarser sediments than the seaward slope facies, and by cross-stratified sets produced by the migration of lunate megaripples. Examples of these structures can be seen in the sequence of cores shown in Figures 5c–e and 6d, e.

Individual sets of plane beds may exceed 20 cm in thickness and have a considerable lateral extent. The dip varies from gently seaward to gently landward, and is controlled by the bar slope at the point of deposition. Landward or seaward migration of the crest may result in both landward and seaward dipping plane bed units being superimposed on one another.

The lunate megaripple cross-stratification can be as much as 15–20 cm in thickness (Fig. 6e) but more frequently is 5–10 cm thick. Each unit extends less than 1 m laterally along the bar and only a few meters normal to the shoreline. Individual laminae within the units frequently have long, curving toesets (Figs. 5e, 6d) but wedge shaped units are also common (Fig. 6e). Most of the megaripple cross-stratification indicates migration nearly perpendicular to the shoreline, but some units indicating oblique movement have been observed.

Seaward dipping units of cross-stratification occur near the center of crescentic bars in both the inner and the outer system (Figs. 6f and 7a, b). Many of the individual foreset laminae are lenticular (Fig. 6f), probably reflecting the pulsations in sediment supply together with lateral transport across the crescentic face under extremely turbulent flow conditions. Fluctuations in the relative strength of waves and seaward flowing currents, and thus in the location of the zone of interaction between them, results in complex interbedding of landward and seaward dipping sets (Fig. 7a).

When the depth of water over the inner bar crest is very shallow, such as during spring low tides, the movement of surf bores across the crest may produce a low accretionary ridge several tens of meters long with a landward slip face 10–20 cm high. It builds landward across the crest and produces a tabular unit of cross-bedding with considerable lateral extent (Davidson-Arnott and Greenwood, 1974).

A few shell fragments may be incorporated in the bar crest sediments but bioturbation structures are almost absent. The bar crest sands have a 'clean' appearance and are generally somewhat coarser and better sorted than those of the seaward slope (Table 3).

LANDWARD SLOPE FACIES

There are numerous differences between the landward slope facies of the inner and outer systems. In the inner system relatively large volumes of sediment are transported over the bar crest during periods of high wave activity and

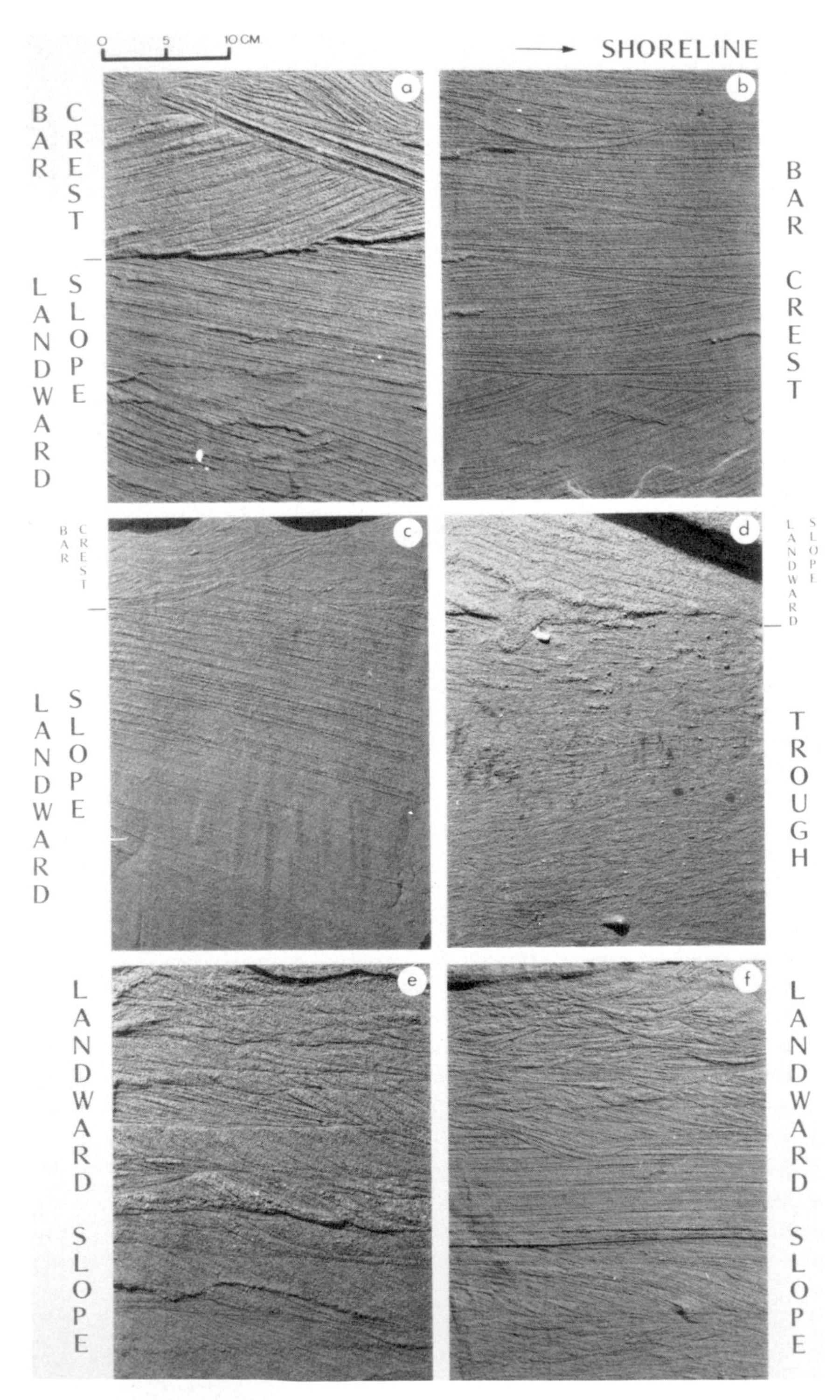

FIG. 7.—Box cores: *a*, crest, outer bar; *b*, crest, inner bar; *c*, landward slope, inner bar; *d*, landward slope—trough junction, inner bar; *e*, *f*, landward slope, outer bar.

TABLE 3.—MEAN AND RANGE OF SIZE-FREQUENCY STATISTICS OF SEDIMENTS FROM THE SEAWARD SLOPE, BAR CREST, LANDWARD SLOPE AND TROUGH FACIES OF THE OUTER SYSTEM. SAMPLES FROM THE LANDWARD SLOPE FACIES HAVE BEEN DIVIDED INTO TWO GROUPS IN ORDER TO SHOW THE CHARACTERISTICS OF SEDIMENTS ACCUMULATING AT THE FOOT OF THE SLOPE.

	$m\phi$			$s\phi$			$sk\phi$		
Facies	min.	av.	max.	min.	av.	max.	min.	av.	max.
Seaward slope	2.19	2.47 (0.18 mm)	2.83	0.36	0.48	0.62	−1.18	−0.59	−0.12
Bar crest	1.84	2.15 (0.23 mm)	2.44	0.39	0.51	0.56	−1.10	−0.15	+0.24
Landward slope	1.30	2.02 (0.25 mm)	2.51	0.39	0.47	0.62	−0.45	−0.09	+0.40
Foot of slope	0.83	1.27 (0.41 mm)	1.75	0.40	0.61	0.86	−0.61	+0.05	+0.42
Trough	1.28	1.92 (0.26 mm)	2.51	0.39	0.64	1.10	−0.87	−0.31	+0.42

are deposited by gravitational avalanching on the landward slope. This is because of the shallower water depths and greater incidence and intensity of wave breaking. The slope is maintained at a steep angle (up to 25°) by the strong longshore currents present in the landward trough which transport much of the sediment delivered over the crest alongshore and eventually offshore. As a result, the landward slope is characterized by plane bedding developed on a steep avalanche slope which produces medium scale planar cross-stratification in cross-section (Fig. 7c). This unit is nearly as thick as the bar is high (up to 1 m) and extends for the full length of the bar (Davidson-Arnott and Greenwood, 1974). The junction with the trough sediments is usually abrupt with little sign of toesets, probably because of transport of sediment by the longshore currents developed in the trough. This can be seen clearly in Figure 7d where the steeply dipping bar sands contrast with the darker trough sands characterized by small scale ripple lamination. The contact is not, however, erosional.

Wave breaking is generally confined to the crest of the bar in the outer system. Strong translatory currents are rarely present and rates of sediment transport are much lower. The landward slope is much more gentle (4.5° average) and there is no evidence of avalanche bedding. In cross-section the bar crest is gently curved and lacks the sharp landward break which is often present in the inner bars. There is a gradual reduction in the effects of oscillatory currents at the bed as water depth increases down the slope. This leads to a reversal of the sequence found on the seaward slope.

The upper part of the landward slope of the outer bar is thus characterized by units of low-angle, landward dipping plane beds and megaripple cross-bedding, the latter often filling the core (Fig. 7e). Toward the bottom of the slope the units of plane bed decrease in thickness, and are interbedded with units of small scale ripple cross-stratification (Fig. 7f).

Sediment size increases down the landward slope, which commonly contains a 10–15 m wide zone of coarse sand and gravel (Table 3) covered with large oscillation ripples. The junction with the trough sediments is usually distinct (Fig. 8a) and is often enhanced by the incorporation of organic material that accumulates at the foot of the bar, and distinctly darkens the trough sediments. In the deeper parts of the outer trough there may be considerable bioturbation.

In the inner system, near the center of crescentic bars, the steep foreset bedding may be interbedded with, or replaced entirely by, seaward dipping cross-stratification resulting from the migration of dunes or megaripples generated by seaward flowing rip currents (Fig. 8b). These units are rare in the outer system, probably because the currents there are much weaker than in the inner system.

TROUGH FACIES

The trough is a zone where longshore currents predominate and wave influence becomes of secondary importance. Under moderate or high wave conditions, waves lose much of their energy on the bar. When they move into the deep water of the trough, the oscillatory currents near the bed are very weak. Once the waves reform in the trough, they begin to shoal again and to control bedform generation.

The characteristic bedforms in the trough are current ripples, primarily oriented normal to the shore under the influence of the longshore currents. The range of orientations does vary somewhat depending on the effectiveness of wave action. The trough facies is thus characterized by small scale ripple lamination similar to that of the seaward slope facies but generally oriented at right angles to the shore (Figs. 7d, 8c, d).

After a storm, organic material such as kelp and eel grass may accumulate to a depth of several cm in the trough. When this is incorporated into the sediments it gives a distinct dark, mottled color which contrasts sharply with the bar sedi-

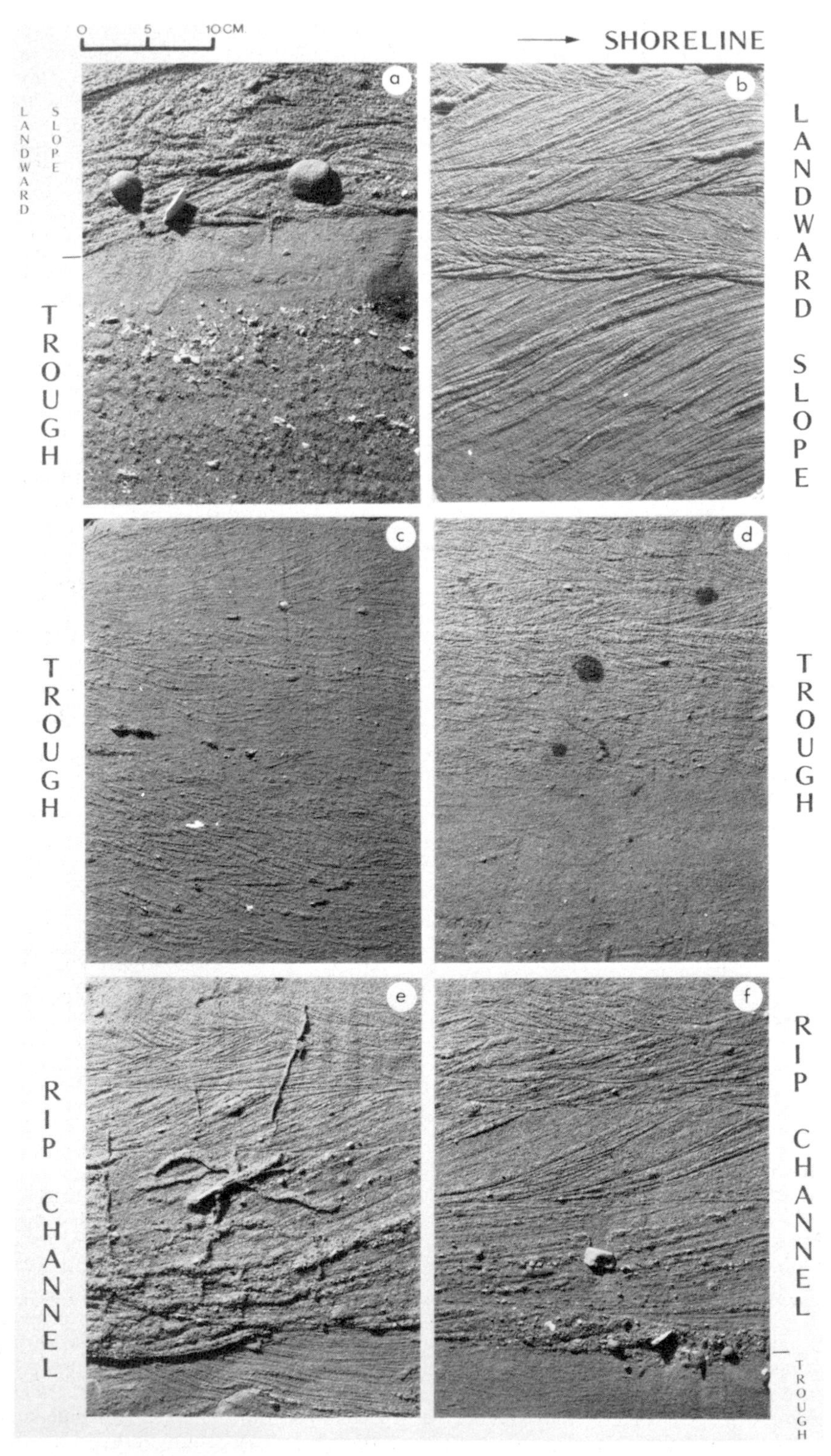

FIG. 8.—Box cores: *a*, landward slope—trough junction, outer bar; *b*, landward slope, inner crescentic bar; *c*, trough, outer bar; *d*, trough, inner bar; *e*, *f*, rip channel, inner system.

ments. Shell fragments which have been swept over the bar are also concentrated in the trough sediments. Although mean sediment size resembles that of the bar, sorting is much poorer and there is a distinct fine component which is not present in the bar sands.

RIP CHANNEL FACIES

The rip channel facies, by definition, is formed in and near the channels that dissect the inner bar system and through which the longshore currents in the trough flow seaward across the breaker line. As in the trough facies, the dominant process controlling bedform generation and movement is the unidirectional current, and wave oscillatory currents are of limited significance. The rip channel facies is distinguished from the trough facies by the presence of seaward dipping cross-bedded units up to 20 cm in thickness. These are produced by migration of megaripples under the influence of unidirectional currents. The crestal form of these bedforms varies between straight and catenary (Allen, 1968) rather than the lunate shape characteristic of the wave-formed megaripples. The distinction between wave-formed and current-formed megaripples in terms of their sedimentary structures is difficult. One common difference, however, is the form of the lower bounding surface to the sets of cross-bedding; the wave-formed lunate megaripples have a curved bounding surface while that of the unidirectional current form is planar. A similar characteristic for ripples has been noted by Boersma (1970).

Because the dominant current is unidirectional, when the longshore current velocities increase near and in the rip channel, the ripples are not washed out to form a plane bed but instead develop into dunes. Highest speeds are usually found in the neck of the rip channel where the volume of flow is greatest and the current is restricted by the sides of the channel. However, if the speeds are high enough, dune formation may extend into part of the trough.

The direction of flow of the rip current can vary considerably, particularly at the seaward side of the rip channel and thus the direction of maximum dip of the cross-bedded units also varies considerably. Examples of the rip channel facies can be seen in Figures 8e and f.

Rip currents transport large volumes of sediment seaward with much of the finer material travelling in suspension. When the rip current flow expands seaward of the breaker zone there is a rapid decrease in speed and much of the sediment is rapidly deposited. In most cases the sediments are probably immediately reworked by wave action. In some areas, however, there is evidence of the effectiveness of this seaward transport preserved in the sedimentary structures.

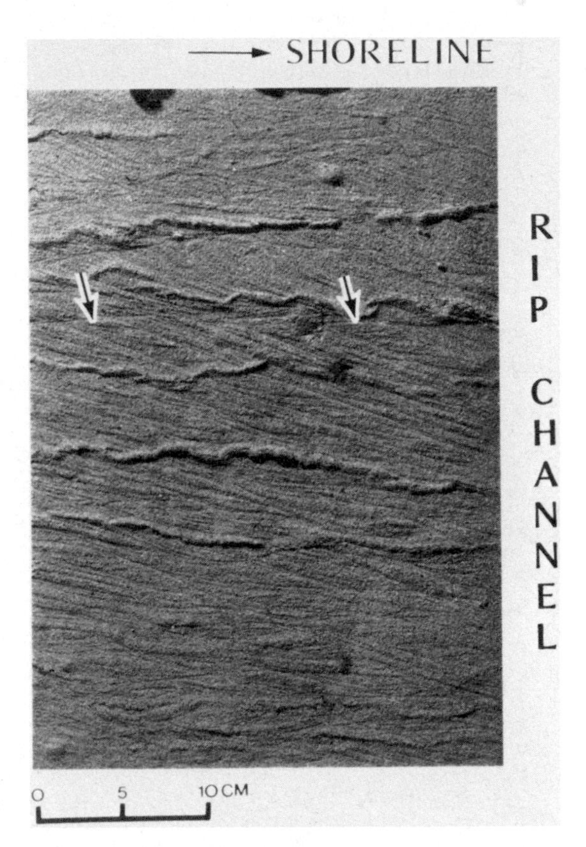

FIG. 9.—Box core from seaward slope, inner bar, marginal to rip channel illustrating sets of climbing ripple cross-stratification (noted by arrows).

One such example is shown in Figure 9, which shows climbing ripples resulting from the rapid deposition of sediment at the seaward edge of a rip channel by the seaward flowing currents.

Sediments in the rip channel facies are often coarser than those of the trough or bar, particularly near the seaward margin of the facies. Coarse material washed over the bar or moved seaward from the step area during a storm is transported along the bed by the longshore and rip current. It is deposited when the rip current velocity is checked on the seaward side of the rip channel. A fine component derived from sediment settling out of suspension as current speed decreases after a storm may also be present. As in the trough facies, the sediments may have a mottled color due to the presence of organic material.

BEACH FACE

Landward of the inner trough, the waves shoal and break again near the bottom of the swash slope. The general sequence of bedforms found on the seaward slope and bar crest is once more repeated with a complex zone of surf-swash interaction landward of the breaker zone and plane bed on the swash slope itself. Details of the bedforms and structures found in this area have been documented already by Clifton and others (1971) and Davidson-Arnott and Greenwood (1974) and need not be reiterated here.

The effects of the shifting of the bedform zones can be seen in Figure 6a in which the lower half of the core consists of structures of the seaward slope facies and the upper part is lunate megaripple cross-bedding formed at the edge of the bar crest facies. The core was taken across the slip face of an inactive megaripple and both landward and seaward dipping foresets are present. In Figure 6b, plane-to-ripple bed units of the seaward slope facies overlie the coarser plane bedding of a rip channel deposit. The latter has been buried by changes in the inner system topography resulting from the migration of a bar over the old rip channel.

FLOW REGIME CONCEPT AND BEDFORM GENERATION UNDER WAVE-DOMINATED CONDITIONS

The flow regime concept, developed primarily by Simons and co-workers (Simons and others, 1961; Simons and Richardson, 1961, 1962; Simons and others, 1965) has found a wide application in sedimentology. The basis of the concept is that there exists a definite sequence of bedforms generated as bed current velocities are increased in an alluvial channel with cohesionless bed material. The flow conditions can be divided into a lower flow regime where the bedforms are not in phase with the water surface and an upper flow regime where they are in phase. These are separated by a transition zone where the bed surface is usually flat. The sequence of bedforms in the lower regime is dependent on the size of the material. For fine sand (<0.2 mm) the sequence is: *no movement—ripples—dunes—plane bed.* As grain diameter increases however, the zone of dunes expands and that of ripples decreases until, with coarse sand (1.0 mm or greater), ripples are no longer formed and the sequence becomes: *no movement—plane bed—dunes—plane bed.* Recently, Southard and Boguchwal (1973) have shown that there is a transitional area between the two sequences in which the succession of bedforms is more complex: *no movement—ripples—plane bed—dunes—upper flat bed.*

The usefulness of the flow regime concept in the examination of sedimentary structures lies in the information it can provide on flow and sediment transport conditions. It has been shown by Southard (1971) that if conditions such as sediment density, fluid density and viscosity are held constant, the bed configuration can be described by an unique combination of sediment size, flow velocity and depth of flow. Because, the size and density of sediment composing the structures can be measured, useful information on depth of flow and flow velocity can then be derived by comparison with measured values. The rate of sediment transport increases from the lower towards the upper flow regime and thus identification of the bedform within the flow regime sequence can also provide a rough estimate of sediment transport rates.

The only field application of the flow regime concept to bedforms and structures produced by wave action appears to be that of Clifton and others (1971) and Clifton herein. In the initial study of bedforms and structures in the high energy nearshore along a non-barred coastline, Clifton and others recognized a sequence of five facies in a landward direction from the zone of shoaling waves (Fig. 10). They suggested that the bedforms in these zones could be compared to those in alluvial channels and that the sequence from ripples through lunate megaripples to plane bed, as the wave moved from deep water to the breaker and surf zones, was similar to that from ripples through dunes to plane bed under unidirectional flow. Thus, the zone of asymmetric ripples corresponds to the lower part of the lower flow regime, the lunate megaripples to the upper part of the lower flow regime and the plane bed to the transition zone.

The sequence of structures found on the seaward slope and bar crest facies of both the inner and outer systems in Kouchibouguac Bay does not appear to follow the same sequence as that described by Clifton and others (1971). Instead of the zone of lunate megaripples succeeding the asymmetric ripples, the ripples are succeeded first by plane bed and then by lunate megaripples. This is obviously in conflict with the sequence documented by Clifton and others (1971) and thus deserves further investigation. Landward of the zone of lunate megaripples in Kouchibouguac Bay the sequence is similar to that found in Oregon. In the outer system bar crest, the units of plane beds and cross-bedding are interbedded making identification of the sequence difficult. In the inner system, particularly on the broad shoal areas where a surf zone is developed, the lunate megaripples are succeeded landward by plane bed.

Results of early laboratory studies by Evans and Ingram (1943), Bagnold (1946), Manohar (1955), Kennedy and Falcon (1965), and more recent work using oscillatory flow tunnels by Carstens and others (1969) and Mogridge and Kamphuis (1972) indicate that under symmetric oscillatory flow conditions ripples are washed out to form plane bed as bed velocities and orbital diameters are increased (Figs. 11a, b). A sequence

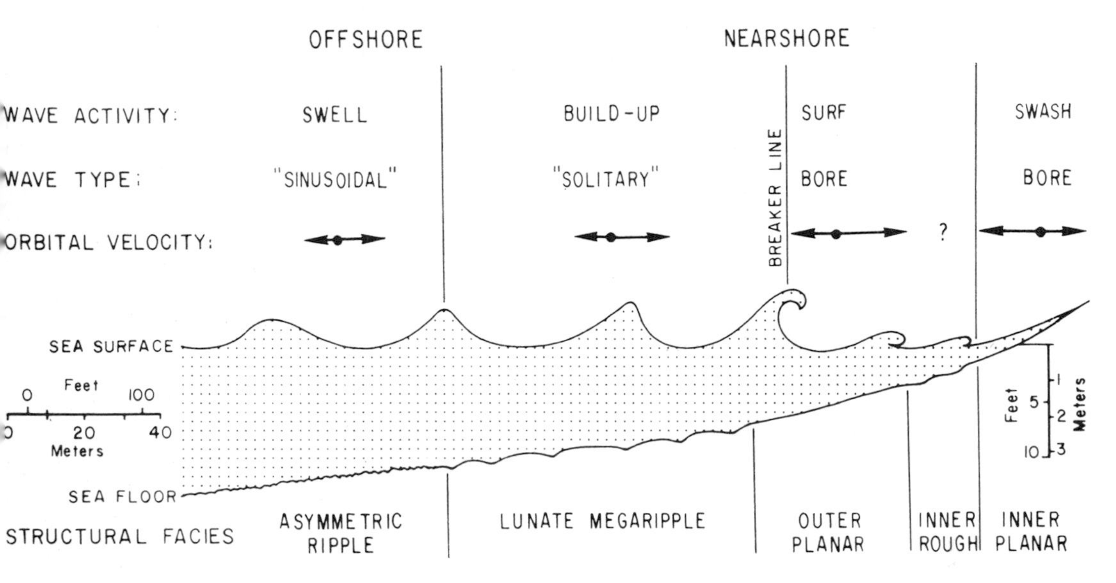

FIG. 10.—Relationship of depositional structures to wave type and activity on a non-barred, high energy coastline, (after Clifton *et al.*, 1971).

similar to that in Kouchibouguac Bay was observed by Inman (1957).

In the oscillatory tunnel experiments, bedform height is compared to the amplitude or diameter of horizontal water motion, which is the most convenient parameter to measure. It seems likely, however, that the most important factor controlling the transition from ripples to plane bed is the maximum orbital velocity at the bed. In Figure 11b the curves for bedform height can be extrapolated to zero bedform height as in Figure 11a. If this is done and all other factors are held constant, it can be seen that for any one orbital diameter, the progression from maximum bedform height to zero bedform height (plane bed) is accompanied by decreasing values for wave period. For a constant orbital diameter this must be accompanied by increasing maximum orbital velocities.

In order to determine whether the results of the laboratory studies in oscillatory flow tunnels were compatible with the observed field data, the results of Carstens and others were used to predict the critical orbital diameter at the bed needed to produce plane bed conditions on the seaward slope of the outer bar. For a wave period of 6.0 seconds and wave heights ranging from 1.0 to 2.0 m (typical of storm waves in the bay) the critical orbital diameter (110 cm) occurs in water depths ranging from 5.2 to 8.7 m assuming a sediment diameter of 0.2 mm. Thus, during a storm, plane bed conditions would be expected to occur for part of the time over the whole of the seaward slope of the outer bar, with ripples occurring during lower wave heights.

Both the observed sequence in Kouchibouguac Bay and that reported from oscillatory flow tunnel experiments indicate that under symmetric oscillatory flow (Clifton, this volume) the sequence corresponding to the lower flow regime for unidirectional flow is: *no motion—ripples—plane bed.*

The problem then arises of accounting for the zones of lunate megaripples and plane bed found in the breaker and surf zones in Oregon (Clifton and others, 1971) and on the bar crest in Kouchibouguac Bay. Although they occur in shallower water and under higher current velocities than the lower flow regime sequence previously defined, it seems unlikely that the lunate megaripples and plane bed form part of the same sequence. If this were the case they would correspond to the upper flow regime and it is difficult to conceive of conditions under wave action corresponding to the upper flow regime under unidirectional flow. Clifton and others (1971) found that the zone of lunate megaripples coincided with the zone of wave transformation where there was a rapid increase in the landward asymmetry of the orbital velocities at the bed. On this basis they distinguished between the offshore area where orbital velocities are near symmetrical and the nearshore area where they are highly asymmetric landward. It is probable, therefore, that the lunate megaripples result from this change in flow conditions and are not part of the sequence

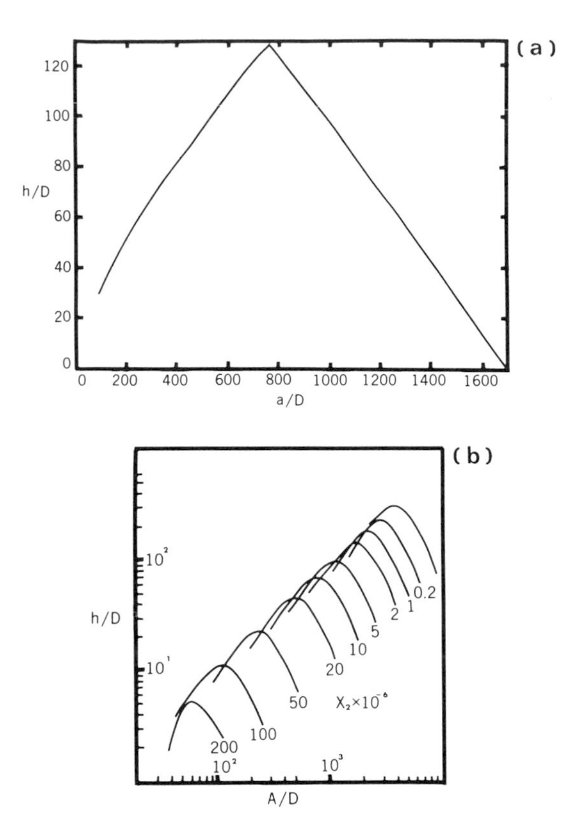

FIG. 11.—*a*, Dimensionless bedform amplitude versus dimensionless orbital amplitude (simplified after Carstens *et al.*, 1971); *b*, dimensionless bedform amplitude versus dimensionless orbital diameter (after Mogridge and Kamphuis, 1972): h = bedform height; D = grain median diameter; a = orbital amplitude; A = orbital diameter (2a); $X_2 = \rho D/\gamma_s T^2$ where ρ = fluid density; γ_s = submerged unit weight; T = period.

generated under nearly equal oscillatory currents. Additional support for this lies in the fact that lunate megaripples are also found occasionally in the lower swash zone (Clifton and others, 1971; Davidson-Arnott and Greenwood, 1974) which is also an area of strong translational currents. As suggested by Clifton and others, the lunate megaripples correspond to dunes, or the upper part of the lower flow regime, and the plane bed in the surf zone corresponds to the transition zone or lower part of the upper flow regime.

It seems, therefore, that there are two distinct flow regime sequences for bedforms in areas of wave domination: (1) a symmetric oscillatory flow regime controlled by near-symmetric oscillatory currents; and (2) a nearshore or asymmetric oscillatory flow regime controlled by highly asymmetric oscillatory and translational currents. In the offshore zone the sequence of bedforms developed in response to increasing orbital diameter and maximum orbital velocities is: *no movement—ripples—plane bed.* The nearshore sequence, controlled by increasing shoreward asymmetry and bed current velocity is: *no movement—ripples—lunate megaripples—plane bed.*

The actual sequence of bedforms from the offshore zone to the breaker zone will depend on the maximum velocities in the area of change from the symmetric oscillatory flow regime to the asymmetric oscillatory flow regime. The results of Mogridge and Kamphuis (1972) indicate that as wave period increases the initiation of plane bed occurs at greater wave orbital diameters and ripples persist into shallower water (Fig. 11b). With steep, short period waves, bed current velocities high enough to generate plane bed may occur before rapid wave transformation results in the transition to bedforms dominated by highly asymmetric flow. In this case, the sequence of bedforms in a landward direction would probably be: *ripples—plane bed (symmetric flow)—plane bed (asymmetric flow)*. Under long period waves, however, the transition may occur before velocities high enough to generate a plane bed are reached. In this case, the symmetric ripples would be succeeded by asymmetric ripples and by lunate megaripples, producing a sequence: *ripples (symmetric flow)—ripples (asymmetric flow)—lunate megaripples—plane bed (asymmetric flow)*.

The structural sequence observed in Kouchibouguac Bay (ripples—plane bed on seaward slope; lunate megaripples—plane bed on the crest) could, therefore, be achieved either by: (1) the superimposition of the two observed sequences, the former produced by very high short period wave conditions during the height of a storm and the latter by lower, longer period waves as the storm abates; or (2) a critical combination of increasing grain size toward the bar crest and wave transformation resulting in a spatially contiguous bedform sequence of: ripples (symmetric)—plane bed (symmetric)—lunate megaripples (asymmetric)—plane bed (asymmetric). This would then provide an explanation for the differences in the sequence observed by Clifton and others and that observed in Kouchibouguac Bay. A more detailed theoretical consideration of the relationships of bedform type to flow conditions and grain size is given by Clifton (this volume).

The application of the flow regime concept in the nearshore area is made even more complex by the presence of a third flow regime sequence, dominated by unidirectional longshore and rip currents, and by the interaction between unidirectional and oscillatory currents. The sequence found under unidirectional currents in the nearshore appears to be similar to that for flow in a stream channel: *no movement—ripples—*

SYMMETRIC OSCILLATORY FLOW
NO MOVEMENT – RIPPLES – PLANE BED
ASYMMETRIC OSCILLATORY FLOW
NO MOVEMENT – RIPPLES – LUNATE MEGARIPPLES – PLANE BED
UNIDIRECTIONAL FLOW
NO MOVEMENT – RIPPLES – MEGARIPPLES – PLANE BED

FIG. 12.—Suggested flow regime sequences in the nearshore area.

megaripples—plane bed.

A further complicating factor is that, although each flow regime sequence is associated with increasing bed velocities, rates of sediment transport vary considerably between similar bedforms in each flow regime. Thus, under symmetric oscillatory flow conditions, there is little or no net sediment transport even with a plane bed, and net rates of transport seaward of the zone of rapid wave transformation are low. Rates of sediment transport are highest in the unidirectional currents where sediment is being transported continuously in one direction, and are intermediate between the two under highly asymmetric or translatory flow because of the occurrence of zero or seaward flow for part of the time.

In conclusion, bedforms in the nearshore area are controlled by three different types of flow conditions, each of which produces a distinct flow regime sequence of bedforms in response to increasing bed velocities (Fig. 12). In the examination of sedimentary structures, care must be taken, therefore, to determine the type of flow under which the structure was formed, particularly in connection with the interpretation of energy levels and rates of sediment transport. Because of the difficulty of distinguishing flow conditions from the individual sedimentary structures, identification of flow conditions probably necessitates identification of the facies type first.

DISCUSSION

The study of bedforms and structures in the barred nearshore environment serves several useful functions. In the first place it provides basic information on the sedimentological characteristics of a depositional environment that has not previously been studied intensely. Also the analysis of bedforms and structures found in each subfacies provides an indication of flow conditions and the direction and rate of sediment transport which control bar dynamics. Further, the subfacies relationships provide a modern analogue for the identification of nearshore barred environments in ancient sedimentary sequences.

In the absence of detailed observations on bedforms under all conditions, the structures provide useful information on bed conditions in the different topographic zones, particularly on the levels and directions of sediment transport and also differences between inner and outer systems. The structures indicate differences between the two systems in terms of sediment transport over the seaward slope, bar crest and landward slope. Spilling breakers on the outer bar fail to produce large scale surf bores characteristic of the more complete wave collapse in the inner system. The landward slope of the inner bar is therefore characterized by a steep avalanche slip face whereas a lower angle slope in the outer system reflects continual sediment transport by oscillatory wave currents.

The structures also give an indication of the predominant direction of sediment transport. In the trough areas the structures indicate that longshore sediment transport takes place under the influence of longshore currents. In general the seaward slope of the bar (both inner and outer) is an area of landward sediment transport by waves. The existence of seaward dipping cross-bedded units in both the inner and outer landward slope and/or bar crest facies of crescentic bars, however, shows that seaward transport of sediment does occur across the bar. This could provide a possible mechanism to explain the maintenance of an equilibrium position and explain why crescentic bars do not migrate continuously shoreward. In contrast, the straight inner bars have structures suggesting continual landward transport over a sharp crest and deposition on an advancing bar front by gravitational slumping. Greenwood and Davidson-Arnott (1975) have shown that straight inner bars do migrate landward but structures in the trough and rip channel facies indicate a seaward flow of sediment which is of significantly higher magnitude than in the outer bar. This acts to reduce the rate of landward migration and trough infilling. Regarding the stability of the crescentic bars, it is interesting to note that the zone of seaward-migrating megaripples is found in the deeper crescent area and not on the shallower shoal or lower areas where landward dipping units occur almost without exception. If this is indeed the mechanism of maintaining equilibrium, then similar facies characteristics should occur in straight bars which also maintain an equilibrium position. Hunter and others (1972), however, note only plane bedding or small scale ripple cross-lamination in the bar crests of straight parallel bars along Padre Island, Texas.

The increasing asymmetry of flow produced by wave shoaling followed by energy loss through spilling breakers suggests that the outer bar is indeed maintained by increasing sediment transport to the point of breaking after which there is a gradual decrease as the wave reforms and water deepens. In contrast the straight inner bar is formed in a distinct surf zone of horizontal translatory currents resulting from greater energy loss on breaking.

The seaward slope facies with composite bedsets of "plane-to-ripple" bedding provides an indicator of storm-induced sedimentation. The sharp erosional contact between successive units is provided by truncation of the ripple cross-bedding. In some seaward slope sequences the total unit is composed of plane bedding suggesting regular removal of the low energy ripple bedding by each storm.

The nearshore bars of Kouchibouguac Bay illustrate certain geometrical and structural characteristics which may provide environmental indices for the interpretation of ancient sandstone bodies. To date no clearly identified wave-formed bars have been documented in the rock record, although "marine bars" have been cited by Masters (1967), Van der Lingen and Andrews (1968), Weidie (1968), Clifton (1972), Hobday and Reading (1972), Brenner and Davies (1973), Exum (1973) and Stricklin and Smith (1973). Although the preservation potential of sediments in the bar form is not very high because the bar itself tends to migrate in response to changing water levels (Saylor and Hands, 1970), preservation is possible under a prograding shoreline situation. The units most likely to be preserved are the lowermost part of the bar slope facies (landward and seaward), together with the trough and rip channel facies. These will be buried under beach face sediment as the progradation proceeds. The bar systems are maintained by sediment transport under storm wave conditions, particularly the outer bar. It is becoming increasingly apparent that the results of higher energy events are the ones most likely to be preserved.

Identification of wave-formed bars in a sedimentary sequence will allow specific environmental interpretations because the bars are only found under restricted, offshore slope, bed material, tidal and wave conditions. It is hoped that the presentation of this detailed facies model for a nearshore barred shoreline will aid in identifying ancient examples.

ACKNOWLEDGEMENTS

This study forms part of a larger project on coastal sedimentation supported by a grant to Greenwood from the National Research Council of Canada (NRCA7956). The basis for the paper forms a part of a doctoral dissertation written by the senior author, which was further supported by a National Research Council of Canada Postgraduate Fellowship. Thanks go to P. Hale, P. Keay and D. McGillivray for their assistance in the field and to P. Hale for photographs of cores from his B.Sc. Research Paper. The staff of the Academic and Electronic workshops helped to design and construct the equipment used in coring and wave and current measurement. We would like to thank the staff of the Graphics and Photography Department at Scarborough College for producing the diagrams and particularly David Harford for the photographs of the peels.

Drs. H. E. Clifton, R. A. Davis Jr. and G. V. Middleton read drafts of the manuscript and we are grateful to them for their constructive criticism. We further benefitted from discussions with Dr. A. V. Jopling. However, the results and interpretations presented are solely the responsibility of the authors.

REFERENCES

Allen, J. R. L., 1968. Current ripples: North-Holland Pub. Co., Amsterdam, 433 p.

Andrews, P. B., and van der Lingen, G. J., 1969, Environmentally significant sedimentological characteristics of beach sands: New Zealand Jour. Geol. and Geophys., v. 12, p. 119-137.

Bagnold, R. A., 1946, Motion of waves in shallow water—interaction between waves and sand bottoms: Royal Soc. London Proc., Ser. A, v. 187, p. 1-15.

Boersma, J. R., 1970, Distinguishing features of wave-ripple cross-stratification and morphology: Ph.D. Thesis, Univ. Utrecht, 65 p.

Brenner, R. L., and Davies, D. K., 1973, Storm-generated coquinoid sandstone: Genesis of high-energy marine sediments from the Upper Jurassic of Wyoming and Montana: Geol. Soc. America Bull., v. 84, p. 1685-1698.

Bryant, E. A., and McCann, S. B., 1973, Long and short term changes in the barrier islands of Kouchibouguac Bay, southern Gulf of St. Lawrence: Canadian Jour. Earth Sci., v. 10, p. 1582-1590.

Burger, J. A., Klein, G. D., and Sanders, J. E., 1969, A field technique for making epoxy relief-peels in sandy sediments saturated with salt-water: Jour. Sed. Petrology, v. 39, p. 338-341.

Campbell, C. V., 1967, Lamina, laminaset, bed and bedset: Sedimentology, v. 8, p. 7-26.

Carstens, M. R., Neilson, F. M., and Altinbilek, H. D., 1969, Bed forms generated in the laboratory under an oscillatory flow: Analytical and experimental study: U.S. Army, Corps of Engineers, Coastal Eng. Res. Cent. Tech. Memo. 28, 39 p.

Clifton, H. E., 1969, Beach lamination—Nature and origin: Marine Geol., v. 7, p. 553-559.

——, 1972, Miocene marine to nonmarine transition in southern coast ranges of California [abs.]: Am. Assoc. Petroleum Geologists Bull., v. 56, p. 609.

——, Hunter, R. E., and Phillips, R. L., 1971, Depositional structures and processes in the non-barred high energy nearshore: Jour. Sed. Petrology, v. 41, p. 651-670.

COOK, D. O., 1969, Sand transport by shoaling waves: Ph.D. Dissertation, Univ. Southern California, 148 p.
DAVIDSON-ARNOTT, R. G. D., 1971, An investigation of patterns of sediment size and sorting in the beach and nearshore area, Kouchibouguac Bay, New Brunswick: M. A. Thesis, Univ. Toronto, 109 p.
——, AND GREENWOOD, B., 1974, Bedforms and structures associated with bar topography in the shallow-water wave environment, Kouchibouguac Bay, New Brunswick, Canada: Jour. Sed. Petrology, v. 44, p. 698–704.
DAVIS, R. A., JR., 1965, Under water study of ripples, southeastern Lake Michigan: Jour. Sed. Petrology, v. 35, p. 857-866.
——, FOX, W. T., HAYES, M. O., AND BOOTHROYD, J. C., 1972, Comparison of ridge and runnel systems in tidal and non-tidal environments: Jour. Sed. Petrology, v. 42, p. 413-421.
EMERY, K. O., AND STEVENSON, R. E., 1950, Laminated beach sand: Jour. Sed. Petrology, v. 20, p. 220–223.
EVANS, O. F., 1941, The classification of wave-formed ripple marks: Jour. Sed. Petrology, v. 11, p. 37-41.
——, AND INGRAM, R. L., 1943, An experimental study of the influence of grain size on oscillation ripple marks: Jour. Sed. Petrology, v. 13, p. 117–120.
EXUM, F. A., 1973, Lithologic gradients in a marine bar, Cadeville Sand, Calhoun Field, Louisiana: Am. Assoc. Petroleum Geologists Bull., v. 57, p. 301-320.
GREENWOOD, B., AND DAVIDSON-ARNOTT, R. G. D., 1975, Marine bars and nearshore sedimentary processes, Kouchibouguac Bay, New Brunswick, Canada: *In* J. R. Hails and A. Carr (eds.), Nearshore sediment dynamics and sedimentation: John Wiley and Sons, New York, p. 123-150.
HAYES, M. O., (ED.), 1969, Coastal environments, northeastern Massachusetts and New Hampshire: Univ. Massachusetts, Dep. Geol. Coastal Res. Group Contr. 1, 462 p.
HOBDAY, D. K., AND READING, H. G., 1972, Fairweather versus storm processes in shallow marine sand bar sequences in late Precambrian of Finnmark, north Norway: Jour. Sed. Petrology, v. 42, p. 318-324.
HOWARD, J. D., 1972, Trace fossils as criteria for recognizing shorelines in stratigraphic record: *In* J. K. Rigby and W. K. Hamblin (eds.), Recognition of ancient sedimentary environments: Soc. Econ. Paleontologists and Mineralogists, Spec. Pub. 16, p. 215-225.
HUNTER, R. E., WATSON, R. L., HILL, G. W., AND DICKINSON, K. A., 1972, Modern depositional environments and processes, northern and central Padre Island, Texas: Gulf Coast Assoc. Geol. Socs., Padre Island Natl. Seashore Field Guide, p. 1-27.
INMAN, D. L., 1957, Wave generated ripples in nearshore sands: U.S. Army Corps of Engineers, Beach Erosion Board Tech. Memo. 100, 42 p.
KENNEDY, J. F., AND FALCON, M., 1965, Wave generated sediment ripples: Massachusetts Inst. Technology, Hydrodynamics Lab. Rep. 86, 55 p.
KLOVAN, J. E., 1964, Box-type sediment-coring device: Jour. Sed. Petrology, v. 34, p. 185–189.
KRANCK, K., 1967, Bedrock and sediments of Kouchibouguac Bay, New Brunswick: Fisheries Res. Board of Canada Jour., v. 24, p. 2242-2265.
MANOHAR, M., 1955, Mechanics of bottom sediment movement due to wave action: U.S. Army Corps of Engineers, Beach Erosion Board Tech. Memo. 75, 121 p.
MASTERS, C. D., 1967, Use of sedimentary structures in determination of depositional environments, Mesaverde Formation, Williams Fork Mountains, Colorado: Am. Assoc. Petroleum Geologists Bull., v. 51, p. 2033-2043.
MCCANN, S. B., AND BRYANT, E. A., 1972, Beach changes and wave conditions, New Brunswick: Am. Soc. Civil Engineers, Proc. 13th Conf. on Coastal Eng., p. 1293-1304.
MCKEE, E. D., 1957, Primary structures in some recent sediments: Am. Assoc. Petroleum Geologists Bull., v. 41, p. 1704–1747.
MOGRIDGE, G. R., AND KAMPHUIS, J. W., 1972, Experiments on bed form generation by wave action: Am. Soc. Civil Engineers, Proc. 13th Conf. on Coastal Eng., p. 1123-1142.
NEWTON, R. S., 1968, Internal structure of wave-formed ripple marks in the nearshore zone: Sedimentology, v. 11, p. 275-292.
REINECK, H. E., 1963, Sedimentgefüge im Bereich der südlichen Nordsee: Senckenb. Naturforsch. Ges. Abh., v. 505, 138 p.
—— AND SINGH, I. B., 1971, Der Golf von Gaeta/Tyrrhenisches Meer: 3. Die Gefüge von Vorstrand- und Schelf-sedimenten. Senckenbergiana Maritima, v. 3, p. 185-201.
—— AND ——, 1973, Depositional sedimentary environments: Springer-Verlag, New York, 439 p.
RISK, M. J., 1967, Shallow water ripple marks in Lake Huron: M.S. Thesis, Univ. Western Ontario.
SAYLOR, J. H., AND HANDS, E. B., 1970, Properties of longshore bars in the Great Lakes: Am. Soc. Civil Engineers, Proc. 12th Conf. on Coastal Eng., p. 839-853.
SIMONS, D. B., AND RICHARDSON, E. V., 1961, Forms of bed roughness in alluvial channels: Am. Soc. Civil Engineers Proc., Jour. Hydraulics Div., HY3, p. 87-105.
—— AND ——, 1962, Resistance to flow in alluvial channels: Am. Soc. Civil Engineers. Trans., v. 127, p. 927-953.
——, ——, AND ALBERTSON, M. L., 1961, Flume studies using medium sand (0.45 mm): U.S. Geol. Survey, Water Supply Paper 1498-A, 76 p.
——, ——, AND NORDIN, C. F., 1965, Sedimentary structures generated by flow in alluvial channels: *In* G. V. Middleton (ed.), Primary sedimentary structures and their hydrodynamic interpretation: Soc. Econ. Paleontologists and Mineralogists, Spec. Pub. 12, p. 34-52.
SOUTHARD, J. B., 1972, Representation of bed configurations in depth-velocity-size diagram: Jour. Sed. Petrology, v. 41, p. 903-915.

—— and Boguchwal, L. A., 1973, Flume experiments on the transition from ripples to lower flat bed with increasing sand size: Jour. Sed. Petrology, v. 43, p. 1114-1121.

Stricklin, F. L., Jr., and Smith, C. I., 1973, Environmental reconstruction of a carbonate beach complex Cow Creek (Lower Cretaceous) Formation of central Texas: Geol. Soc. America Bull., v. 84, p. 1349-1368.

Tanner, W. F., 1959, Shallow water ripple marks varieties: Jour. Sed. Petrology, v. 30, p. 481-485.

——, 1963, Origin and maintenance of ripple marks: Sedimentology, v. 2, p. 307-311.

——, 1971, Numerical estimates of ancient waves, water depth and fetch: Sedimentology, v. 16, p. 71-88.

Thompson, W. Q., 1937, Original structures of beaches, bars and dunes: Geol. Soc. America Bull., v. 48, p 723-751.

van der Lingen, G. J., and Andrews, P. B., 1968, Grain size parameters and sedimentary structures of a last interglacial marine sand body near Westport, New Zealand: New Zealand Jour. Marine and Freshwater Res., v. 2, p. 447-471.

Weidie, A. E., 1968, Bar and barrier island sands: Gulf Coast Assoc. Geol. Socs. Trans., v. 18, p. 405-415.

INTERACTION OF BIOLOGICAL AND GEOLOGICAL PROCESSES IN THE BEACH AND NEARSHORE ENVIRONMENTS, NORTHERN PADRE ISLAND, TEXAS

GARY W. HILL AND RALPH E. HUNTER[1]
U.S. Geological Survey, Corpus Christi, Texas 78411

ABSTRACT

Padre Island, a barrier island off the southern Texas coast, experiences a low tidal range, is affected by waves of moderate height, and the northern part of the island is composed largely of fine sand. The seaward part of the island and the adjacent nearshore can be differentiated geomorphically into a beach (backshore and foreshore) and a bar-trough system.

Macrobenthos zonation is related to geomorphic features and can be defined in terms of species distribution and density, diversity-equitability, and niche classification. Dominant species characterize each landform: *Ocypode quadrata* (ghost crab) is concentrated in the beach, and *Callianassa islagrande* (ghost shrimp) and *Mellita quinquiesperforata* (sand dollar) are mainly in the nearshore bar-trough system. Diversity is lowest in the backshore and highest in the foreshore and the bar-trough system. The backshore has a homogeneous trophic nucleus (scavengers) and the foreshore and bar-trough system have a mixed trophic nucleus of suspension and deposit feeders.

Characteristic depositional structures are common to each landform: (1) beach, gently seaward-dipping planar laminations produced by wave swash; (2) bar-trough system, small-scale lenticular crossbedding usually without any well-defined preferred orientation of dip direction and produced by deposition on sand ripples, planar lamination produced when bar crests are planar because of high waves, and medium-scale cross-bedding produced by migrating megaripples and dipping in the direction of the longshore current. The depositional structures commonly are destroyed by organisms. Characteristic biogenic sedimentary structures are found in each geomorphic element. Variation in the areal density, size, morphology, and orientation of *O. quadrata* burrows can be used to define subenvironments of the beach. Deep oblique to vertical burrows of *C. islagrande* and shallow horizontal intrastratal trails of *M. quinquiesperforata* characterize the bar-trough system.

INTRODUCTION

Studies of modern marine environments have long been used to provide an insight into the processes that created the sedimentary rocks found in ancient basins. In particular, elements of the modern beach and littoral environments are detailed to serve as models in documenting the facies changes from marine to nonmarine. This paper describes the results of a detailed study of biological and geological features in the Texas Gulf Coast beach and nearshore environments and discusses the nature of, and balance between, biological and physical processes.

Although much has been written about the geology (King, 1972) and biology (Pearse, Humm, and Wharton, 1942) of beaches and their adjacent environments, relatively little work has been directed toward the interaction of physical and biological factors in controlling the sedimentary features of beach and nearshore environments. Studies in the United States on animal-sediment relationships have largely been restricted to the Atlantic coast, especially Georgia (Howard and Frey, 1973). In Texas, animal-sediment relationships in the neritic environment have received meager study. Biogenic structures common to modern depositional environments of northern Padre Island were briefly described by Hunter and others (1972), and the relationship of variations in morphology, distribution, and density of ghost crab *Ocypode quadrata* (Fabricius) burrows to the physical zonations of the beach environment was described by Hill and Hunter (1973).

ENVIRONMENTAL SETTING

Geographic setting.—Padre Island is the southernmost part of a gently curving chain of barrier islands and spits that stretches some 322 km along the Texas coast from the Brazos River delta on the northeast to the Rio Grande delta in the southeast. Padre Island itself extends a distance of 177 km without a permanently open natural pass and is divided only by Mansfield Pass, an artificial cut which is maintained by dredging. The beach and nearshore section of the South Bird Island 7.5-minute quadrangle, which is typical of northern Padre island, served as the principal study area (Fig. 1).

Climate.—Padre Island has a subhumid to semiarid warm-temperature climate. The mean annual rainfall at Corpus Christi is 66.9 cm; the mean monthly temperature varies from 14.1° C in January to 29.0° C in August, and the mean annual temperature is 22.1° C (NOAA, 1974). The mean annual wind speed is 5.3 m/sec, and the resultant wind direction is 121°. The climate becomes more

[1]Present address: U.S. Geological Survey, Menlo Park, Calif.

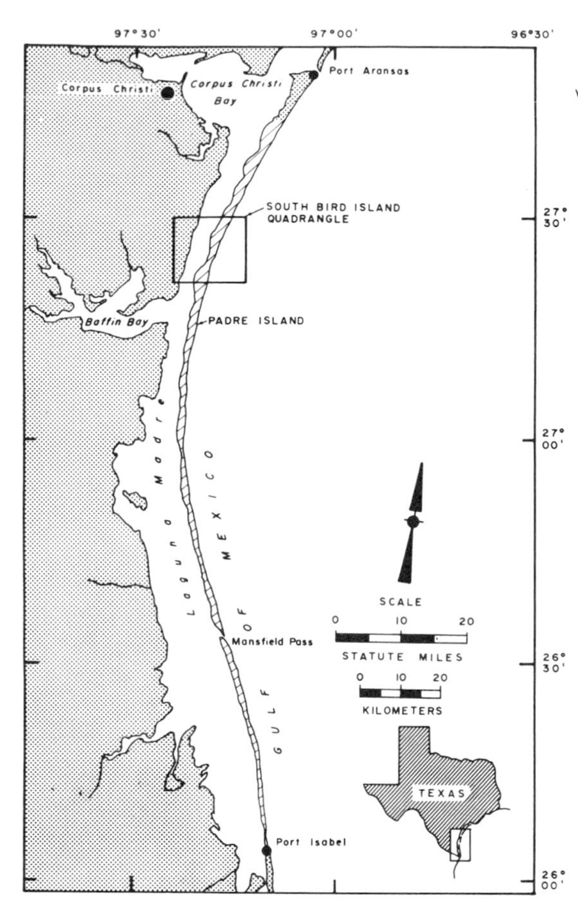

FIG. 1.—Index map of study area. Padre Island is obliquely hachured.

humid and less windy along the coast to the northeast; for those reasons, the role of wind in shaping landforms and transporting sand is and has been greater on Padre Island than on the northeastern part of the Texas barrier chain.

Tides, water properties, and wave climate.—The Padre Island shore is microtidal, having a mean diurnal range of 51.8 cm at Port Aransas (NOAA, 1975). The mean monthly water temperature at Port Aransas varies from 13.6° C in January to 30.0° C in August, and the mean annual water temperature is 22.7° C (NOAA, 1973). The mean monthly salinity, calculated from the water density at Port Aransas, varies from 29.5‰ in May to 36.6‰ in August, and the mean annual salinity is 32.0‰ (NOAA, 1973).

Few data exist regarding wave climate. The heights of breaking waves are normally 0.3 to 1.0 m by visual observation. Breakers higher than 2 m occur several days per year, largely during storms in the fall, winter, and spring. Waves approach the coast from the southeast during the summer and dominantly from the northeast during the winter.

Sediment.—The mean grain size of shell-free beach sand along northern Padre Island is in the range 2.5 to 3.0 ϕ and is well sorted (Hayes, 1964; Dickinson, 1971). Shells and shell fragments other than *Donax* are not abundant on the beach and nearshore. However, shells of many different molluscs have formed surficial layers in the outer trough at times when it was cut to an unusually great depth, and pipe cores show that shell layers occur at depths greater than 0.7 m beneath the beach surface. These shells are derived from shelly sands that underlie the more recent, nearly shell-free sands.

Geomorphology.—Northern Padre Island and the adjacent inner shelf are marked by a set of elongate, shore-parallel geomorphic units (Fig. 2). The geomorphic units with which this paper is concerned are the beach and inner part of the shoreface.

The beach along most of northern Padre Island is bounded landward by a vegetated foredune ridge. Where vehicular traffic is permitted on the beach, the beach-foredune boundary is sharply defined. Where vehicular traffic has been banned on a section of beach within Padre Island National Seashore, patches of vegetation have become established on the backshore, and since 1971 have caused the growth of small dune mounds that tend to obscure the beach-foreshore boundary. The permanence of this dune growth remains to be tested by a major hurricane. Beach studies for this report were carried out largely within this untrafficked zone, where the influence of man is minimal. Here the beach normally consists of a gently sloping (<2°-5°) foreshore, 25-45 m wide, and a nearly horizontal backshore, 65-115 m wide, at an elevation of 1.5-2.0 m above low water (Fig. 2).

The beach along northern Padre Island has not migrated measurably in any systematic way from the position surveyed in 1877 and shown on the first reliable map published in 1882 (Hunter and Dickinson, 1970). Hurricanes, however, cause striking temporary erosion, producing a gently sloping, nearly planar hurricane beach (Hayes, 1964). Sand eroded from the beach during hurricanes is gradually carried back by normal wave activity and results in the rebuilding of the normal backshore and foreshore (Fig. 2). The backshore may persist for several years between hurricanes because it is outside the influence of normal wave action after its full development.

A low-tide terrace often makes up the seaward part of the intertidal zone. Nearly as often, however, this terrace is replaced by an ephemeral swash bar (ridge) and runnel. A topographic step

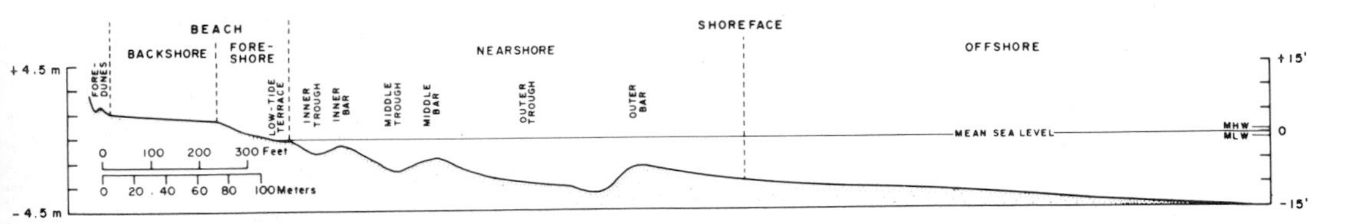

FIG. 2.—Profile of northern Padre Island beach and shoreface. The morphology is typical of northern Padre Island during summer conditions. An ephemeral bar and runnel commonly replace the low-tide terrace.

usually occurs at the seaward edge of the low-tide terrace.

The shoreface, which is the relatively narrow, steep, inner part of the continental shelf, can be divided into nearshore and offshore parts. Only the former is part of this study. The nearshore part of the shoreface consists of a series of bars, typically three along this section of coast, having intervening troughs (Fig. 2). The bars on this part of the Texas coast are generally straight, continuous, and aligned parallel to shore. The bars are remarkably stable in a dynamic sense. They shift back and forth and change shape as wave conditions change, but seldom are destroyed, migrate onto the beach, or grow from a previously planar bottom. In unusually calm weather, however, the inner bar migrates shoreward and becomes irregular in plan view as tongues of sand are built from the bar into the inner trough. In the late summer of 1974, this migration proceeded to the extent that the inner trough was filled and the inner bar was planed off. In contrast, during the passage of Hurricane Fern in September 1971, the outer bar was planed off and the outer trough was filled.

Hydrodynamic processes.—The swash zone, the dynamic zone dominated by wave runup and backswash, coincides with the foreshore during normal wave and tide conditions. The backshore, in contrast, is inactive during fair weather except for the generally landward transport of sand by wind. Because of the dampness of the sand due to its low elevation, wind erosion seldom lowers the level of the backshore by more than a few centimeters. Wave swash reaches the backshore several times a year, mostly during winter storms, but only hurricanes cause severe erosion.

The nearshore is the dynamic zone dominated by breaking waves and by wave-induced currents. Breaking waves and surf are confined mainly to the bars and to the edge of the beach with the waves tending to reform in the troughs. The waves are seldom so small that they do not break on the inner bar, but breaking is commonly absent on the outer bars.

The prevailing longshore currents vary seasonally, being mostly to the south in winter and to the north in summer. The strongest currents, more than 1 m/sec, are mainly to the south and occur several times a year, mostly during winter storms. Rip currents commonly flow seaward across the bar crests when waves are breaking on the bars. Aerial photographs taken about a half-hour apart show that the rip currents migrate with the longshore current. The absence of rip-current channels across the bars is probably related to the migration of the rip currents.

The strongest onshore- and offshore-directed water motions in the nearshore are the oscillatory motions due to waves. The oscillations are markedly asymmetric: the onshore pulses reach higher maximum velocities but are of shorter duration than the offshore pulses (Clifton, this volume). This asymmetric oscillatory water motion can transport sand landward even in the absence of superimposed net currents.

Observations of slightly negatively buoyant drifters and of plumes from anchored dye packets show that net onshore near-bottom currents are common on the seaward sides of the bars and on the bar crests. Similar observations show that net offshore, near-bottom currents occur on the landward sides of the bars and in the troughs when waves are breaking on the bars. Although the sand-transporting power of these offshore bottom currents has not been documented, they may be the main factor in promoting bar stability by counteracting the tendency of the asymmetric oscillatory water motion to transport sand landward.

METHODS

Biota.—From January 1971 to February 1975, numerous transects were made perpendicular to the shoreline across northern Padre Island from the foredune ridge to the outer bar in the bar-trough system. Bulk sediment samples for biota were taken along each transect, and a constant volume of sediment, to a depth of 15 cm and having a surface area of 0.01 m^2, was recovered by pipe coring. The bulk samples were washed through a 0.5 mm mesh sieve. Organisms recovered were fixed in 10 percent formalin, preserved in 45 percent isopropyl alcohol, and later identified and counted in the laboratory.

Biota of the beach and nearshore was quantitatively sampled in the summer of 1972 and in the winter and spring of 1973 and 1975. Burrows of the larger crustaceans, particularly *Ocypode quadrata*, were examined during all seasons from 1971 to 1975.

The spacing of sample stations along each transect varied according to the subenvironment being sampled. Sample stations on the backshore and foreshore were spaced 8 m and 3 m apart, respectively. In the nearshore, divers using SCUBA took samples in the deepest part of each trough and on each bar crest.

To ensure representative sampling of large deep-burrowing organisms, different sampling techniques were used. The local distribution and density of the ghost crab *Ocypode quadrata* were determined by marking out 3 m wide strips perpendicular to the shoreline and divided into sections 3 m square from the waterline to the foredune ridge. Within each section, the number of ghost crab burrows was counted and their diameter, length, morphology, and orientation were measured. To determine more accurately the distribution-density of *Callianassa islagrande* Schmitt, two methods of counting burrow openings were used. During spring low tides, a tape measure was placed across the foreshore perpendicular to the shoreline, and *C. islagrande* burrow openings were counted in each square meter along the tape. In the surf zone, a diver dropped a metal loop, 30 cm in diameter, onto the bottom in each trough and on each bar crest and counted the burrow openings inside the loop. A benthic suction sampler modified from Brett (1964) was also used to collect organisms quantitatively from sediment depths as great as 1 m.

Sediments.—Relatively undisturbed box cores (10 × 14 × 20 cm) and pipe cores as much as 1.3 m long were collected from the beach and bar-trough system. In the laboratory, the cores were opened and X-ray radiographs and epoxy casts were made for the analysis of physical and biogenic sedimentary structures.

A modification of burrow-casting techniques (Mayou, Howard, and Smith, 1969), and trenching were used to determine burrow morphologies. Burrow casts were made using epoxy and fiberglass resins.

Laboratory techniques.—X-ray radiographic examinations of biogenic sedimentary structures were made using a 3 cm wide plexiglas aquarium having a continuous water-circulation system (Fig. 3). Sediment erosion or deposition in the aquarium was maintained at various rates. Radiographs were made using a portable industrial X-ray unit equipped with a beryllium window hotshot head and type-M industrial X-ray film.

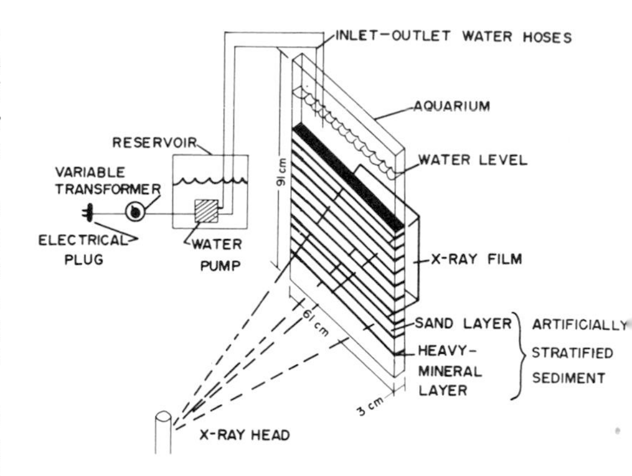

FIG. 3.—Schematic diagram of aquarium, X-ray, and continuous water-circulating system used to study species-specific biogenic sedimentary structures under laboratory conditions.

BIOTA

Community structure.—The 660 short pipe cores taken at 110 sample stations along four transects on the northern Padre Island beach and bar-trough system yielded 4,546 individuals belonging to 27 macrobenthic species representing several taxonomic groups (Table 1). The distribution and density of *Ocypode quadrata*, as indicated by their burrows, were determined at 597 sample stations (3 m^2 each) along 27 transects which contained 2,800 burrows. An additional 387 *O. quadrata* burrows were examined to determine burrow orientation. Burrows of *Callianassa islagrande* were studied at 76 stations along four transects.

Most of the species are represented by few individuals. Seventy-seven percent (3,443 individuals) belong to the single most frequently occurring taxon (haustoriid amphipods), and 99 percent (4,450 individuals) belong to the five most frequent species. In contrast, 1 percent (96 individuals) of the biota comprises 81.5 percent of the species (Tables 1 and 2).

TABLE 1.—NUMBER OF SPECIES AND SPECIMENS COLLECTED FOR ALL SAMPLES FROM ALL STATIONS

Taxa	No. of species	No. of specimens
Crustacea	9 (33.5%)	3,550 (78%)
Polychaeta	9 (33.5%)	691 (15%)
Mollusca	4 (15%)	194 (4%)
Echinodermata	3 (11%)	99 (2%)
Others	2 (7%)	12 (1%)
	27 (100%)	4,546 (100%)

TABLE 2.—COMPARISON OF NUMBER OF SPECIES AND SPECIMENS COLLECTED IN THE BEACH AND BAR-TROUGH SYSTEM OF NORTHERN PADRE ISLAND

	Beach				Bar-trough system	
	Backshore		Foreshore			
No. of stations	40		46		24	
Total no. of species	6		21		19	
Total no. of specimens	335		2,337		1,874	
No. of species:						
Crustacea	2 (33.3%)		9 (43%)		5 (26%)	
Polychaeta	2 (33.3%)		9 (43%)		7 (37%)	
Mollusca	0		2 (10%)		4 (21%)	
Echinodermata	0		1 (10%)		3 (16%)	
Others	2 (33.3%)		0		0	
Most abundant:						
2 species	322 (96%)		2,076 (89%)		1,656 (88%)	
5 species	334 (99.7%)		2,272 (97%)		1,844 (98%)	
No. of specimens:						
Crustacea	272 (81%)		1,712 (73%)		1,551 (83%)	
Polychaeta	51 (15%)		579 (25%)		76 (4%)	
Mollusca	0		45 (2%)		149 (8%)	
Echinodermata	0		1 (1%)		98 (5%)	
Others	12 (4%)		0		0	
Most common species:	Haustoriidae	272	Haustoriidae	1,656	Haustoriidae	1,514
	Scolelepis squamata	50	*Scolelepis squamata*	420	*Donax variabilis*	142
	Staphylinidae	9	*Lumbrineris* sp.	133	*Mellita quinquiesperforata*	93
			Donax variabilis	39	*Lumbrineris* sp.	65
			Lepidopa websteri	24	*Pinnixa chacei*	30

The taxonomic group comprising the greatest number of individuals collected was Crustacea (78%) followed by Polychaeta (15%), Mollusca (4%), and Echinodermata (2%). Other taxonomic groups accounted for less than one percent of the total specimens collected (Table 1).

Two macrozoobenthic communities and an intervening ecotone, each related to a specific geomorphic feature (Table 3), are defined by the zonation in distribution and density of the most common and/or conspicuous species (characteristic species, Table 4; Fig. 4).

The backshore contains the fewest zoobenthic species and individuals compared with the other landforms in this study (Table 2). Some of these species, among which the haustoriid amphipods and the polychaete worm *Scolelepis squamata* (Müller) are most common, are restricted to the seaward edge of the backshore (Fig. 4). Because these species represent landward extensions of populations more abundant in other subenvironments and occupy a restricted zone, they are not characteristic of the backshore in general. The ghost crab *Ocypode quadrata* is a large conspicu-

TABLE 3.—COMMUNITIES AND RELATED GEOMORPHIC FEATURES OF THE NORTHERN PADRE ISLAND BEACH AND BAR-TROUGH SYSTEM

Community	Geomorphic feature
Ocypode	Backshore
Ecotone	Foreshore
Callianassa-Donax-Haustoriidae	Bar-trough system

TABLE 4.—CHARACTERISTIC MACROZOOBENTHIC SPECIES OF COMMUNITIES FOUND IN THE BEACH AND BAR-TROUGH ENVIRONMENTS OF NORTHERN PADRE ISLAND

	Sand beach community	Bar-trough community
1st order	*Ocypode quadrata*	*Callianassa islagrande* *Donax variabilis*
2nd order	*Scolelepis squamata*	*Mellita quinquiesperforata*
3rd order	—	*Haustoriidae*
Influents	Staphylinidae	*Lumbrineris sp.*

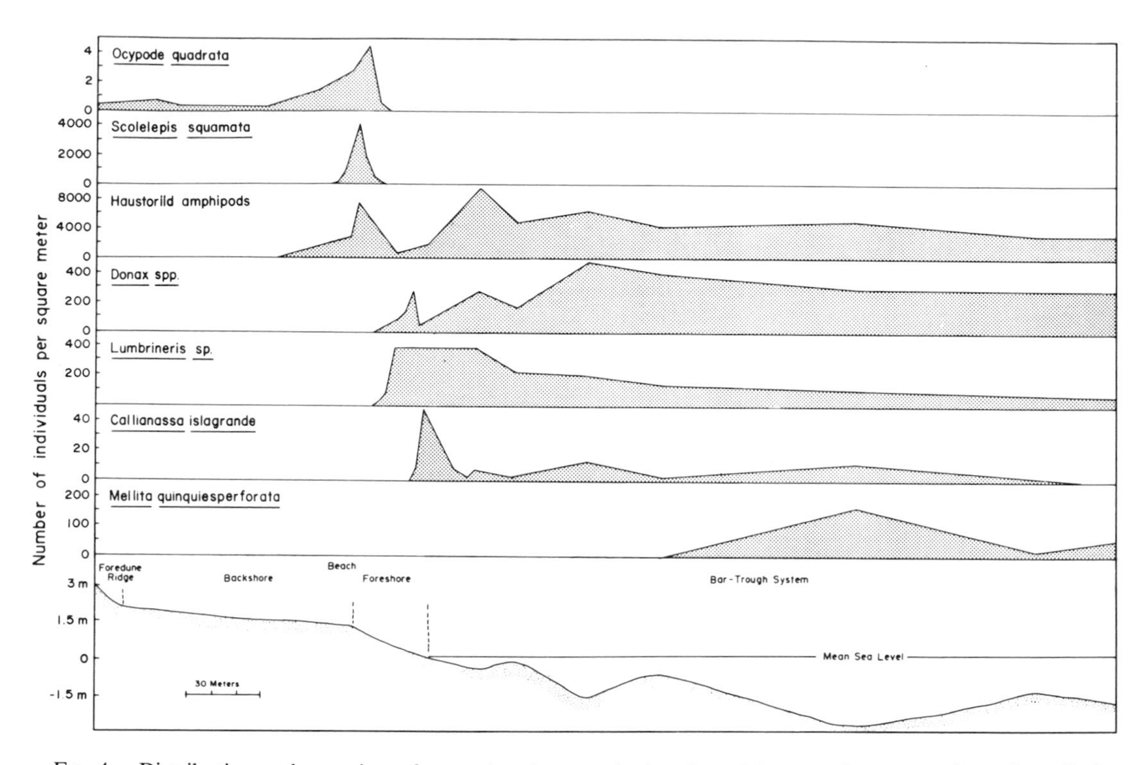

FIG. 4.—Distribution and zonation of macrobenthos on the beach and bar-trough system of northern Padre Island.

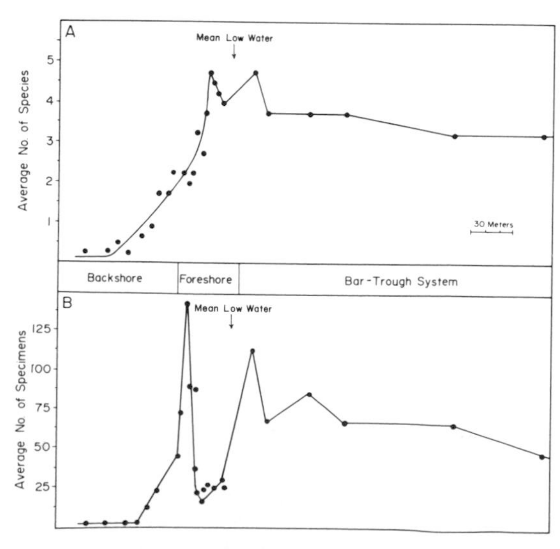

FIG. 5.—Cross section of the northern Padre Island beach and bar-trough system. *A*. Distribution of species. *B*. Distribution of individuals.

ous organism that is found everywhere in the backshore in relatively small numbers (Fig. 4) and is considered a first-order characteristic species of the backshore (Table 4). The backshore community was named after this species—*Ocypode* community. Rove beetles (Staphylinidae) are abundant in the backshore, but the 0.5 mm mesh size of the sieves used in collecting samples was not small enough to collect them quantitatively.

The second largest number of zoobenthic species and individuals is found in the bar-trough system (Table 2). Characteristic species that are distributed throughout the bar-trough system (Table 4) include haustoriid amphipods, the bivalve *Donax variabilis* Say, the polychaete worm

TABLE 5.—DIVERSITY AND EQUITABILITY VALUES OF MACROZOOBENTHIC ASSEMBLAGES RELATED TO SPECIFIC GEOMORPHIC FEATURES COMMON TO NORTHERN PADRE ISLAND

Landform	Diversity	Equitability
Backshore	0.6280	0.2950
Foreshore	.9327	.1070
Bar-trough System	.8054	.1047

TABLE 6.—PERCENTAGES OF BIOMASS FOR EACH TROPHIC TYPE FOUND IN DIFFERENT NERITIC MACROZOOBENTHIC COMMUNITIES ON NORTHERN PADRE ISLAND

Trophic type	*Ocypode* community	*Callianassa-Donax*-Haustoriidae community
Suspension feeders	< 10%	28%
Detritus feeders	> 90%	70%
Carnivores	< 1%	2%
Trophic nucleus	Homogeneous	Mixed

Lumbrineris sp., and the ghost sprimp *Callianassa islagrande* (Fig. 4). The sand dollar *Mellita quinquiesperforata* (Leske) is not found landward of the crest of the middle bar. Because of the large numbers of haustoriid amphipods and *Donax*, and the restricted distribution of the large conspicuous *Callianassa islagrande*, the macrozoobenthic community of the bar-trough system was named the *Callianassa-Donax*-Haustoriidae community.

Samples from the foreshore yielded the greatest number of species and individuals (Table 2). The foreshore represents an ecotone—a boundary between two adjacent communities. Species common to the backshore and foreshore include *Ocypode quadrata* and staphylinid beetles. Species common to the bar-trough system and foreshore include haustoriid amphipods, *Scolelepis squamata*, *Callianassa islagrande*, *Donax variabilis*, and *Lumbrineris sp.* (Fig. 4, Table 2). To further distinguish the ecotone from either adjacent community, the densities of some characteristic species found in the two communities are greatest in the foreshore. Both *Ocypode quadrata* and *Callianassa islagrande* reach their greatest density in the foreshore, but their distribution patterns generally do not overlap (Fig. 4). The polychaete *Scolelepis squamata* is almost totally limited in its distribution to the upper foreshore, where it attains very high densities (Fig. 4).

Diversity-equitability.—As a measure of processes operating in an ecologic system, diversity reflects external environmental stresses as well as internal stability and productivity (Beerbower and Jordan, 1969). Diversity was determined by simply counting the number of species and individuals in each sample and also by calculating the Shannon diversity index.

Both the number of species (Fig. 5a) and the number of individuals (Fig. 5b) increases seaward across the backshore, rises sharply in the foreshore, and then gradually declines across the bar-trough system. A slight reduction in the number of species and a large drop in the number of total individuals seems to occur just landward of the mean low water line. Howard and Dörjes (1972) noticed a similar reduction of species and individuals landward of the low water line on Sapelo Island, Georgia. On Padre Island, this zone of reduced density generally corresponds to the plunge point of waves on the lower foreshore, suggesting that extreme turbulence near the bed may play a role in causing this reduction.

The Shannon diversity index, H (Shannon and Weaver, 1963), is an approximation of the Brillouin index.

$$H = -\sum_{i=1}^{N} P_i \ln P_i \qquad (1)$$

FIG. 6.—X-radiograph of box-core peel from crest of inner bar. Planar lamination probably formed by deposition on planar bed.

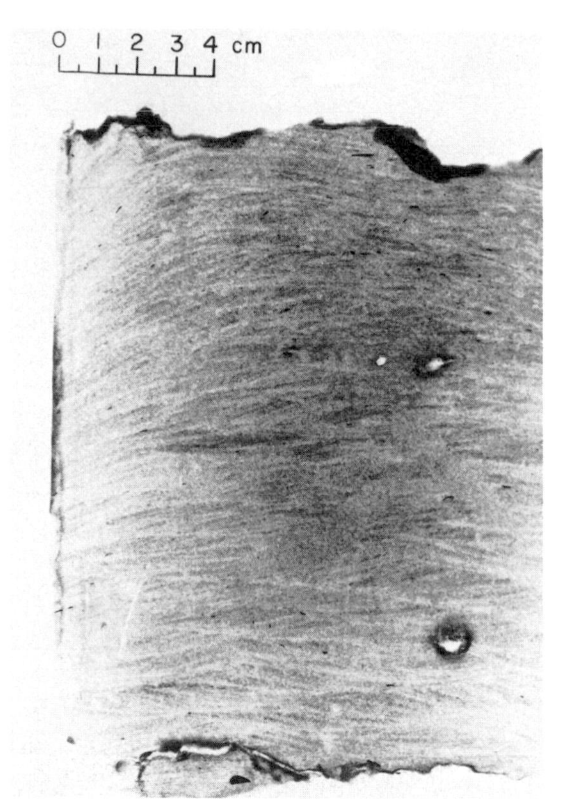

FIG. 7.—X-radiograph of box-core peel from inner trough. Small-scale cross-bedding formed by ripple migration; contortion of bedding at edges of peel is artificial.

where P_i = the number of individuals of the ith species divided by the number of individuals of all species in a sample, and n = the number of species in the sample. It is influenced by two components: the total number of species present (species-richness component) and the evenness of distribution of the individuals among the different species (equitability component). Whereas the number of species depends primarily on the structural diversity of a habitat, equitability is more sensitive to the stability of physical conditions. To separate the two components of the Shannon diversity index, Lloyd and Ghelardi's (1964) equitability index, E, was used:

$$E = si/s \qquad (2)$$

where s = observed number of taxa, and si = calculated number of taxa whose random properties yield the diversity H.

Equitability decreased seaward from the backshore to the bar-trough system (Table 5). The bar-trough system has a higher diversity index (Table 5) than the backshore (Table 2, Fig. 5a), but the greatest diversity index is found in the foreshore or intertidal zone. Such an abundance of species is explained in the biological concept of the ecotone or "edge effect"; when two habitats come together, the edge between the two will be more favorable as a habitat then either type considered alone.

Niche classification.—According to Scott (1972), the two most important parameters of the niche hypervolume are the feeding habit and the position of each species relative to the sediment-water interface (its substrate niche).

Trophic groupings are formed by listing species (trophic types) in order of decreasing biomass. The dominant species, whose combined biomass is greater than 80 percent of the total sample, form the trophic nucleus, which may be classified as either homogeneous (consisting of species of a similar feeding type) or mixed (consisting of species of different feeding types) (Rhoads, Speden and Waage, 1972).

FIG. 8.—X-radiograph of box-core peel from inner trough. Medium-scale cross-bedding formed by migration of megaripples; the cross-bedding dips in the direction of the longshore current.

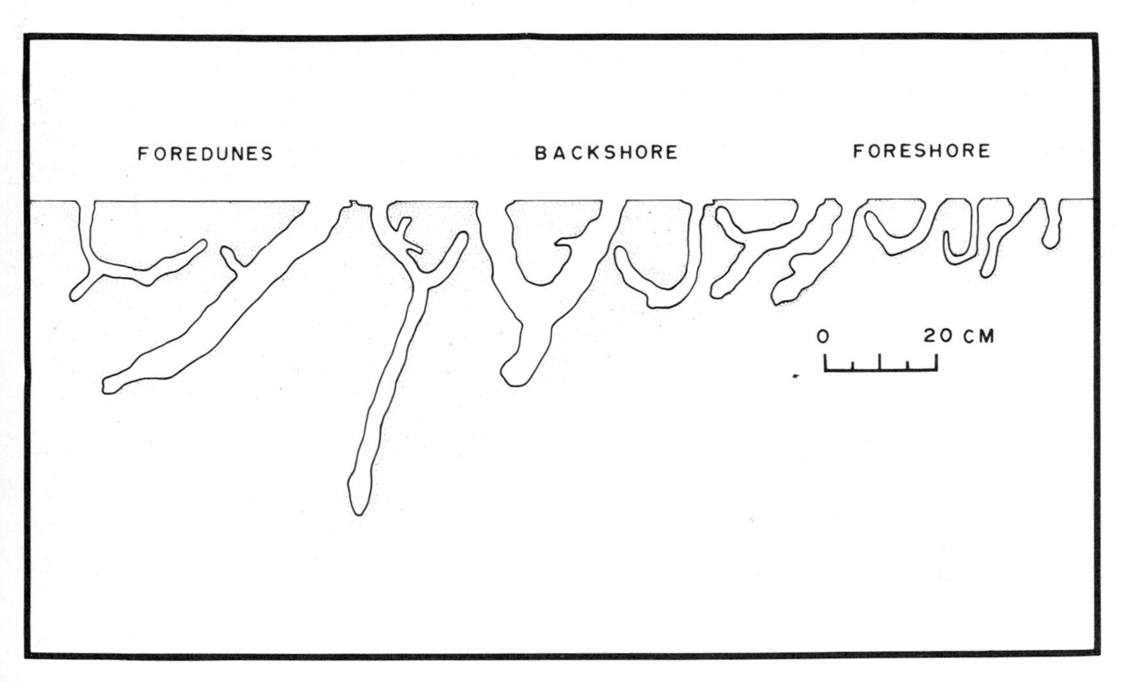

FIG. 9.—Representative *Ocypode quadrata* burrows of a beach cross section, northern Padre Island (from Hill and Hunter, 1973).

Samples within each community were grouped together, and species were listed by trophic types to determine the composition of the trophic nucleus. On the basis of percentages of biomass, the trophic nucleus of the *Ocypode* community is homogeneous, and the *Callianassa-Donax*-Haustoriidae community is mixed (Table 6). Detritus feeders dominate both communities. Scavengers, however, are prevalent on the backshore, whereas deposit feeders abound in the bar-trough system.

Closely associated with the substrate and its condition is the mode of movement or attachment of benthic animals. Thorson's (1957, p. 461) classification of three "ecologically different groups of benthic marine animals" was used: (1) epifauna, animals occupying area upon the substrate; (2) infauna, animals occupying sediment volume; and (3) vagrant invertebrates.

The majority of benthic organisms found in the beach and surf environments are infaunal forms. The environments are too harsh for epifauna to exist. Animals such as *Donax* and haustoriid amphipods, living near the sediment-water interface without the protection of an open burrow for downward escape, are streamlined and possess the ability to burrow rapidly when dislodged from the sediment.

SEDIMENTARY STRUCTURES

Physical sedimentary structures.—The dominant internal structure produced by wave swash on the beach is gently seaward-dipping planar lamination (McKee, 1957; Milling and Behrens, 1966; Moiola and Spencer, 1973). Some irregular stratification is produced on the backshore by scour and deposition around stranded debris and by the growth of incipient dunes. Cores suggest that such irregularities are less common at depths below 0.5 m. Most of the beach sand body was probably deposited on foreshores, and most of

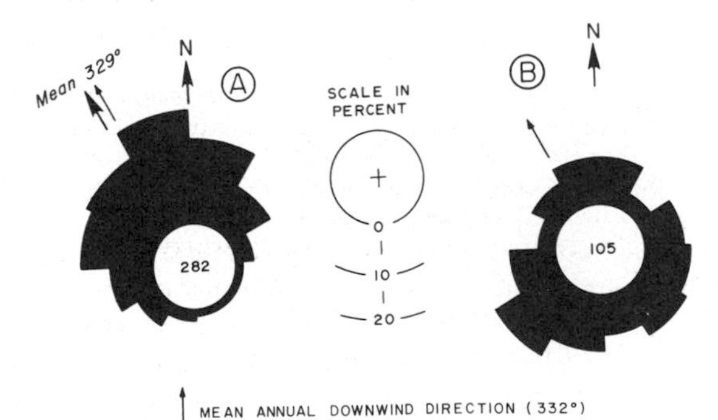

FIG. 10.—Orientation of *O. quadrata* burrows, defined as the direction of burrow descent. Number in the center of each circle indicates the number of burrows measured. Data plotted in 30° classes. *A*. Orientation of backshore burrows. *B*. Orientation of burrows in the interdune flats.

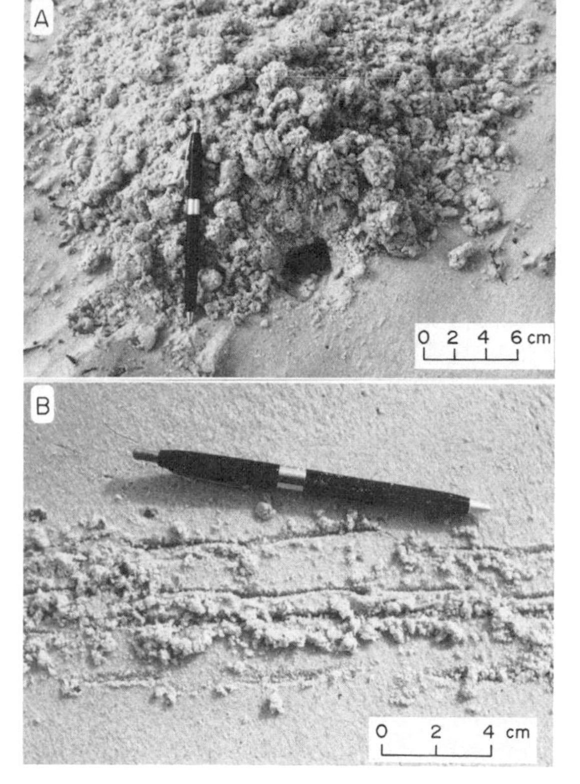

FIG. 11.—Lebensupurren produced by *O. quadrata*. *A*. Mound of sand around burrow opening resulting from digging activities. *B*. Trackway.

the sand was probably deposited rapidly during post-hurricane recovery periods.

Three kinds of bed configurations and corresponding internal structures produced by wave and current action on fine-grained sand have been found in the bar-trough system of northern Padre Island. These are plane beds with planar lamination, ripples with small-scale cross-stratification, and megaripples with medium-scale cross-stratification.

Plane beds occur on the bar crests whenever the waves are large enough to break and extend down the sides of the bars. Deposition sometimes occurs on such surfaces, producing planar lamination (Fig. 6). The laminae tend to be similar in thickness to adjacent laminae, suggesting that each lamina is the product of a single wave in a series of similar waves.

Ripples occur at depths greater than those in which plane beds are generated under equivalent conditions and as a result tend to be restricted to the troughs between bars. In the transition from rippled to plane beds, the ripples are destroyed during the peak of each passing wave surge and form again during the deceleration of the surge. In somewhat deeper water, the ripples become permanent, but their asymmetry reverses as wave surge oscillates. These very active ripples, of the type called "nonorbital" by H. E. Clifton (this volume), range from 5 to 10 cm in spacing, are less than 1 cm high, and have flat, wide troughs and long straight crests. They are aligned parallel to the waves or, if a longshore current is present, at some angle between the waves and the direction normal to the current. The internal structure produced by the migration of these ripples is poorly known; if the ripples climb at a very low angle, the resulting structure may closely resemble planar lamination.

In the deeper parts of the troughs, during low and moderate wave activity, the ripples are less active. These less active ripples tend to be larger and steeper than the "nonorbital" ripples, have spacings as much as 15 cm, have rounded troughs and no distinct asymmetry, are short-crested, and form an irregular branching pattern. The complexity of the plan-view pattern is probably due to the interference of wave surge, the net offshore flow, and the longshore current. The internal

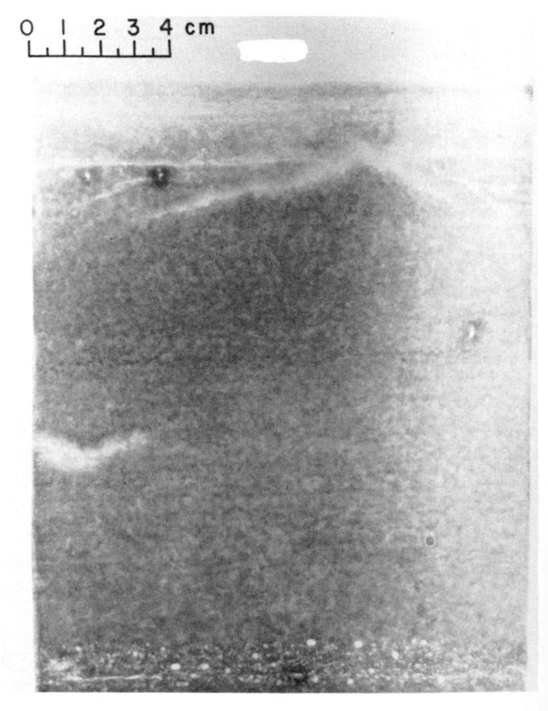

FIG. 12.—X-radiograph of box-core peel from the backshore. Sand intensely bioturbated by rove beetles (Staphylinidae). Because of the small size of the beetles, the sand was not displaced very far, and stratification is still visible. Gas-bubble cavities are still preserved near the base.

structure produced by deposition on rippled surfaces in the interbar troughs is small-scale lenticular cross-bedding without any well-defined preferred onshore or offshore component of foreset dip direction but at times having preferred longshore components, which, however, are subject to reversals (Fig. 7).

Longshore currents exceeding about 0.5 m/sec produce megaripples ranging from 10 to 60 cm in height. Megaripples produced by both north-flowing and south-flowing currents have been observed. The internal structure produced by migration of the megaripples is medium-scale (defined here as units 4 to 100 cm thick) cross-bedding that dips in the direction of the current (Fig. 8). This cross-bedding commonly dips at angles less than the angle of repose and in places at angles as low as 10°. Currents that flow along the leeward slopes of megaripples, rather than currents that flow down these slopes, as suggested by Imbrie and Buchanan (1965), are probably the most common cause of the low dip angles that are typical in shallow marine cross-bedding.

Ripples having spacings of 0.3 m or more can exist in shelly sand adjacent to small ripples in fine-grained sand. Sand that is sufficiently shelly for the existence of large ripples is rarely found off northern Padre Island but is common off central Padre Island. The orientation and symmetry of these ripples varies. Onshore-facing lunate megaripples have been seen on the landward side of the inner trough, and offshore-facing megaripples were observed at the same time on the seaward side of the inner trough. Symmetric, shore-parallel, large wave ripples and longshore-facing, current-generated megaripples have been observed in shelly sand in the middle and outer troughs.

Biogenic sedimentary structures.—Distinctive biogenic sedimentary structures are produced by many of the macrozoobenthic organisms found in the beach and bar-trough system of northern Padre Island. These biogenic sedimentary structures help to define the macrobenthos zonation and may be related to specific geomorphic features.

Variation in areal density, size, morphology, and orientation of *Ocypode quadrata* burrows can be used to define subenvironments of the beach (Hill and Hunter, 1973). An increase in burrow diameter, length, and complexity of shape (Fig. 9) coupled with a decrease in burrow density (Fig. 4) from the upper foreshore to the back edge of the beach, differentiates beach subzones. Burrows in the backshore show a preferred orientation, descending in a northwest direction (Fig. 10a). Laboratory experiments and field observations under varying wind conditions show that this orientation is controlled by the direction of

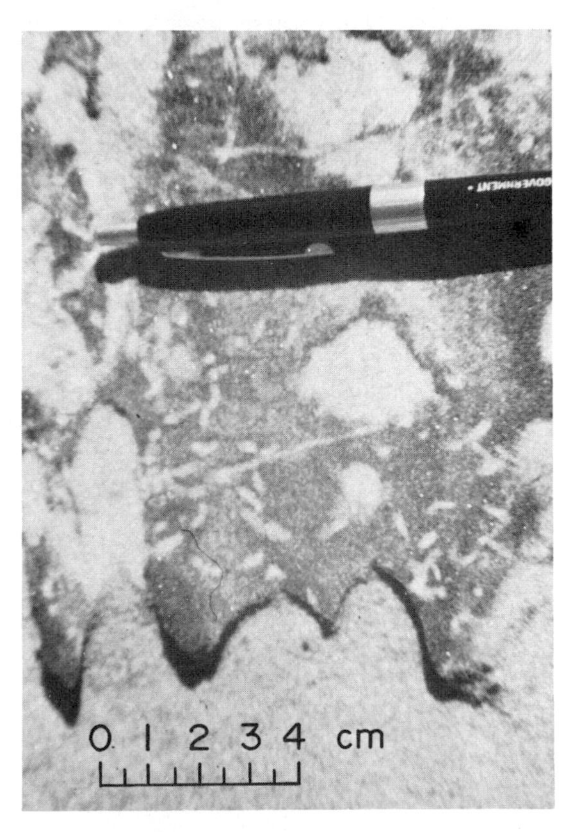

Fig. 13.—Filled rove-beetle burrows penetrating a heavy mineral layer exposed by wind scour on the backshore.

onshore winds. In contrast, burrow orientation is random in the interdune flats of the foredune ridge, where wind shadows occur (Fig. 10b).

Other lebenspurren produced by *O. quadrata*, such as mounds of excavated sand around burrow openings, trackways, and feeding pellets (Fig. 11), are normally destroyed by wind and wave activity. The ghost crabs will inhabit any large burrow found on the backshore, such as those made by the south Texas pocket gopher *Geomys personatus* (True).

Intense burrowing by rove beetles (Staphylinidae) on the backshore produces bioturbated sediments (Fig. 12). The presence of rove beetles on the backshore is marked by small burrow openings (1 mm) and stringers of loose sand pushed from the burrows. Filled rove beetle burrows can be observed on the beach when a heavy-mineral layer is exposed by wind scour (Fig. 13).

The south Texas pocket gopher, *G. personatus*, which is found in great numbers on the eolian flats of Padre Island, produces a large distinct burrow found occasionally on the beach. The

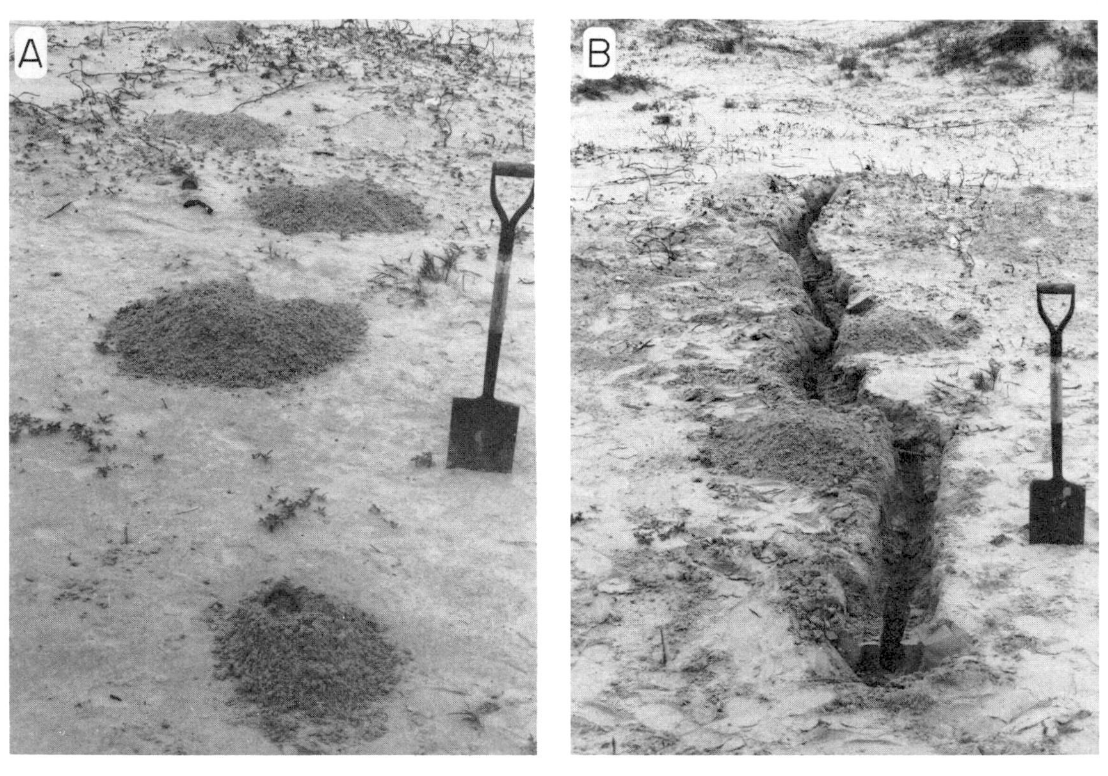

FIG. 14.—Burrow of the south Texas pocket gopher on the backshore of northern Padre Island. *A*. Mounds of sand excavated from the burrow. *B*. Excavated burrow revealing a somewhat sinuous morphology.

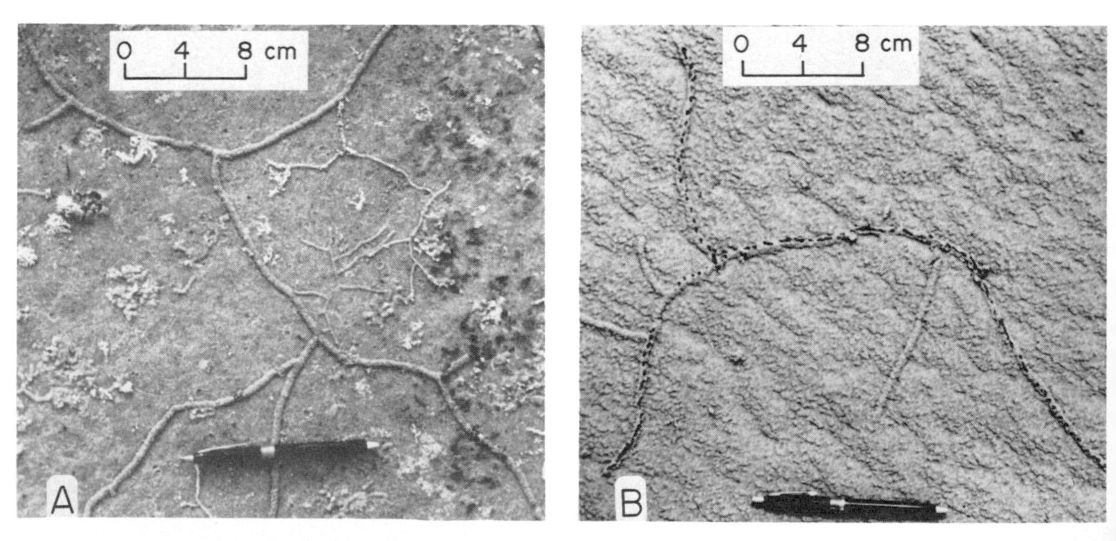

FIG. 15.—Horizontal burrow network of *Omophron* beetle larvae on the backshore of northern Padre Island. *A*. Undisturbed burrow. *B*. Burrow showing peck marks of birds searching for food.

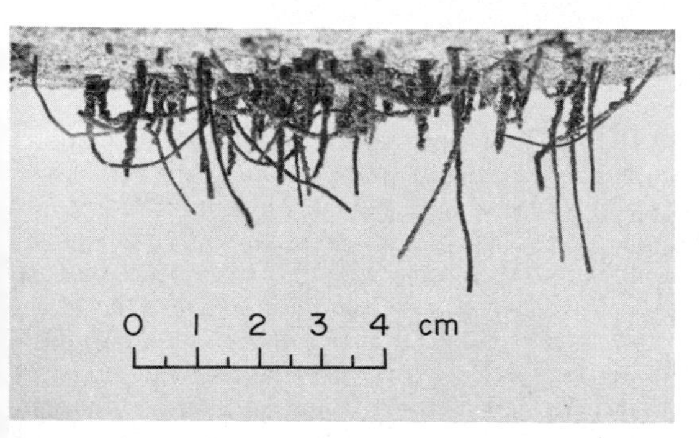

FIG. 16.—Epoxy burrow casts of the polychaete worm *Scolelepis squamata.*

burrows are horizontal, slightly sinuous, average 10 cm in horizontal diameter, 13 cm in vertical diameter, and lie about 25 cm below the sand surface (Fig. 14). Most of the burrows terminate near the seaward edge of the backshore, where the water level is very close to the sediment surface. One burrow system excavated on the backshore was more than 35 m long and had short side branches. These side branches normally contain a food cache or fecal pellets. The fecal pellets are capsule shaped and about 19 mm in length and 7 mm in diameter (Davis, 1966). A typical mound of sand measured 10 to 15 cm in height, 48 to 60 cm in diameter, and consisted of 4.5 to 6.5 kg of sand.

A horizontal burrow network produced by the larvae of the beetle *Omophron sp.* (Fig. 15a) is occasionally found along the landward edges of the backshore. This distinctive burrow system is a feeding target for birds which methodically peck the burrow when searching for the larvae (Fig. 15b).

The upper foreshore is characterized by the dense vertical burrows of the polychaete worm *Scolelepis squamata* (Fig. 16). The burrows have a thin mucoid lining. Lebenspurren, such as footprints, produced by shorebirds feeding along the upper foreshore are superficial and easily destroyed by wave swash. Pecking marks made by shorebirds when searching for polychaetes, mole crabs, or *Donax* reach depths of as much as 30 mm and could be preserved.

Lebenspurren found in both the foreshore and the bar-trough system include the small body-sized burrows of the coquina clam *Donax variabilis* just below the substrate surface and a bioturbated sediment produced by numerous haustoriid

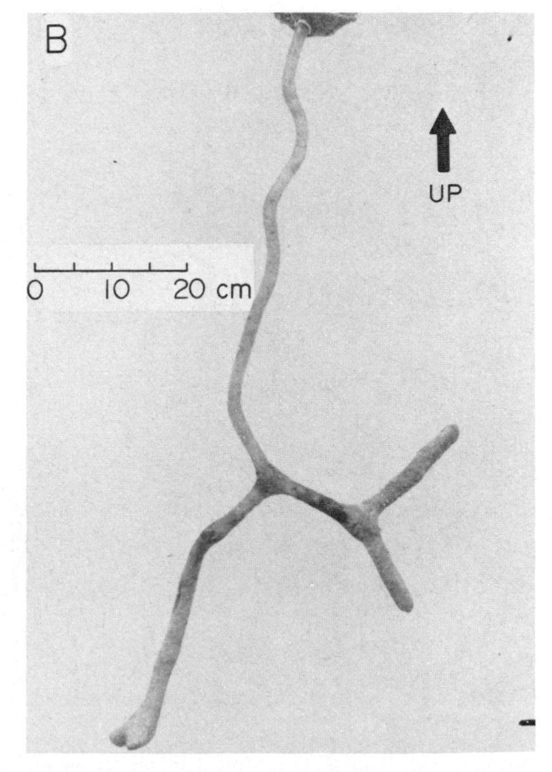

FIG. 17.—*Callianassa islagrande* burrow. *A.* Narrow upper section of burrow. *B.* Fiberglass cast of main burrow.

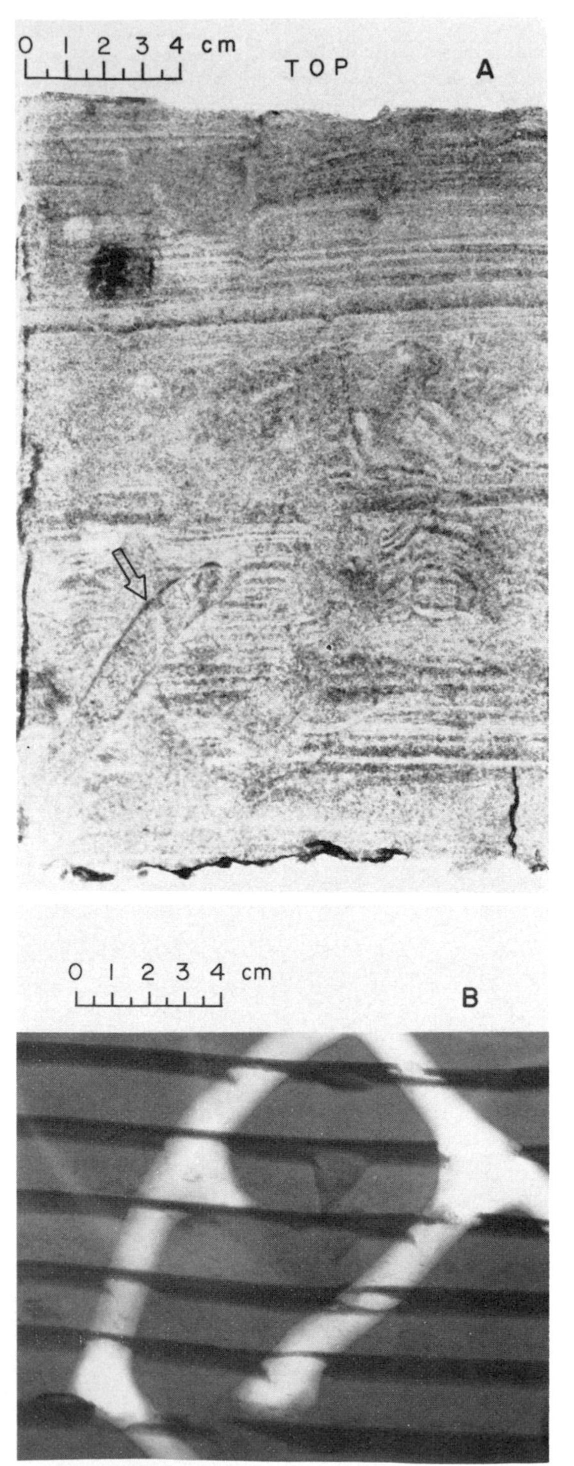

amphipods. The subvertical burrows of the polychaete *Lumbrineris* sp. are common in the lower foreshore and inner trough.

In the bar-trough system, the large conspicuous burrowing animals can be divided into two broad groups based on gross burrow orientation. Species that produce deep vertical burrows are dominant in the inner surf zone, whereas species that burrow horizontally are more common in the outer nearshore. The ghost shrimp *Callianassa islagrande* constructs deep, nearly vertical burrows (Fig. 17) to depths of a meter or more and is the dominant burrowing species throughout the nearshore. The length of the narrow (6–7 mm) upper section of the *C. islagrande* burrow depends on the rate of sediment deposition. The narrow section of the burrow is almost never present on bar crests. Burrows of *C. islagrande* are well defined in cores by their morphology, size, and cemented walls (Fig. 18). Surface expressions of ghost shrimp burrows include chimney structures, mounds of sediment, conical depressions, and simple burrow openings (Fig. 19).

C. islagrande produce large numbers of fecal pellets that measure approximately 0.75 to 1.0 mm in diameter and 2 to 3 mm in length. These fecal pellets consist largely of clay-size material (97%) and are found locally in rings around the burrow openings of the ghost shrimp burrows at low tide (Fig. 20a). Dense accumulations of *C. islagrande* fecal pellets are commonly found along strand lines on the beach (Fig. 20b). These fecal pellets are also concentrated in troughs between ripples where they become covered by sediment (Fig. 20c). Although the fecal production rate was not determined for *C. islagrande* on Padre Island, *C. major* Say was found to produce about 456 pellets per burrow per day in the sand-beach community at Sapelo Island, Georgia (Frankenberg, Coles, and Johannes, 1967).

Sediment excavated by *C. islagrande* during its deposit-feeding process may be redeposited at a level higher or lower than the level from which it was removed. Under laboratory conditions, *C. islagrande* was found to dispose of excavated sand by redepositing it in abandoned feeding burrows at all levels in the sediment column and by venting it out the burrow opening at the sediment surface (Fig. 21).

Although the burrows of *C. islagrande* are lined to increase the wall strength, the burrows occasionally collapse, thereby producing a variety of

FIG. 18.—X-radiographs of *C. islagrande* burrows. *A*. Box-core peel from the lower foreshore showing a well defined *C. islagrande* burrow and other irregularly shaped burrows, and an intensely bioturbated interval between intervals having partially preserved bedding; the planar bedding was formed by wave swash. *B*. Burrow made in a narrow aquarium under laboratory conditions.

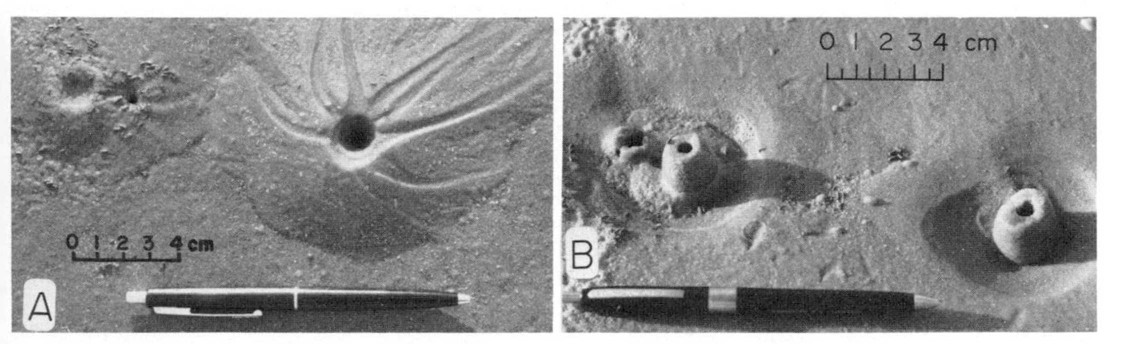

FIG. 19.—Surface openings to *C. islagrande* burrows. *A*. Mound produced by excavated sand vented by the organism from burrow opening. *B*. Chimney structure produced when sand is eroded away from burrow near surface opening.

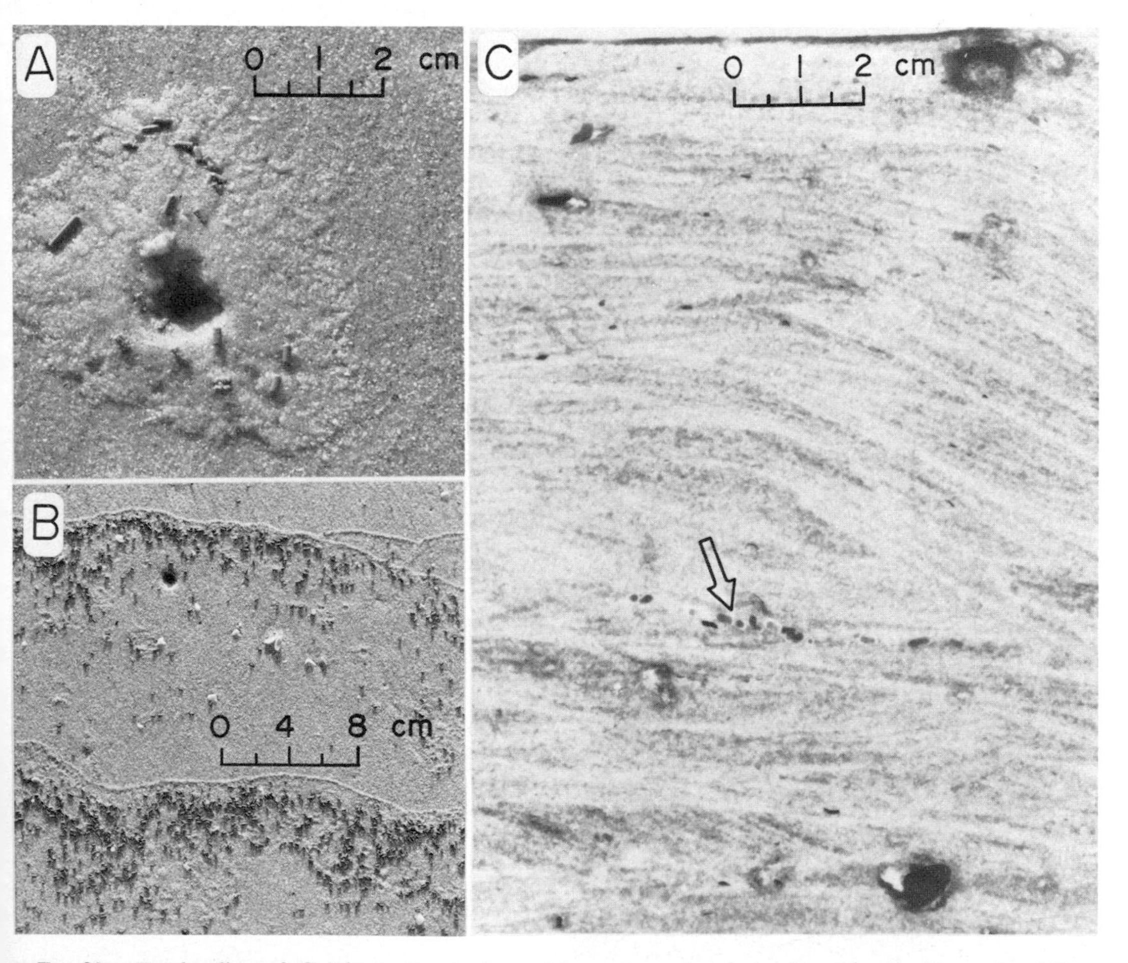

FIG. 20.—Fecal pellets of *C. islagrande*. *A*. Around burrow opening during low tide. *B*. Along strand lines on the upper foreshore. *C*. X-radiograph of box-core peel showing fecal pellets buried along a depositional surface in the inner trough.

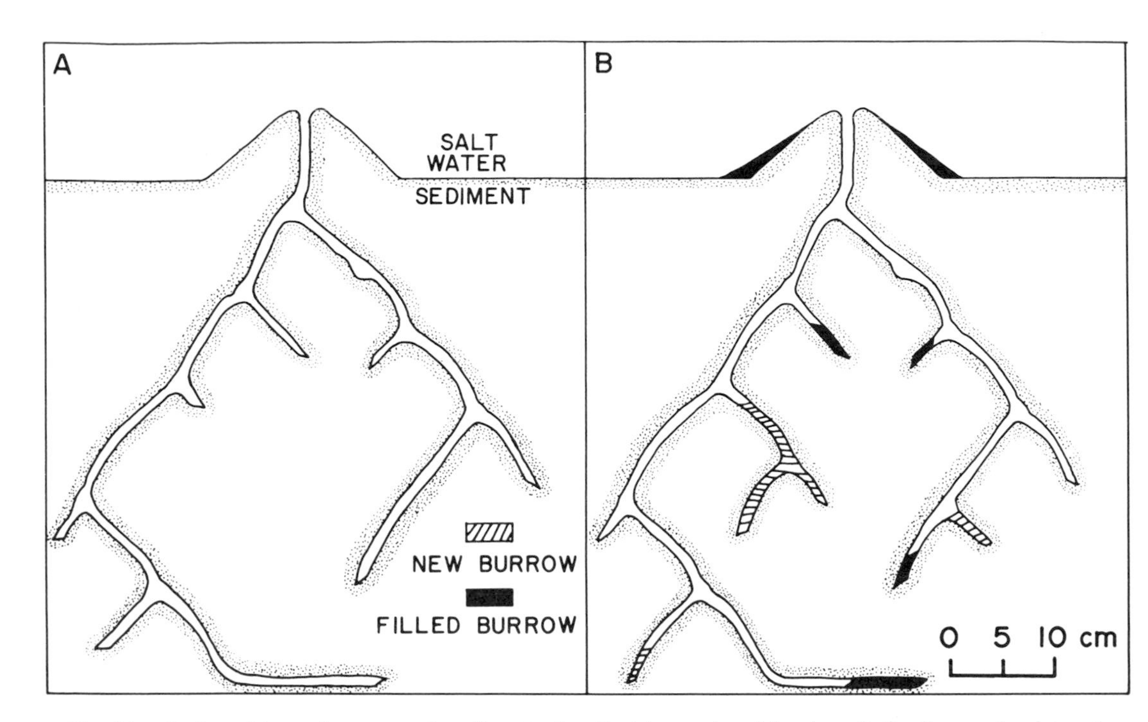

Fig. 21.—Redeposition of excavated sediments by *C. islagrande* while deposit feeding under laboratory conditions, as determined by time-lapse X-ray radiography. *A*. Morphology of the burrow at the start of the experiment. *B*. Morphology of the burrow 24 hours into the experiment.

sedimentary structures (Fig. 22). Nodular wall structure is common to callianassids found on the East Coast of the United States (Weimer and Hoyt, 1964). A general lack of obviously nodular structure in the burrow wall of recently constructed ghost-shrimp burrows has been observed in this study, both in the field and in the laboratory.

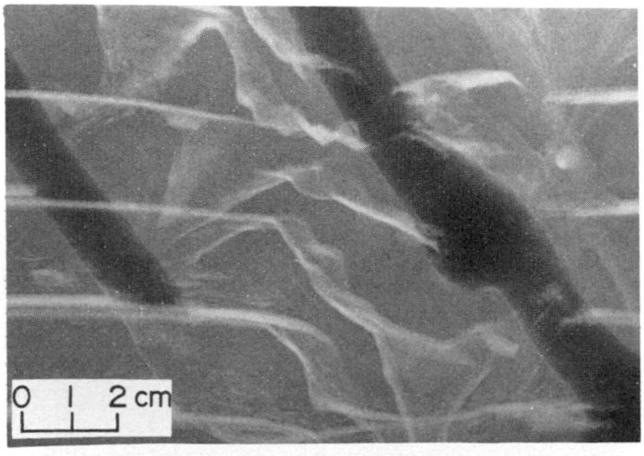

Fig. 22.—Sedimentary structures produced by the collapse of *C. islagrande* burrows under laboratory conditions. Heavy-mineral layers (light areas) bounded on each side by fine sand layers (dark areas).

Rarely, the thicker, stronger, somewhat nodular walls of older(?) *C. islagrande* burrows have been seen on the middle and outer bar crests.

Intrastratal trails probably produced by the horizontal burrowing of sand dollars *Mellita quinquiesperforata* (Fig. 23) are common seaward of the middle bar in the surf zone. Less common but conspicuous horizontal burrowers and surface crawlers include the olive shell *Oliva sayana* Ravenel, moon shell *Polinices duplicatus* (Say), and the auger snail *Terebra salleana* Deshayes.

Although distinct biogenic structures are common in the beach and nearshore sands of northern Padre Island, the burrowing activities of the organisms are even more strikingly manifest in the essentially complete homogenization of the nearshore sands below a thin surficial layer. Pipe cores as long as 1.5 m show that lamination is almost completely absent at depths greater than 0.3 m below the sediment-fluid interface in the nearshore. In fact, seaward of the middle bar, lamination seldom extends more than 2 cm below the sediment surface.

On the beach, lamination is found down to a sharp contact at a level near mean low water. Below this contact, the sand is completely homogenized and contains shells and shell fragments of Gulf origin. This shelly sand is apparently a nearshore deposit that predates the beach. The lower foreshore deposits above this older deposit are alternating beds of highly bioturbated sand and relatively undisturbed laminated sand. Farther landward on the beach, the sands above the older deposit are well laminated, having thin shelly layers in the lower part. The shelly layers were probably derived from the older deposit. Some *Callianassa* burrows occur in the lower part, at levels corresponding to the lower foreshore. Burrows typical of the upper foreshore and backshore occur near the top.

Although Moore and Scruton (1957) suggested that the absence of visible physical sedimentary structures in shoreface sands of the Texas coast might be the result of rapid deposition of well-sorted sand, this hypothesis must be ruled out. Downward increases in the degree of bioturbation, remnant wispy inhomogeneities of irregular shape, and the random orientation of shell fragments are definitive evidence that the absence of lamination is due to bioturbation. The degree of bioturbation is related to changes in the balance between biogenic and physical sedimentary structures.

The balance between biogenic and physical sedimentary structures at depth changes significantly over relatively short periods of time. As the beach and nearshore are rebuilt following erosion caused by severe storms such as hurricanes, the newly deposited sediment is well laminated down to depths as great as one meter, and little biogenic activity is evident. As biological populations destroyed by a major storm become reestablished, distinct biogenic sedimentary structures and bioturbated sediments become more common. Within a few years, the sediment at depths greater than a few centimeters has been worked and reworked by organisms, resulting in a high degree of homogenization.

Given the fact that the uppermost part of the inner nearshore sand is often well laminated, the general absence of lamination at greater depths must be largely the result of the activities of deep-dwelling burrowers. *Callianassa islagrande* is the only organism that seems capable of performing this work. Laboratory experiments to determine burrowing rates and patterns under varying environmental conditions showed that *C. islagrande* is capable of reworking 17 percent of the upper one meter of sediment in the inner nearshore each year.

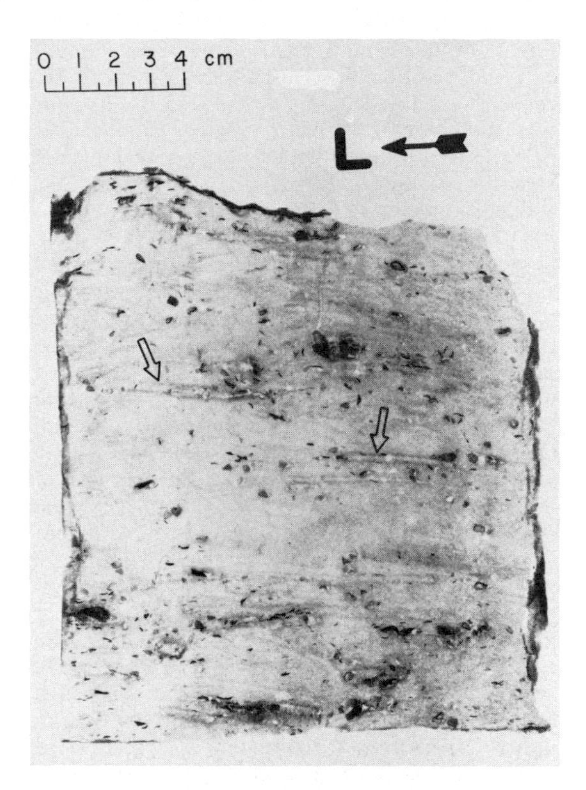

FIG. 23.—X-radiograph of box-core peel from the outer bar in the surf zone showing abundant horizontal burrows in shelly sand, probably produced by sand dollars *Mellita quinquiesperforata*. Orientation of peel is perpendicular to shoreline (L).

SUMMARY

Macrobenthos zonation and characteristic sedimentary structures relate to geomorphic features and dynamic zones along the coast. Perhaps the most striking aspect of the interaction of biologic and geologic processes in this area is the destruction of such a large fraction of the depositional structures by burrowing organisms. Macrobenthos zonation can be defined in terms of species distribution and density, diversity-equitability, and niche classification. Characteristic sedimentary structures common to each landform include depositional structures and biogenic structures. However, biogenic sedimentary structures are most effective in defining beach subzones, whereas physical sedimentary structures are most useful in defining subzones of the inner nearshore.

The *Ocypode* community is found primarily on the backshore and upper foreshore and is characterized by low species diversity, low density of individuals, and infaunal forms which are primarily detritus feeders, mainly scavengers. The dominant internal depositional structure of the backshore is gently seaward-dipping planar lamination produced by wave swash. The burrows of the ghost crab *Ocypode quadrata* are the dominant biogenic sedimentary structures on the backshore. *O. quadrata* burrows are larger in diameter, extend deeper into the sediment, and have a more complex morphology than those in the upper foreshore. In the backshore, a substantial amount of the depositional structures are preserved because the only deep-burrowing form, *Ocypode quadrata*, does not occur in great numbers, is not active during winter months, and does not construct extensive burrows.

Specific to the bar-trough system, the *Callianassa-Donax*-Haustoriidae community is characterized by comparatively high diversity of infaunal species, most of the many individuals belonging to a few species. The trophic nucleus of this community is mixed, although detritus feeders, mainly deposit feeders, are dominant. Three kinds of sedimentary structures produced by waves and currents are common in the bar-trough system: (1) small-scale lenticular cross-bedding, usually without any well-defined preferred orientation of dip direction produced by deposition on sand ripples; (2) planar lamination produced when bar crests are planar because of surge caused by large waves; and (3) medium-scale cross-bedding dipping in the direction of the longshore current produced by migrating megaripples. *Callianassa islagrande* burrows are the dominant biogenic sedimentary structure throughout the bar-trough system. Horizontal intrastratal trails of *Mellita quinquiesperforata* are abundant seaward of the middle bar. Depositional structures in this zone are preserved in only the upper 30 cm because the deeper sediment is reworked continually by the deep-burrowing species, *Callianassa islagrande*, which occurs in relatively dense population, is active during all seasons, and continually reworks its extensive burrow system.

The ecotone or boundary between the communities of the backshore and the bar-trough system is in the foreshore. The greatest diversity and density of macrozoobenthic organisms, all of which are infaunal forms, are found in the ecotone. Species common to both adjacent communities are abundant but locally restricted in the foreshore. The trophic nucleus of the ecotone is mixed. As in the backshore, the dominant internal depositional structure in the foreshore is gently seaward-dipping planar lamination produced by wave swash. Highest densities of the large conspicuous burrows of *O. quadrata* and *C. islagrande* occur in the upper and lower foreshore, respectively. In the sediment column of the foreshore, alternating zones of bioturbated sediment and well preserved depositional structures are common. Considering the high densities of organisms in this landform, it is surprising that any depositional structure is preserved.

ACKNOWLEDGEMENTS

We are indebted to Drs. J. W. Tunnell, Texas A&I University at Corpus Christi, Texas; H. E. Clifton, U.S. Geological Survey, Menlo Park, California; and L. E. Garrison, U.S. Geological Survey, Corpus Christi, Texas, for their critical reading of the manuscript. Ronald J. Miller, Michael E. Dorsey, Kenneth A. Roberts, and Jack L. Kindinger of the U.S. Geological Survey, Corpus Christi, Texas, are thanked for their invaluable assistance in field sampling and laboratory work.

REFERENCES

Beerbower, J. R., and Jordan, D., 1969, Application of information theory to paleontologic problems: Taxonomic diversity: Jour. Paleontology, v. 43, p. 1184–1198.

Brett, C. W., 1964, A portable hydraulic diver operated dredge-sieve for sampling subtidal macrofauna: Jour. Marine Res., v. 22, p. 205–209.

Davis, W. B., 1966, The mammals of Texas: Texas Parks and Wildlife Dept. Bull., v. 41, 267 p.

Dickinson, K. A., 1971, Grain-size distribution and the depositional history of northern Padre Island, Texas: U.S. Geol. Survey Prof. Paper 750-C, 6 p.

Frankenberg, D., Coles, S. L., and Johannes, R. E., 1967, The potential trophic significance of *Callianassa major* fecal pellets: Limnology and Oceanography, v. 12, p. 113–120.

Hayes, M. O., 1964, Grain size modes in Padre Island sands: Gulf Coast Assoc. Geol. Socs., Field Trip Guidebook 1965 Ann. Meeting, p. 121–216.

Hill, G. W., and Hunter, R. E., 1973, Burrows of the ghost crab *Ocypode quadrata* (Fabricius) on the barrier islands, south-central Texas: Jour. Sed. Petrology, v. 43, p. 24–30.

Howard, J. D., and Dörjes, Jurgen, 1972, Animal-sediment relationships in two beach-related tidal flats: Sapelo Island, Georgia: Jour. Sed. Petrology, v. 42, p. 608–623.

——, and Frey, R. W., 1973, Characteristic physical and biogenic sedimentary structures in Georgia estuaries: Am. Assoc. Petroleum Geologists Bull., v. 57, p. 1169–1185.

Hunter, R. E., and Dickinson, K. A., 1970, Map showing landforms and sedimentary deposits of the Padre Island portion of the South Bird Island 7.5 minute quadrangle, Texas: U.S. Geol. Survey, Misc. Geol. Inv. Map I-659.

——, Watson, R. L., Hill, G. W., and Dickinson, K. A., 1972, Modern depositional environments and processes, northern and central Padre Island, Texas: Gulf Coast Assoc. Geol. Socs., Field Trip Guidebook 1972 Ann. Meeting, p. 1–27.

Imbrie, John, and Buchanan, Hugh, 1965, Sedimentary structures in modern carbonate sands of the Bahamas:

In G. V. Middleton (ed.), Primary sedimentary structures and their hydrodynamic interpretation: Soc. Econ. Paleontologists and Mineralogists, Spec. Pub. 12, p. 149–172.

King, C. A. M., 1972, Beaches and coasts: St. Martin's Press, New York, 570 p.

Lloyd, M., and Ghelardi, R. J., 1964, A table for calculating "equitability" components of species diversity: Jour. Animal Ecol., v. 33, p. 217–225.

Mayou, T. V., Howard, J. D., and Smith, K. L., 1969, Techniques for sampling tracks, trails, burrows, and bioturbate textures in unconsolidated sediments: Geol. Soc. America Spec. Paper 121, p. 665–666.

McKee, E. D., 1957, Primary structures in some recent sediments: Am. Assoc. Petroleum Geologists Bull., v. 41, p. 1704–1747.

Milling, M. E., and Behrens, E. W., 1966, Sedimentary structures of beach and dune deposits—Mustang Island, Texas: Univ. Texas, Inst. Marine Sci. Pub., v. 11, p. 135–148.

Moiola, R. J., and Spencer, A. B., 1973, Sedimentary structures and grain-size distribution, Mustang Island, Texas: Gulf Coast Assoc. Geol. Socs. Trans., v. 23, p. 324–332.

Moore, D. G., and Scruton, P. C., 1957, Minor internal structures of some recent unconsolidated sediments: Am. Assoc. Petroleum Geologists Bull., v. 41, p. 2723–2751.

Pearse, A. S., Humm, H. J., and Wharton, G. W., 1942, Ecology of sand beaches at Beaufort, N.C.: Ecol. Mon., v. 12, p. 136–180.

Rhoads, D. C., Speden, I. G., and Waage, K. M., 1972, Trophic group analysis of Upper Cretaceous (Maestrichtian) bivalve assemblages from South Dakota: Am. Assoc. Petroleum Geologists Bull., v. 56, p. 1100–1114.

Scott, R. W., 1972, Preliminary ecological classification of ancient benthic communities: 24th Internat. Geol. Congr. Proc., Sec. 7, p. 103–110.

Shannon, C. E., and Weaver, W., 1963, The mathematical theory of communication: Univ. Illinois Press, Urbana, 117 p.

Thorson, G., 1957, Bottom communities (sublittoral or shallow shelf): *In* J. W. Hedgpeth (ed.), Treatise on marine ecology and paleoecology: Geol. Soc. America Mem. 67, v. 1, p. 461–534.

U.S. National Oceanic and Atmospheric Administration, 1973, Surface water temperature and density, Atlantic coast, North and South America: Natl. Ocean Survey Pub. 31-1.

——, 1974, Local climatological data, annual summary with comparative data, 1974, Corpus Christi, Texas: Natl. Environmental Data Service.

——, 1975, Tide tables, 1975, east coast of North and South America: Natl. Ocean Survey.

Weimer, R. J., and Hoyt, J. H., 1964, Burrows of *Callianassa major* Say, geologic indicators of littoral and shallow neritic environments: Jour. Paleontology, v. 38, p. 761–767.